W0099526

Genetics of Bacterial Polysaccharides

Genetics of Bacterial Polysaccharides

Edited by

Joanna B. Goldberg, Ph.D.

Department of Microbiology
University of Virginia
Charlottesville, Virginia

CRC Press

Taylor & Francis Group
Boca Raton London New York

CRC Press is an imprint of the
Taylor & Francis Group, an **informa** business

CRC Press
Taylor & Francis Group
6000 Broken Sound Parkway NW, Suite 300
Boca Raton, FL 33487-2742

ISBN-13: 978-0-8493-0021-9 (hbk)

Library of Congress Cataloging-in-Publication Data

Genetics of bacterial polysaccharides / edited by Joanna B. Goldberg.
    p. cm.
Includes bibliographical references and index.
ISBN 0-8493-0021-5 (alk. paper)
    1. Microbial polysaccharides. 2. Bacterial genetics.
I. Goldberg, Joanna B.
QR92.P6G46 1999
572'.566293—dc21

99-13548
CIP

Library of Congress Card Number 99-13548

Visit the Taylor & Francis Web site at
http://www.taylorandfrancis.com

and the CRC Press Web site at
http://www.crcpress.com

About the Editor

Joanna B. Goldberg, Ph.D., has been an Associate Professor in the Department of Microbiology at the University of Virginia Health Sciences Center since 1996. She received her B.A. in Biology with Distinction from Boston University in 1979 and her Ph.D. from the Department of Microbiology and Immunology at the University of California, Berkeley, in 1986. In 1989, Dr. Goldberg joined the faculty at the Channing Laboratory, Brigham and Women's Hospital, Harvard Medical School, as an instructor and then assistant professor. Her research has focused on bacterial pathogenesis with an emphasis on polysaccharides as virulence factors and vaccine candidates. She has published more than 30 research papers on these subjects.

Contributors

Michael A. Apicella, M.D.
Department of Microbiology
University of Iowa
Iowa City, Iowa

Russell W. Carlson, Ph.D.
Complex Carbohydrate Research
  Center
University of Georgia
Athens, Georgia

L. Scott Forsberg, Ph.D.
Complex Carbohydrate Research
  Center
University of Georgia
Athens, Georgia

Joanna B. Goldberg, Ph.D.
Department of Microbiology
University of Virginia
Charlottesville, Virginia

Emil C. Gotschlich, M.D.
Laboratory of Bacterial
  Pathogenesis and Immunology
The Rockefeller University
New York, New York

Elmar L. Kannenberg, Ph.D.
Mikrobiologie/Biotechnologie
Universität Tübingen
Tübingen, Germany

Chia Y. Lee, Ph.D.
Department of Microbiology,
  Molecular Genetics, and
  Immunology
University of Kansas Medical
  Center
Kansas City, Kansas

Jean C. Lee, Ph.D.
Channing Laboratory
Department of Medicine
Brigham & Women's Hospital and
  Harvard Medical School
Boston, Massachusetts

Michael McNeil, Ph.D.
Department of Microbiology
Colorado State University
Fort Collins, Colorado

Paul A. Manning, Ph.D.
Astra Research Center Boston
Cambridge, Massachusetts

Susie Y. Minor, B.S.
Laboratory of Bacterial
  Pathogenesis and Immunology
The Rockefeller University
New York, New York

Andrew Preston, Ph.D.
Department of Clinical Veterinary
  Medicine
University of Cambridge
Cambridge, England

Peter R. Reeves, Ph.D.
Department of Microbiology
University of Sydney
Sydney, Australia

Bradley L. Reuhs, Ph.D.
Complex Carbohydrate Research
  Center
University of Georgia
Athens, Georgia

Mikael Skurnik, Ph.D.
Department of Medical
  Biochemistry
University of Turku
Turku, Finland

Uwe H. Stroeher, Ph.D.
Mikrobiologie/Membranphysiologie
Universität Tübingen
Tübingen, Germany

Janet Yother, Ph.D.
Department of Microbiology
University of Alabama at
  Birmingham
Birmingham, Alabama

Preface

Bacterial surface or secreted polysaccharides are molecules that can function as barriers to protect bacterial cells against environmental stresses, as well as act as adhesins or recognition molecules. In some cases, these molecules are immunodominant antigens eliciting a vigorous immune response, while in other cases, the expression of polysaccharides camouflages the bacteria from the immune system. In a historical sense, investigations of bacterial polysaccharides have paved the way for the advent of modern molecular genetics: examination of the expression of *Streptococcus pneumoniae* capsules early in this century identified bacterial gene transfer and recognized DNA as the genetic material for this transfer.

Bacterial polysaccharides can be considered tertiary gene products. The pathways for biosynthesis of bacterial polysaccharides involve multiple enzymatic steps for the synthesis of precursor molecules, the assembly of these molecules, and their subsequent transport to the cell surface. These processes may be highly regulated, including at the transcriptional level, where specific environmental stimuli may induce signal transduction events leading to the selective expression of genes for the synthesis of polysaccharides. Until recently, most studies on the enzymatic steps and regulation of these molecules were performed on the enteric Gram-negative bacteria, *Escherichia coli* and *Salmonella typhimurium*.

With the advent of modern bacterial genetics, techniques such as construction and characterization of polysaccharide mutants, cloning of genes and complementation of these mutations, and expression of polysaccharides in heterologous bacterial hosts have prompted investigations into the roles and functions of these molecules for many different bacteria. Nucleotide and inferred amino acid sequence analysis of these various polysaccharide loci in conjunction with polysaccharide structural comparisons has helped to define functions for polysaccharide genes and their products and has increased the body of knowledge related to the synthesis and regulation of these complicated structures. In addition, the biochemical analysis of protein products encoded by genes from many different bacterial genera involved in polysaccharide expression has advanced the characterization of enzymes and regulatory molecules involved in the synthesis of these complex molecules. Here, we present the genetic analysis of polysaccharides from a number of bacteria pathogenic to humans and one symbiotic with plants in hopes that similarities in the experimental approaches as well as findings from such investigations may lead to a general understanding of polysaccharide synthesis and regulation in various bacteria.

Joanna B. Goldberg
Charlottesville, Virginia
January 1999

Contents

Introduction xiii
Peter R. Reeves

Chapter 1   Genetics of *Pseudomonas aeruginosa* polysaccharides 1
Joanna B. Goldberg

Chapter 2   Molecular genetics of *Yersinia* lipopolysaccharide 23
Mikael Skurnik

Chapter 3   Rhizobial cell surface carbohydrates:
Their structures, biosynthesis, and functions 53
Russell W. Carlson, Bradley L. Reuhs, L. Scott Forsberg,
and Elmar L. Kannenberg

Chapter 4   The genetics of capsule and lipooligosaccharide
biosynthesis in *Haemophilus influenzae* 91
Andrew Preston and Michael A. Apicella

Chapter 5   The genetics of LPS synthesis by the gonococcus 111
Susie Y. Minor and Emil C. Gotschlich

Chapter 6   Genetics of *Vibrio cholerae* O1 and O139 surface
polysaccharides 133
Uwe H. Stroeher and Paul A. Manning

Chapter 7   Common themes in the genetics of streptococcal
capsular polysaccharides 161
Janet Yother

Chapter 8   Capsular polysaccharides of *Staphylococcus aureus* 185
Jean C. Lee and Chia Y. Lee

Chapter 9   Arabinogalactan in mycobacteria: Structure, biosynthesis,
and genetics 207
Michael McNeil

Index 225

Introduction

Bacterial polysaccharides were in at the beginning of bacterial genetics. A 1928 paper by Griffith[1] on capsules of the Pneumococcus (now *Streptococcus pneumoniae*) is often seen as the first paper on bacterial genetics, though there was prior literature on bacterial variation and it was only later that the paper was seen as being on bacterial genetics. It describes transformation, later shown to be due to DNA: it was the first in a chain of events that led to the study of genetics at the molecular level. The capsules of *S. pneumoniae* occur in a variety of forms detected first by their antigenic specificity, but now also by their underlying chemical differences. The first transformation experiments showed that the ability to make capsules could be restored by transformation and also that the newly smooth (encapsulated) organism could have capsule specificity of the donor. Work on capsule genetics was pursued strongly in the 1950s and 1960s. With the nature of transformation now understood as transfer of genes, it became clear that observations made in the 1928 paper and confirmed in other situations, that transformation could confer the ability to make a capsule of the donor strain type, meant that the genes for capsule synthesis must be clustered. This has turned out to be a very general phenomenon.

This book includes discussion of the genetics of those same *S. pneumoniae* capsules, but for the genetics of the day *S. pneumoniae* was not the easiest organism to work with. *Escherichia coli, Salmonella enterica,* and *Bacillus subtilis* became the major foci for bacterial genetics, and *S. enterica* in particular for study of the genetics of polysaccharide synthesis, with the underlying form of the genetics of its lipopolysaccharide (LPS) resolved in the 1960s. This era is well documented in a 1969 review by Mäkelä and Stocker,[2] which summarizes the situation at that time for biochemistry and genetics of both *S. pneumoniae* capsules and *S. enterica* lipopolysaccharide. The pattern of separate gene clusters for O antigen and lipid A/core of LPS had been established by that time, together with the recognition that such clusters include genes for nucleotide sugar synthesis pathways and sugar transferases. Crosses between strains showed that the genes for the O antigens were located at the same locus in strains of different serogroups. These characteristics have been found to be quite general with several species shown to have a locus for O antigen genes, with variation in the chemical nature of the O antigen reflected in variation in the genes of the cluster. In the case of LPS, there is a separate locus for genes of the lipid A/core, and by 1969 there were 8 lipid A/core genes recognized in one cluster and 10 genes concerned with synthesis of the O antigen in another gene cluster. The contrast with the level of understanding at that time of the genetics of *S. pneumoniae* capsules is striking because for these capsules there is no list of individual genes, although the gene for UDP-glucose dehydrogenase is defined. This difference did not reflect the number or ability of the people working on the two species so much as the genetic tools available for this area of work.

The discrepancy continues to the present day, but the gap is narrowing. The genes for O antigens and lipid A/core for *E. coli* K-12 and *S. enterica* LT2 are now in general known, although in some cases in one species only. There are still significant gaps in the case of lipid A. There has also been extensive work on the genetics of variation in O antigens in the two species. The complete sequences are known for 9 O antigen gene clusters of *S. enterica* and 7 of the closely related *E. coli* (if one includes Shigella strains as one should). In several cases the function of all genes is also known, whereas for most other species at most 2 or 3 sequences are known and rarely is there much information on gene function, although there are some striking exceptions of very detailed studies. The work on *E. coli, S. enterica,* and close relatives is reviewed quite frequently (for example, see References 3 and 4), often under a more general heading but with the two related species getting most of the coverage.

However the development of cloning, sequencing, and PCR means that one no longer needs a well-developed genetic manipulation system for a species in order to do genetics of specific gene clusters and, in this more democratic world, there is a resurgence of interest in polysaccharides of many species, including the capsules of *S. pneumoniae*. The editor of this book has chosen to focus on the excellent work on these other species. It is timely to do this because there is now a large body of knowledge that begins to address the enormous diversity of bacterial polysaccharides. This book celebrates the growing interest in a wide variety of bacterial polysaccharides.

The emphasis on genetics is justified because this is the area of most obvious advance, with complete sequences for many gene clusters. We must not forget, however, the need for reliable structures, because it is the combination of good sequence and good structural data that enables us to study the genetics of biosynthesis and assembly of the polysaccharides.

The enormous diversity of bacterial polysaccharides is impressive. There are 90 known capsule structures for *S. pneumoniae*[5] and 80 for *E. coli,*[6] with three separate loci involved in *E. coli*. There are approximately 187 and 50 O antigens found in *E. coli* and *S. enterica*, respectively. (The 187 in *E. coli* includes the Shigella O antigens, as Shigella and *E. coli* are clearly one species.) Of these, only three are common to both species. Many other species have a large repertoire of O antigens and capsules. These numbers make the ABO blood group polymorphism look puny, and even more so if one remembers that blood groups A and B differ chemically at only one sugar residue of an oligosaccharide, the residue being either N-acetylgalactosamine or galactose for blood groups A and B, respectively. This major polymorphism is due to relatively minor differences in the allelic sugar transferase genes, which differ at only a few bases that affect only four amino acid residues.[7] The bacterial polymorphisms commonly involve presence or absence of several genes.

There is not only variation in the structures but also in the mechanisms for assembly and processing of the polysaccharides. The lipid A/core component of LPS is synthesized by sequential addition of sugars and fatty acids, and this also applies to the complete structures of some of the rather larger lipooligosaccharides. However, most O antigens are synthesized as a small repeat unit on the inner surface of the cytoplasmic membrane, flipped to the periplasmic face, and then polymerized — but then some, like the O9 antigen of E. coli and some of those of Klebsiella pneumoniae, are synthesized by sequential addition of individual sugars to make a complete polysaccharide, which is then exported by a quite different mechanism. (Neither is well understood, but the genes involved encode quite different proteins.)

I now refer to problems that arise for nomenclature with such a rich diversity of structures and genes. Nomenclature of the structures themselves is not too bad. Chemistry has very well-defined nomenclature for the description of individual structures, and the trivial names of different forms have generally followed fairly consistent rules, which nonetheless differ a little from case to case.

The major problem has arisen in naming genes. The Demerec system of genetic nomenclature[8] is widely used and provides for four-letter gene names. The system was set up to handle nomenclature when genetic analysis was focused on mutations: the first 3 letters together with an allele number are sufficient to define a mutation. The fourth letter is capitalized and differentiates genes of a given pathway or other group.

For instance, his-1 defines a mutation in the his group of genes for histidine biosynthesis. The mutation can be named hisA1 or hisB1, if one knows which his gene is involved in the mutation. There have been a few cases where the number of genes in a pathway exceeded 26, and special steps were taken (e.g., use of fla* and flb* etc.) for flagella genes of E. coli and S. enterica. The Demerec system continues to work well in most cases, but because of their extreme diversity it gives major problems when applied to genes for bacterial polysaccharides.

If we look just at O antigen genes, the first gene cluster described was that for S. enterica strain LT2, which we now know has 16 genes. They were named rfbA, rfbB, etc. The number of genes in O antigen gene clusters varies from about 6 to 19, but because each O antigen form has its own set of genes, the total number for species like S. enterica will be many times 26, and the grand total for all species will be very large indeed. This is important because, although genetic nomenclature has grown up independently for each species, there is now a strong incentive to use common nomenclatures as we try to compare genomes for many species.

This problem was discussed recently and a proposal put forward to adopt a new nomenclature which allows a very large number of gene names. The proposal, described in detail in the original paper by Reeves et al.[9] and on a web site (http://www.angis.su.oz.au/BacPolGenes/welcome/html) has been adopted by several groups in addition to the original authors. I will illustrate the bacterial polysaccharide gene nomenclature (BPGN) scheme by reference to the O antigen gene cluster of *E. coli* Dysenteria (nee *Bacterium dysenteriae* and commonly known as *Shigella dysenteriae*), chosen because it is short, sequenced, and all its genes have a known function based either on direct experiment or homology.[10]

rmlA, *rmlB*, *rmlC*, and *rmlD* are the four genes of the TDP-L-rhamnose pathway and represent the first class of polysaccharide synthesis genes. Under the BPGN scheme each nucleotide sugar pathway has its own name, as for other biosynthetic pathways, and the name gets used wherever that pathway occurs. The four *rml* genes were previously known as *rfbA,B,C,D*, but these names were also used in other gene clusters, not only for TDP-rhamnose pathway genes but also for genes of quite unrelated function.

The second class of genes are those for transferases. *wbbR* and *wbbQ* are the two rhamnose transferase genes of Dysenteriae. Transferase genes are specific for the linkage and tend to be specific to the cluster so there are going to be many different rhamnose transferase genes. All such genes in a cluster get a w*** name, and those in a given cluster in general get the same first three letters, *wbb* in this case. There are 26^3 w*** names, so they should last a while.

Note that in some less well-documented gene clusters, presumptive sugar pathway genes have been given w*** names as a temporary measure until function is established. This is likely to be a short-term problem because in general one can identify pathway genes by homology once one example of that gene has been characterized. The time is not far off when almost all pathway genes in a newly sequenced cluster will be identifiable from the sequence. However, the same is not true for the transferases, which can be extremely difficult to identify with certainty without experimental evidence for each.

The third class of genes are those for export and processing, which often occur in many clusters with different specificity. The Dysenteriae cluster has *wzx* and *wzy*, best known for their occurrence in gene clusters for O antigens which have the LT2 type O antigen export and assembly system. They are named *wzx* and *wzy* in all cases, although they will often have different specificities, particularly in the case of *wzy*. There are similar suggestions for genes of other export and processing genes which have been given wz* names as a subset of the w*** names.

There is not the space to give much detail on nomenclature, but there are examples in the presentations in this book and a database of bacterial polysaccharide genes is available at the BPGD web site given above. We worked longer than I like to remember to develop such a simple system, and those involved hope it serves you well.

It is a pleasure to launch this book on developments in bacterial polysaccharide genetics.

Peter R. Reeves
University of Sydney
Sydney, Australia

References

1. Griffith, F. Significance of pneumococcal types. *Journal of Hygiene.* 27, 113-159, 1928.
2. Mäkelä, P. H. and B. A. D. Stocker. Genetics of polysaccharide synthesis, *Annual Review of Genetics.* 3, 291-322, 1969.
3. Reeves, P. R. Biosynthesis and assembly of lipopolysaccharide, in *Bacterial Cell Wall, New Comprehensive Biochemistry,* A. Neuberger and L. L. M. van Deenen, Eds., Elsevier Science Publishers, Amsterdam. 1994.
4. Whitfield, C., P. A. Amor and R. Koplin. Modulation of the surface architecture of gram-negative bacteria by the action of surface polymer-lipid A-core ligase and determinants. *Molecular Microbiology.* 23, 629-638, 1997.
5. Henrichsen, J. Six newly recognized types of *Streptococcus pneumoniae. Journal of Clinical Microbiology.* 33, 2759-2761, 1995.
6. Ørskov, F. and I. Ørskov. *Escherichia coli* serotyping and disease in man and animals. *Canadian Journal of Microbiology.* 38, 699-704, 1992.
7. Yamamoto, F., H. Clausen, T. White, J. Marken and S. Hakomori. Molecular genetic basis of the histo-blood group ABO system. *Nature.* 345, 229-233, 1990.
8. Demerec, M., E. Adelberg, A. Clark and P. Hartman. A proposal for a uniform nomenclature in bacterial genetics. *Genetics.* 54, 61-74, 1966.
9. Reeves, P. R., M. Hobbs, M. Valvano, M. Skurnik, C. Whitfield, D. Coplin, N. Kido, J. Klena, D. Maskell, C. Raetz and P. Rick. Bacterial polysaccharide synthesis and gene nomenclature. *Trends in Microbiology.* 4, 495-503, 1996.
10. Klena, J. D. and C. A. Schnaitman. Function of the *rfb* gene cluster and the *rfe* gene in the synthesis of O antigen by *Shigella dysenteriae* 1. *Molecular Microbiology.* 9, 393-402, 1993.

chapter one

Genetics of Pseudomonas aeruginosa *polysaccharides*

Joanna B. Goldberg

Contents

Introduction 2
P. aeruginosa polysaccharides 2
Alginate 3
    Genes encoding alginate biosynthetic enzymes 4
    Alginate genes encoding regulatory functions 4
    Role of alginate 8
O antigen 8
    Genes for O-antigen synthesis 8
    Role of the O antigen 10
LPS core 10
    LPS core genes 11
    Role for the LPS core 12
Common antigen 12
    Common antigen genes 12
    Role of the common antigen 13
Lipid A 13
    Genes for lipid A synthesis 13
    Role of the lipid A 14
Link between alginate and LPS synthesis 14
    Role for the LPS-rough, mucoid phenotype associated
      with CF isolates 14
Conclusions 15
Acknowledgments 15
References 15

Introduction

Pseudomonas aeruginosa is an opportunistic pathogen that is ubiquitous in the environment and is capable of infecting patients whose health is compromised; in most cases these infections begin with some breakdown of normal host defenses. The fourth most common nosocomial pathogen, *P. aeruginosa* is the most frequent cause of pneumonia in hospitalized patients. As people with human immunodeficiency virus (HIV) disease are living longer, bacterial pneumonia is becoming more common, and *P. aeruginosa* is being isolated with increasing frequency. In addition, this bacterial species is a common cause of urinary tract infections and a frequent isolate in surgical infections. Patients with burns are particularly susceptible to *P. aeruginosa* infections. Septicemia caused by *P. aeruginosa* is difficult to treat because of the inherent resistance of this bacterium to many antibiotic regimens. *P. aeruginosa* is the most common cause of bacterial corneal ulcers or keratitis, infections that usually begin after injury to the eye. *P. aeruginosa* is also responsible for most of the mortality in patients with cystic fibrosis (CF) who die as a consequence of chronic lung infections with this bacterium.[1]

P. aeruginosa *polysaccharides*

P. aeruginosa produces at least two polysaccharides, lipopolysaccharide (LPS) and alginate.[2] The general structure of *P. aeruginosa* LPS is similar to that of other Gram-negative bacteria: an O-polysaccharide side chain, an LPS core, and lipid A. Common antigen is a polymer composed of rhamnose that is also attached to the LPS core. Alginate is a high-molecular-weight, capsule-like exopolysaccharide. The structures of these polysaccharides are shown in Figure 1.1.

P. aeruginosa strains isolated from various clinical sources differ with respect to the polysaccharides they produce. Isolates from acute infections and those from the environment express a smooth LPS with many long O side chains that protects the organism from complement-mediated killing. These strains typically have a nonmucoid phenotype due to low-level production of alginate.

Strains that initially infect the lungs of CF patients have the same phenotype with respect to polysaccharide expression as strains in the environment and from acute infections. However, as the clinical course in the CF patient progresses, chronic *P. aeruginosa* isolates colonizing the respiratory tract express a rough LPS with few or no long O-polysaccharide side chains. This characteristic renders the bacteria nontypable (unable to agglutinate with O-antigen typing serum) or polyagglutinable (reacting with more than one serum) and sensitive to killing by normal human serum.[3-5] In addition, these strains are often mucoid, producing large amounts of alginate,[6] which surrounds the bacterium as a loosely associated capsule.

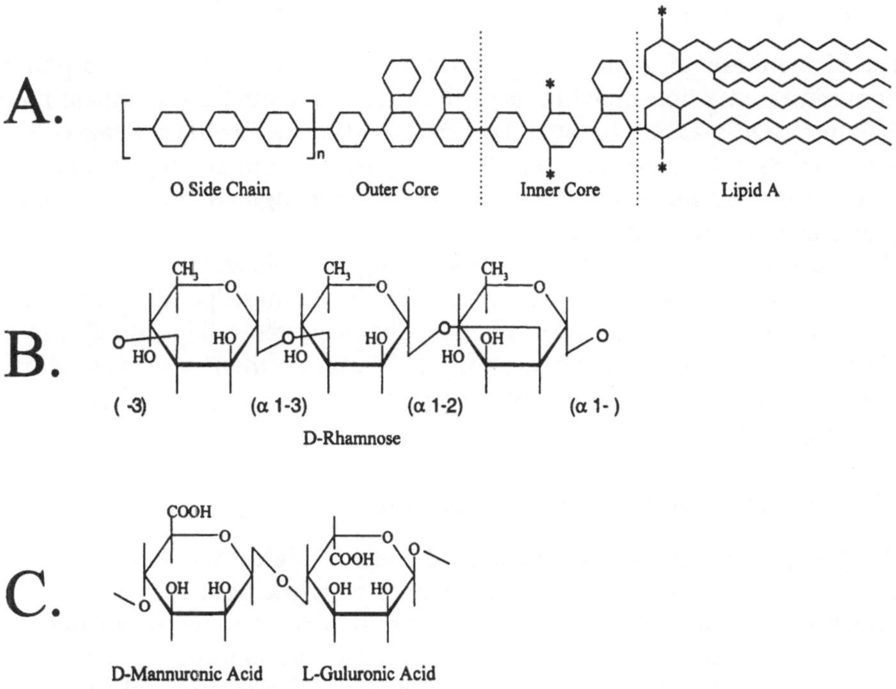

Figure 1.1 *P. aeruginosa* polysaccharides. (A) Lipopolysaccharide (LPS) includes O polysaccharide (also called O antigen, O side chain, or B band), composed of trisaccharide or tetrasaccharide repeating units, is attached to the LPS core, which has an inner core region, composed of ketodeoxyoctonate (KDO) and heptose, and an outer core region composed of hexoses. The LPS core is attached to lipid A inserted into the outer membrane. The hexagons represent monosaccharides; the * represent phosphates; and "n" indicates an O-antigen repeating unit. There are 20 different serogroups of *P. aeruginosa* that differ in their monosaccharide components and linkages. (B) Common antigen (also called A band) is a regular homopolymer D-rhamnose that is also attached to the LPS core. (C) The exopolysaccharide alginate is a long, linear, nonrepeating polymer composed of D-mannuronic acid and L-guluronic acid.

Alginate

Alginate (also known as mucoid exopolysaccharide or MEP) is a high-molecular-weight, linear, O-acetylated polymer composed of D-mannuronic acid and its C5-epimer, L-guluronic acid (Figure 1.1). The block structure and degree of acetylation give alginate its physicochemical properties of flexibility and water-binding capacity. Overexpression of this polysaccharide gives *P. aeruginosa* a mucoid appearance when grown on solid medium in the laboratory. In liquid growth conditions, alginate is not attached to the bacterium but is released into the extracellular medium, where it can accumulate to levels as high as 2 g/L. Overproduction of alginate can also be detected in expectorated sputum from CF patients with *P. aeruginosa* lung infections.[7]

Genes encoding alginate biosynthetic enzymes

The biosynthesis of alginate (Figure 1.2) begins with fructose 6-phosphate from central metabolism, which is further converted into GDP-mannose and then into GDP-mannuronic acid. This last step, directed by the enzyme GDP-mannose dehydrogenase, is considered the first committed step in alginate production. This nucleotide sugar is a precursor in alginate polymerization, epimerization, and transport.

The genes encoding the enzymes for alginate biosynthesis have been identified and cloned; the products of many of these genes have been expressed and their enzymatic activities demonstrated[8,9] (Figure 1.2). Most of the alginate biosynthetic genes map to 34 min on the *P. aeruginosa* chromosome and form an operon. The exception is the *algC* gene, which maps at 10 min (Figure 1.3).

Alginate genes encoding regulatory functions

The pathway for the biosynthesis of alginate is highly regulated; a large number of genes form a complex regulatory network. Some of these genes have been detected by their ability either to repress alginate production in mucoid strains or to promote alginate production in nonmucoid strains.[10,11] These genes map in the 67- to 69-min region of the chromosome (Figure 1.3); nucleotide sequence analysis has revealed their function. *algT* (*algU*) encodes a functional homolog of the extreme heat-shock sigma factor of *Escherichia coli*. Downstream of *algT* is *mucA* (encoding an anti-sigma factor), *mucB* (encoding a negative regulator), *mucC* (encoding a putative regulator), and *mucD* (encoding a serine protease homolog). A distinct *mucD* homolog, *algW*, has also been detected in this region of the chromosome. Each of these genes has a putative role in regulating AlgT.[12]

Another locus contains additional regulatory genes at 9 to 10 min (Figure 1.3): *algR* (*algR1*) encodes a response regulatory protein, *algQ* (*algR2*) encodes a regulator of NDP-kinase, and *algP* (*algR3*) encodes a histone-like product.[12] A sensor-like gene, *fimS*, has recently been identified upstream of *algR*.[13] Yu et al. also identified this gene and found that it has a negative impact on alginate production.[14]

Like AlgT, AlgR is thought to control *algD* synthesis directly. Additional regulatory gene products that affect *algD* expression are AlgB, which is a response regulator, and KinB, which is the AlgB-kinase; both map to the 13-min region of the *P. aeruginosa* chromosome (Figure 1.4). Unlike many other two-component regulatory systems, the transcriptional activation of alginate genes by AlgB and AlgR is not mediated by conventional phosphorylation.[15] Another product involved in alginate production, AlgZ, is an AlgT-dependent DNA-binding protein; the *algZ* gene has not yet been identified.[16]

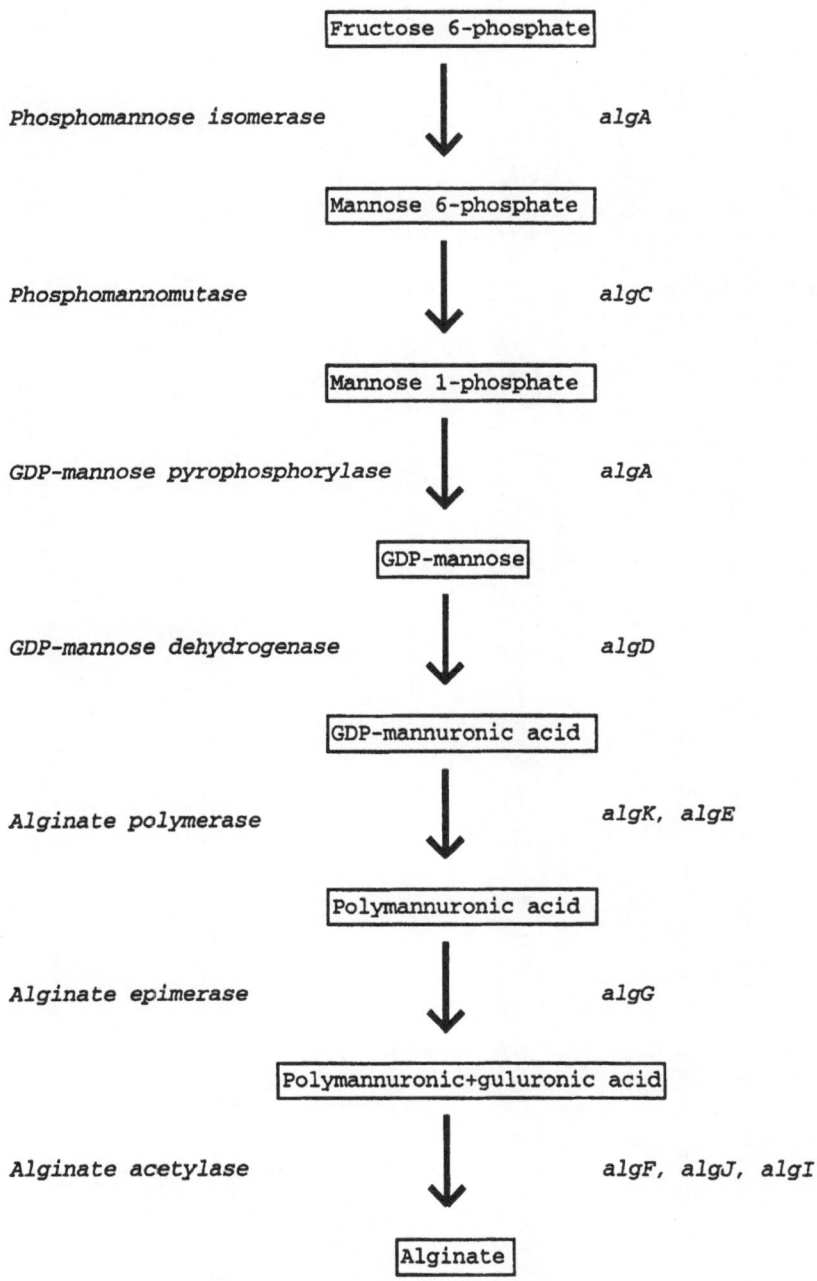

Figure 1.2 Biosynthesis of *P. aeruginosa* alginate. The pathway for alginate biosynthesis and the genes encoding the enzymes for each step are shown.

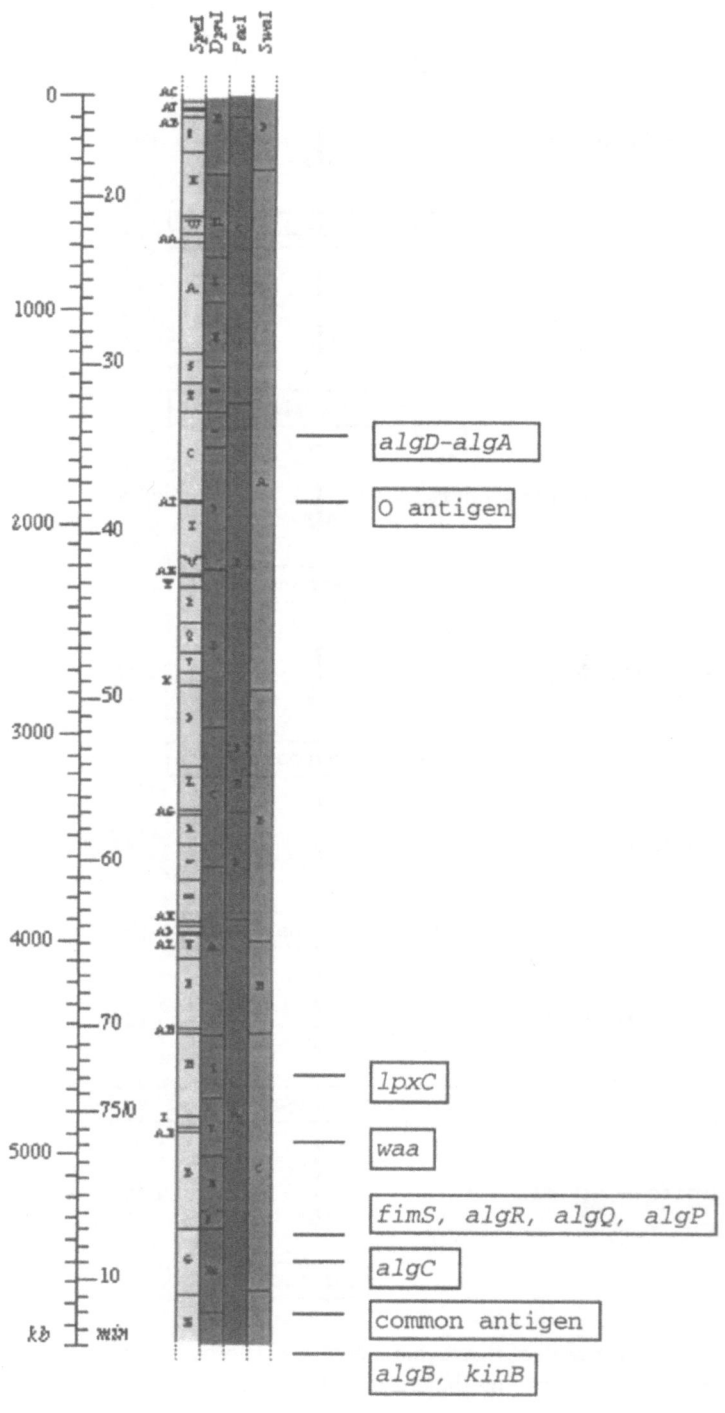

Figure 1.3 Physical map of *P. aeruginosa* strain PAO1 showing genes involved in the production of polysaccharides. Physical map of the *P. aeruginosa* chromosome is derived from a combination of genetic linkage and pulsed-field gel electrophoresis data.[87]

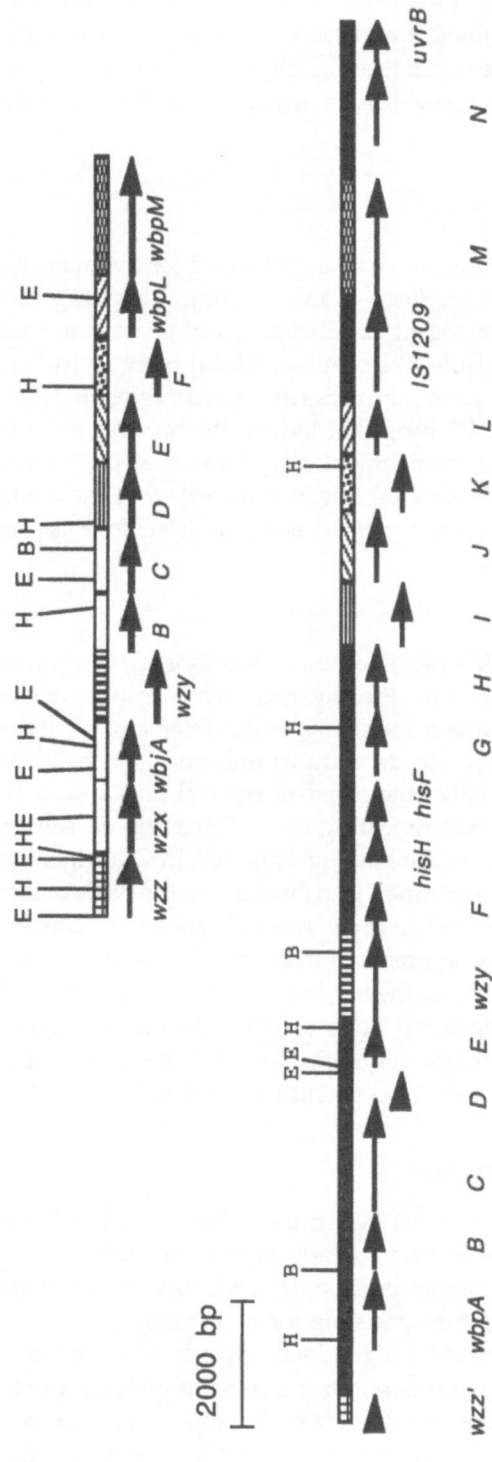

Figure 1.4 Comparison of O antigen gene loci from *P. aeruginosa* serogroup O11 strain PA103 and serogroup O5 strain PAO1. Accession number U44089 corresponds to the locus from PA103 (top); the unique biosynthetic genes in this locus are designated *wbjA-F*. Accession number U50396 corresponds to PAO1 (bottom), where the O antigen genes are referred to as *wbpA-N*. The genes for the common functions of chain length determination, O antigen transport, and O antigen polymerase are referred to by their assigned names *wzz*, *wzx*, and *wzy*, respectively (see the Introduction of this book and Reference 37). ORFs are represented by arrows, with similar inferred amino acid sequences shown by similar fill patterns. Both loci are shown on the same scale. Restriction endonuclease sites for *Bam*HI, B; *Eco*RI, E; and *Hind* III, H are shown.

Additional genes and gene products that are not specific to alginate production but that influence its expression have also been identified: CRP,[17] IHF,[18] glpM,[19] ndk,[20] algH,[21] fumC,[22] and sodA.[22] Finding that these genes and products play a role in the regulation of alginate synthesis suggests that production of this polysaccharide is integral to the physiology of *P. aeruginosa.*

Role of alginate

A number of potential functions have been proposed for the overproduction of alginate by strains of *P. aeruginosa* in the CF lung, including adherence, resistance to antibiotics, scavenging of nutrients, and protection from dehydration. In addition, the capsule-like coat can inhibit phagocytosis and possibly suppress an effective immune response to *P. aeruginosa.* The relevant biological conditions of the CF lung that induce the mucoid phenotype are still unknown; however, the involvement of a stress-related alternate sigma factor (AlgT) in control of alginate production suggests that the environment in the CF lung may be harsh for *P. aeruginosa* and may select for this phenotype.

O antigen

The O-antigen polysaccharide (also known as the B-band) portion of LPS is the immunodominant antigen on *P. aeruginosa* and is responsible for serogroup specificity. The O-antigen repeating units differ among *P. aeruginosa* serogroups in their monosaccharide composition and structural organization. O side chains are usually composed of neutral and amino sugars in trisaccharide or tetrasaccharide repeating units. Currently, 20 serogroups of *P. aeruginosa* are recognized on the basis of variation in O-antigen structure. O-antigen serogroups can be grouped into families on the basis of serological cross-reactivity and further subdivided into subtypes that contain subtle structural variations.[23] It is interesting that only 10 of the 20 identified O-antigen serogroups account for the majority of *P. aeruginosa* strains isolated from infections.[24] The O repeating units extend into the environment and confer upon LPS-smooth strains the ability to resist the bactericidal effects of normal human serum *in vitro* at concentrations of ≥10%.

Genes for O-antigen synthesis

Much of our understanding of the genetics of LPS O-antigen biosynthesis derives from studies of the enteric bacteria *E. coli* and *Salmonella typhimurium*.[25-27] Studies on the O-antigen portion of LPS in these enteric organisms have defined a genetic region responsible for its synthesis. In *E. coli* and *S. typhimurium*, the genes of the O antigen locus encode enzymes for the synthesis of nucleotide sugar precursors, for the transfer of sugars to build the O units, and to carry out the specific assembly steps to convert a single O antigen unit onto the LPS. These loci are usually long, contiguous regions of DNA; typically, 20 to 30 kb of DNA are required to encode the O-antigen biosynthesis genes. The functions of the O antigen genes from strains of various serogroups have been deduced on the basis of similarities in nucleotide sequence, in the inferred amino acid sequence, in the organization of genes, and in the structures of the O antigens encoded by these genes.[28]

To isolate the genes required for *P. aeruginosa* LPS production, the LPS-rough phenotype of isolates from lung infections in CF patients was exploited.[29] These LPS-rough strains, which arise from LPS-smooth isolates found in the environment,[30] are sensitive to normal human serum.[4] The serum-sensitive phenotype of three CF isolates of *P. aeruginosa* was complemented with DNA from a cosmid gene bank from the LPS smooth serogroup O11 *P. aeruginosa* strain PA103, on a plasmid called pLPS2.[29] Serogroup O11 has a relatively simple O-antigen structure consisting of the sugars *N*-acetyl-L-fucosamine, *N*-acetyl-D-fucosamine, and D-glucose.[31] It is the serogroup most commonly found among strains isolated from the environment and from clinical infections.[32]

The recombinant plasmid pLPS2 led to the expression of a second serogroup antigen in two of the three CF isolates tested. This expression represents the restoration of the recipient strain's capacity to synthesize its original O side chain, an ability that presumably is repressed or undergoes mutation during infection of CF patients.[29] Subcloning has revealed that different regions of pLPS2 overcome the mutations in these various CF isolates.[33] Sequence analysis of these genes suggests that they encode enzymes involved in the biosynthesis of LPS sugars rather than regulatory proteins. pLPS2 contains all the genes necessary for synthesis of the serogroup O11 antigen, as evidenced by its ability to elicit expression of this antigen by *E. coli* and *Salmonella*.[29,34]

Nucleotide sequence analysis of pLPS2 has revealed 11 open reading frames (ORFs) with similarities to genes involved in polysaccharide synthesis (Figure 1.4). Use of pLPS2 DNA as a probe to Southern blots of pulsed-field gels revealed that the O antigen region maps at 39 to 40 min on the physical linkage map of the *P. aeruginosa* chromosome (Figure 1.3).[34a]

Lightfoot and Lam isolated and mapped a similar O antigen gene locus from the common laboratory strain PAO1 (serogroup O5).[35] Plasmid pFV100, obtained by complementation of an LPS-defective transposon-insertion mutant of PAO1, induced expression of *P. aeruginosa* serogroup O5 LPS upon transformation into *E. coli* strain HB101.[36] This plasmid also complemented the core-plus-one O-repeat unit phenotype in the PAO1-derived semirough mutant AK1401, indicating that the *wzy* gene, previously called *rfc* (see the Introduction of this book and Reference 37), encoding O-antigen polymerase was also contained on pFV100.[38]

We also identified *wzy* from strain PAO1 by complementing the serum-sensitive phenotype of AK1401.[39] The *wzy* gene sequence has an aberrantly low G+C content (44.6%), which is particularly apparent against the high-G+C background of *P. aeruginosa* DNA (65.2 to 67.2%).[40,41] The discordantly high A+T content of O antigen genes, and in particular of the *wzy* gene, has been cited as evidence of external acquisition for these genes.[28] While the *P. aeruginosa* PAO1 *wzy* gene has similarities to other presumed *wzy* genes, including rare codon usage and a highly hydrophobic inferred amino acid sequence, the O antigen polymerase function can only be determined by making specific nonpolar insertions and observing the diagnostic "core-plus-one" phenotype. O-antigen-polymerase genes from differing organisms cannot complement one another, and no *wzy* genes isolated from serogroups belonging to different families show any obvious similarity to each other.[42,43] Thus, it is not surprising that the *wzy* gene from *P. aeruginosa* serogroup O11 has no homology to the *wzy* gene of serogroup O5.[34a]

Dasgupta and Lam identified a second LPS gene, *rfbA*, now referred to as *wbpL*, on pFV100.[36] Recently, Rocchetta et al. noted that this gene can partially complement the *wecA* (previously called *rfe*) mutation in *E. coli*.[44] WecA catalyzes the transfer of *N*-acetylglucosamine to a lipid carrier undecaprenol phosphate. A comparable gene was found in the O antigen gene cluster of *P. aeruginosa* serogroup O11.[34a]

Burrows et al.[45] reported the complete nucleotide sequence of pFV100, recognized 16 ORFs (Figure 1.4) thought to be involved in O5-antigen biosynthesis, and have proposed a tentative pathway for the synthesis of this polysaccharide.[44-46] A comparison of the ORFs encoding O antigen genes from *P. aeruginosa* serogroups O11 and O5 is shown in Figure 1.4.

Role of the O antigen

The long O side chains on *P. aeruginosa* LPS help the organism avoid complement-dependent host defense mechanisms. The *P. aeruginosa* LPS O side chains are also the immunodominant epitopes, and antibodies to these variant structures are protective.[47,48] In contrast to that of many other Gram-negative organisms, the LPS core of *P. aeruginosa* is relatively exposed in LPS-smooth strains: less than 20% of the LPS-core oligosaccharides contain attached O antigenic side chains.[49,50] Thus, much of the LPS of *P. aeruginosa* may be considered similar to the lipooligosaccharide (LOS) of oral–pharyngeal pathogens such as *Neisseria* spp., *Haemophilus influenzae*, and *Bordetella pertussis*, which completely lack O antigen;[51] this LOS-like structure may give *P. aeruginosa* a selective advantage.

LPS core

The LPS core of *P. aeruginosa* has a structure similar to that found in other Gram-negative bacterial LPSs. It is composed of ketodeoxyoctonate (KDO), heptose, hexosamine, and hexoses. The structure of the LPS core has been determined for four strains of *P. aeruginosa* from different serogroups, and all these structures are distinct.[52]

de Kievit and Lam isolated monoclonal antibodies specific for the outer-core region of 9 of 20 strains (each representing a different serogroup) as well as monoclonal antibodies that reacted with the inner-core region of each of the 20 strains.[53] The reactivity of these monoclonal antibodies suggests that more structural diversity exists in the outer-core region of *P. aeruginosa* LPS than in the inner-core region. Other immunological studies indicate that the heterogeneity of the LPS core region co-varies with the O side chain-defined serogroup structures.[54-56] However, it is not known whether all strains of a particular serogroup have the same LPS core structure or whether additional variability exists.

LPS core genes

A cluster of genes distinct from those encoding the O antigen, known as the *waa* locus (previously known as the *rfa* locus; see the Introduction of this book and Reference 37), is responsible for LPS core synthesis. In order to characterize the *waa* genes of *P. aeruginosa*, a strategy was used based on the structural similarity of the *P. aeruginosa*[2] LPS core to that of *S. typhimurium*.[26] *S. typhimurium waaC* mutant SA1377 is defective in heptosyltransferase I, which is required for the transfer of the first heptose onto the KDO in the LPS core; as a consequence, SA1377 is sensitive to hydrophobic antibiotics such as novobiocin. Novobiocin-resistant transformants were selected after transfer of a gene bank of *P. aeruginosa* DNA from the serogroup O6 strain PAK. Sequence analysis of the insert DNA showed that the recombinant plasmid contained not only the *waaC* gene but also the *waaF* gene (encoding heptosyltransferase II, for the addition of the second linked heptose onto the LPS backbone of the inner core) upstream of *waaC*. Downstream of *waaC* were *waaG* (required for the transfer of the first hexose residue onto the heptose in the LPS core) and *waaP* (which modifies the LPS core, perhaps by the addition of phosphates). The four *waa* genes were homologous (>50 to 53% identical) to the corresponding genes from *E. coli* and *S. typhimurium*, which indicated that, as in *E. coli* and *S. typhimurium*, the *P. aeruginosa waa* genes were clustered on the chromosome. Downstream of *waaP* was another gene with homology to *waaP*. The translational start site for each gene over-lapped with the stop site of the upstream gene, an indication that these genes are likely part of a single operon. As with other *P. aeruginosa* genes, the G+C content of the cloned DNA was approximately 68%. Southern blot hybridization using DNA from this region as a probe revealed that strains of various serogroups contain homologous DNA on similar-sized restriction fragments, an observation suggesting conservation of this region across serogroups.[56a] de Kievit and Lam identified the *P. aeruginosa waaF* and *waaC* genes from the serogroup O5 strain PAO1 using a similar approach and mapped these genes to 0.9 to 6.6 min on the *P. aeruginosa* chromosome[57] (Figure 1.3).

Role for the LPS core

P. aeruginosa reportedly is ingested by corneal cells during experimental eye infection.[58,59] This ingestion is mediated by the outer-core portion of the LPS.[60] Bacterial entry into corneal cells may contribute to eye pathology and may partially explain why this infection is often difficult to treat and why antibiotic therapy is not always effective.

In contrast to the wounded eye, where the epithelial cells ingesting *P. aeruginosa* are buried within the cornea, a mucosal surface may clear bacteria by shedding epithelial cells that have ingested bacteria. This hypothesis comes from studies with derivatives of a transformed airway epithelial cell line originating from a CF patient who was homozygous for the most common mutation in the CF transmembrane conductance regulator (CFTR) gene, ΔF508. These cells were deficient in uptake of *P. aeruginosa* compared with the same cell line expressing wild-type CFTR following transfection; *P. aeruginosa* was the only respiratory bacterial pathogen ingested by airway epithelial cells with wild-type CFTR. Thus, the deficiency in *P. aeruginosa* uptake by the epithelial cells in the airways of CF patients may underlie the hypersusceptibility of these individuals to *P. aeruginosa* infection. In these studies, the *P. aeruginosa* LPS core was found to be the bacterial ligand for this ingestion.[61] Subsequent experiments have identified CFTR as the epithelial cell receptor for *P. aeruginosa*.[62]

Common antigen

P. aeruginosa LPS contains an additional polysaccharide antigen (also known as A band) common to many serogroups. This antigen is a polymer composed principally of D-rhamnose. There is some disagreement as to whether the common antigen is attached to the same LPS core as the O antigen or to a distinct core and lipid A component.[63] The differential migration of D-rhamnan-containing LPS and O antigen-containing LPS upon gel filtration chromatography has suggested to Rivera et al. the presence of two distinct LPSs.[64,65] However, using delipidated O side chains that would not aggregate, and monoclonal and polyclonal antibodies specific for either the O antigens or D-rhamnan antigen, Hatano et al. showed that these two structures could coprecipitate. This finding indicates that *P. aeruginosa* synthesizes only a single lipid A/core substituted by O antigens and/or the D-rhamnan polymer.[66]

Common antigen genes

Lightfoot and Lam identified a clone containing common antigen genes on the basis of its ability to complement a common antigen-deficient mutant of PAO1. They mapped this region to 11 to 13 min on the *P. aeruginosa* chromosome in a locus distinct from the O antigen gene cluster[67] (Figure 1.3). They identified a gene *gmd* (previously referred to as *gca*), whose product converts

GDP-mannose to GDP-rhamnose.[35] Southern blot analysis and polymerase chain reaction (PCR) amplification revealed that *gmd* is conserved across all serogroups of *P. aeruginosa* and within some other *Pseudomonas* species.[68] Further analysis revealed a total of 8 genes involved in common antigen synthesis: two genes (*wzm* and *wzt*) with inferred amino acid sequences similar to an ATP-binding cassette transport system and three other genes (*wbpX, wbpY,* and *wbpZ*) with inferred amino acid similarity to rhamnosyl-transferases.[44,69] It is interesting that these authors showed that one of the genes in the O antigen cluster, *wbpL,* is required for the initiation of common antigen synthesis as well as O antigen synthesis.[44] Thus, this gene may link both of these antigens to the LPS core.

Role of the common antigen

The function of the common antigen in *P. aeruginosa* is unknown; however, it may contribute to the surface hydrophobicity of this bacterium.[70] Unlike the O antigen, the expression of the common antigen does not promote resistance to normal human serum,[71] and its expression remains unchanged in serial isolates from the lungs of CF patients[72] as well as under various growth conditions.[73,74] This invariable expression during all stages of infection may suggest that antibodies to this antigen would be protective against infection; however, either long O-side-chain antigens (found on acute and environmental isolates) or alginate (found on chronic isolates from the CF lung) can block the efficacy of such antibodies.[75]

Lipid A

The lipid A (endotoxin) portion of *P. aeruginosa* LPS differs from lipid A from *E. coli* and *S. typhimurium* in that the lipid substituents acylating the glucosamine disaccharide backbone from *P. aeruginosa* are shorter, and the phosphate groups are attached in a different manner.[76] These differences in structure result in *P. aeruginosa* lipid A being significantly less toxic than that of *E. coli* or *S. typhimurium*. In general, the interaction of lipid A with macrophages results in cellular activation and the release of cytokines, chemokines, and inflammatory mediators such as tumor necrosis factor α and interleukin-1β. In most cases, activation leads to an innate immune response to eliminate the Gram-negative bacteria. However, dysregulation of this response can result in Gram-negative septic shock.

Genes for lipid A synthesis

The *P. aeruginosa* gene *lpxA* encoding UDP-*N*-acetylglucosamine-3-*O*-acyltransferase has been cloned and expressed.[77] This enzyme catalyzes the first step in lipid A biosynthesis. The gene *lpxC* encoding UDP-3-*O*-acyl-glucosamine deacetylase has also been cloned, expressed,[78] and mapped to the *P. aeruginosa* chromosome to about 70 to 74 min[79] (Figure 1.3). This enzyme is the second in the pathway and catalyzes the first irreversible reaction, the likely site of regulation, in the synthesis of lipid A.[80]

Role of the lipid A

Lipid A is apparently critical for the integrity of the bacterial cell surface: in *E. coli* only conditional mutants defective in lipid A synthesis genes have been isolated. The essential role of lipid A for Gram-negative bacteria makes the enzymes required for the synthesis of lipid A attractive targets for the development of antimicrobial agents.[81] Some of these inhibitors are active against *P. aeruginosa* enzymes, but have no effect against *P. aeruginosa* cells, which suggests that these particular inhibitors cannot penetrate into or are extruded from these bacteria.[81] It is likely that additional lipid A synthesis inhibitors that do not have these problems will be developed.

Link between alginate and LPS synthesis

Studies of the *algC* gene have linked the synthesis of the polysaccharide alginate and that of LPS. This gene restored complete LPS synthesis to an LPS mutant of strain PAO1, which is O side chain- and common antigen-deficient because of a lack of glucose residues on the LPS core.[82,83] The *algC* gene, previously identified in the pathway for alginate biosynthesis, encodes phosphomannomutase, which interconverts mannose 6-phosphate and mannose 1-phosphate.[84] The *algC* gene product also has phosphoglucomutase activity that interconverts glucose 6-phosphate and glucose 1-phosphate; this activity was confirmed by successful complementation of an *E. coli pgm* mutation with the *P. aeruginosa algC* gene.[83] Further studies have shown that phosphoglucomutase activity is required for the synthesis of a complete LPS and phosphomannomutase activity for the biosynthesis of alginate. Ye et al.[85] have confirmed the bifunctional nature of the purified *P. aeruginosa algC* gene product and its involvement in the synthesis of both alginate and LPS.

Role for the LPS-rough, mucoid phenotype associated with CF isolates

Strains of *P. aeruginosa* isolated from patients with CF typically have an LPS-rough phenotype that is rarely observed in isolates from other sources. When tested in animal models of acute infection, the LPS-rough isolates from the lungs of CF patients are less virulent than LPS-smooth strains.[86] This finding suggests that in the environment of the CF lung, long LPS O side chains — and thus serum resistance — are not crucial for the virulence of these bacteria.

The various rationales proposed for the appearance of *P. aeruginosa* strains with an LPS-rough phenotype in the CF lung environment have included resistance to host defense and environmental regulation. Results with CFTR airway epithelial cells indicate that the emergence of strains expressing an incomplete LPS during chronic infection of CF patients may enhance the ability of these bacteria to avoid epithelial cell uptake and therefore may contribute to bacterial survival in the mucous secretions in the bronchial lumen.[61] Production of alginate by *P. aeruginosa* and entanglement of bacteria in the mucus of CF patients further protect these bacteria from other host defenses, such as phagocytosis by polymorphonuclear leukocytes and by alveolar macrophages. Thus, the loss of the LPS O side chain (resulting in a rough LPS) and the overproduction of alginate (resulting in mucoidy) may be adaptive changes necessary for the establishment of chronic infection in the lung and may provide *P. aeruginosa* with a means of escaping host immune responses.

Conclusions

P. aeruginosa polysaccharides play an important role in the pathogenesis of this organism. Many genes involved in the biosynthesis and regulation of alginate have been identified; some of them have additional functions in central metabolism. Genetic studies of the production of LPS have thus far revealed only biosynthetic genes, which suggests that the production of LPS is constitutive and less responsive to environmental or genetic regulation. The *P. aeruginosa algC* gene represents an overlap in the synthesis of these polysaccharides. Further identification of genes involved in polysaccharide production should emerge with the completion of the genomic sequence of *P. aeruginosa* strain PAO1.

Acknowledgments

This work was supported by grants from the National Institutes of Health (AI30050, AI35674, and AI37632) and from the Cystic Fibrosis Foundation (G937 and GOLDBE96PO).

References

1. Pollack, M. *Pseudomonas aeruginosa*, in G. L. Mandell, R. G. Douglas, Jr., and J. E. Bennett (Eds.), *Principles and Practice of Infectious Diseases*, pp. 1980-2003. New York: Churchill Livingstone, 1995.
2. Knirel, Y. A. Polysaccharide antigens of *Pseudomonas aeruginosa*, *Microbiology*, 17, 273-304, 1990.
3. Penketh, A., Pitt, T., Roberts, D., Hodson, M. E., and Batten, J. C. The relationship of phenotype changes in *Pseudomonas aeruginosa* to the clinical condition of patients with cystic fibrosis, *American Review of Respiratory Disease*, 172, 605-608, 1983.
4. Hancock, R. E. W., Mutharia, L. M., Chan, L., Darveau, R. P., Speert, D. P., and Pier, G. B. *Pseudomonas aeruginosa* isolates from patients with cystic fibrosis: a class of serum-sensitive, nontypable strains deficient in lipopolysaccharide O side chains, *Infection and Immunity*, 42, 170-177, 1983.
5. Ojeniyi, B., Baek, L., and Hoiby, N. Polyagglutinability due to loss of O-antigenic determinants in *Pseudomonas aeruginosa* strains isolated from cystic fibrosis patients, *Acta Pathol. Microbiol. Scand. Sect. B*, 93, 7-13, 1985.

6. Evans, L. R. and Linker, A. Production and characterization of the slime polysaccharide of *Pseudomonas aeruginosa, Journal of Bacteriology,* 116, 915-924, 1973.

7. Pedersen, S. S., Kharazmi, A., Espersen, F., and Hoiby, N. *Pseudomonas aeruginosa* alginate in cystic fibrosis sputum and the inflammatory response, *Infection and Immunity,* 58, 3363-3368, 1990.

8. May, T. B. and Chakrabarty, A. M. *Pseudomonas aeruginosa*: genes and enzymes of alginate synthesis, *Trends in Microbiology,* 2, 151-156, 1994.

9. Gacesa, P. Bacterial alginate biosynthesis—recent progress and future prospects, *Microbiology,* 144, 1133-1143, 1998.

10. Martin, D. W., Schurr, M. H., Mudd, M. H., Govan, J. R. W., Holloway, B. W., and Deretic, V. Mechanism of conversion to mucoidy in *Pseudomonas aeruginosa* infecting cystic fibrosis patients, *Proceedings of the National Academy of Sciences, U.S.A.,* 90, 8377-8381, 1993.

11. Flynn, J. L. and Ohman, D. E. Use of a gene replacement cosmid vector for cloning alginate conversion genes from mucoid and nonmucoid *Pseudomonas aeruginosa* strains: *algS* controls expression of *algT, Journal of Bacteriology,* 170, 3228-3236, 1988.

12. Govan, J. R. W. and Deretic, V. Microbial pathogenesis in cystic fibrosis: mucoid *Pseudomonas aeruginosa* and *Burkholderia cepacia, Microbiological Reviews,* 60, 539-574, 1996.

13. Whitchurch, C. B., Alm, R. A., and Mattick, J. S. The alginate regulator AlgR and an associated sensor FimS are required for twitching motility in *Pseudomonas aeruginosa, Proceedings of the National Academy of Sciences, U.S.A.,* 93, 9839-9843, 1996.

14. Yu, H., Mudd, M., Boucher, J. C., Schurr, M. J., and Deretic, V. Identification of the *algZ* gene upstream of the response regulator *algR* and its participation in control of alginate production in *Pseudomonas aeruginosa, Journal of Bacteriology,* 179, 187-193, 1997.

15. Ma, S., Selvaraj, U., Ohman, D. E., Quarless, R., Hassett, D. J., and Wozniak, D. J. Phosphorylation-independent activity of the response regulators AlgB and AlgR in promoting alginate biosynthesis in mucoid *Pseudomonas aeruginosa, Journal of Bacteriology,* 180, 956-968, 1998.

16. Baynham, P. J. and Wozniak, D. J. Identification and characterization of AlgZ, an AlgT-dependent DNA-binding protein required for *Pseudomonas aeruginosa algD* transcription, *Molecular Microbiology,* 22, 97-108, 1996.

17. DeVault, J. D., Hendrickson, W., Kato, J., and Chakrabarty, A. M. Environmentally regulated *algD* promoter is responsive to the cAMP receptor protein in *Escherichia coli, Molecular Microbiology,* 5, 2503-2509, 1991.

18. Wozniak, D. J. Integration host factor and sequences downstream of the *Pseudomonas aeruginosa algD* transcription start site are required for expression, *Journal of Bacteriology,* 176, 5068-5076, 1994.

19. Schweizer, H. P., Po, C., and Bacic, M. K. Identification of *Pseudomonas aeruginosa glpM,* whose gene product is required for efficient alginate biosynthesis from various carbon sources, *Journal of Bacteriology,* 177, 4801-4804, 1995.

20. Sundin, G. W., Shankar, S., and Chakrabarty, A. M. Mutational analysis of nucleoside diphosphate kinase from *Pseudomonas aeruginosa*: characterization of critical amino acid residues involved in exopolysaccharide alginate synthesis, *Journal of Bacteriology,* 178, 7120-7128, 1996.

21. Schlictman, D., Kubo, M., Shankar, S., and Chakrabarty, A. M. Regulation of nucleoside diphosphate kinase and secretable virulence factors in *Pseudomonas aeruginosa*: roles of *algR2* and *algH*, *Journal of Bacteriology,* 177, 2469-2474, 1995.

22. Hassett, D. J., Howell, M. L., Sokol, P. A., Vasil, M. L., and Dean, G. E. Fumarase C activity is elevated in response to iron deprivation and in mucoid, alginate-producing *Pseudomonas aeruginosa*: cloning and characterization of *fumC* and purification of native FumC, *Journal of Bacteriology,* 179, 1442-1451, 1997.

23. Knirel, Y. A. and Kochetkov, N. K. The structure of lipopolysaccharides of gram-negative bacteria. III. The structures of O-antigens: a review, *Biochemistry (Moscow)*, 59, 1325-1383, 1994.

24. Pier, G. B. and Thomas, D. M. Lipopolysaccharide and high molecular weight polysaccharide serotypes of *Pseudomonas aeruginosa, Journal of Infectious Diseases*, 148, 217-223, 1982.

25. Mäkelä, P. H. and Stocker, B. A. D. Genetics of lipopolysaccharide, in E. T. Rietschel (Ed.) *Handbook of Endotoxin, Vol. 1: Chemistry of Endotoxin*, pp. 59-137. New York: Elsevier Science Publishers, 1984.

26. Schnaitman, C. A. and Klena, J. D. Genetics of lipopolysaccharide biosynthesis in enteric bacteria, *Microbiological Reviews*, 57, 655-682, 1993.

27. Whitfield, C. Biosynthesis of lipopolysaccharide O antigens, *Trends in Microbiology,* 3, 178-185, 1995.

28. Reeves, P. Evolution of *Salmonella* O antigen variation by interspecific gene transfer on a large scale, *Trends in Genetics*, 9, 17-22, 1993.

29. Goldberg, J. B., Hatano, K., Small-Meluleni, G., and Pier, G. B. Cloning and surface expression of *Pseudomonas aeruginosa* O antigen in *Escherichia coli*, *Proceedings of the National Academy of Sciences, U.S.A.*, 89, 10716-10720, 1992.

30. Mahenthiralingam, E., Campbell, M. E., and Speert, D. P. Nonmotility and phagocytic resistance of *Pseudomonas aeruginosa* isolates from chronically colonized patients with cystic fibrosis, *Infection and Immunity,* 62, 596-605, 1994.

31. Knirel, Y. A., Vinogradov, E. V., Kocharova, N. A., Paramonov, N. A., Kochetkov, N. K., Dmitriev, B. A., Stanislavsky, E. S., and Lanyi, B. The structure of O-specific polysaccharides and serological classification of *Pseudomonas aeruginosa*, *Acta Microbiologica Hungarica*, 35, 3-24, 1988.

32. Farmer 3rd, J. J., Weinstein, R. A., Zierdt, C. H., and Brokopp, C. D. Hospital outbreaks caused by *Pseudomonas aeruginosa*: importance of serogroup O11, *Journal of Clinical Microbiology,* 16, 266-270, 1982.

33. Evans, D. J., Pier, G. B., Coyne Jr., M. J., and Goldberg, J. B. The *rfb* locus from *Pseudomonas aeruginosa* strain PA103 promotes the expression of O antigen by both LPS-rough and LPS-smooth isolates from cystic fibrosis patients, *Molecular Microbiology,* 13, 427-434, 1994.

34. Pier, G. B., Meluleni, G., and Goldberg, J. B. Clearance of *Pseudomonas aeruginosa* from the murine gastrointestinal tract is effectively mediated by O-antigen-specific circulating antibodies, *Infection and Immunity,* 63, 2818-2825, 1995.

34a Dean et al. submitted for publication.

35. Lightfoot, J. and Lam, J. S. Chromosomal mapping, expression and synthesis of lipopolysaccharide in *Pseudomonas aeruginosa*: a role for guanosine diphospho (GDP)-D-mannose, *Molecular Microbiology,* 8, 771-782, 1993.

36. Dasgupta, T. and Lam, J. S. Identification of *rfbA*, involved in B-band lipopolysaccharide biosynthesis in *Pseudomonas aeruginosa* serotype O5, *Infection and Immunity,* 63, 1674-1680, 1995.

37. Reeves, P. R., Hobbs, M., Valvano, M. A., Skurnik, M., Whitfield, C., Coplin, D., Kido, N., Klena, J., Maskell, D., Raetz, C. R. H., and Rick, P. D. Bacterial polysaccharide synthesis and gene nomenclature, *Trends in Microbiology,* 4, 495-502, 1996.

38. de Kievit, T. R., Dasgupta, T., Schweizer, H., and Lam, J. S. Molecular cloning and characterization of the *rfc* gene of *Pseudomonas aeruginosa* (serotype O5), *Molecular Microbiology,* 16, 565-574, 1995.

39. Coyne Jr., M. J. and Goldberg, J. B. Cloning and characterization of the gene (*rfc*) encoding O-antigen polymerase of *Pseudomonas aeruginosa* PAO1, *Gene,* 167, 81-86, 1995.

40. West, S. E. and Iglewski, B. H. Codon usage in *Pseudomonas aeruginosa, Nucleic Acid Research,* 16, 9323-9335, 1988.

41. Palleroni, N. J. Family I. Pseudomonadaceae, in N. R. Kreig and J. G. Holt (Eds.), *Bergey's Manual of Systematic Bacteriology,* Vol. 1, pp. 141. Baltimore, MD.: Williams & Wilkins, 1984.

42. Collins, V. L. and Hackett, J. Molecular cloning, characterization, and nucleotide sequence of the *rfc* gene, which encodes an O-antigen polymerase of *Salmonella typhimurium, Journal of Bacteriology,* 173, 2521-2529, 1991.

43. Morona, R., Mavris, M., Fallarino, A., and Manning, P. A. Characterization of the *rfc* region of *Shigella flexneri, Journal of Bacteriology,* 176, 733-747, 1994.

44. Rocchetta, H. L., Burrows, L. L., Pacan, J. C., and Lam, J. S. Three rhamnosyltransferases responsible for assembly of the A-band D-rhamnan polysaccharide in *Pseudomonas aeruginosa*: a fourth transferase, WbpL, is required for the initiation of both A-band and B-band lipopolysaccharide synthesis, *Molecular Microbiology,* 28, 1103-1119, 1998.

45. Burrows, L. L., Charter, D. F., and Lam, J. S. Molecular characterization of the *Pseudomonas aeruginosa* serotype O5 (PAO1) B-band lipopolysaccharide gene cluster, *Molecular Microbiology,* 22, 481-495, 1996.

46. Burrows, L. L., Chow, D., and Lam, J. S. *Pseudomonas aeruginosa* B-band O-antigen chain length is modulated by Wzz (Rol), *Journal of Bacteriology,* 179, 1482-1489, 1997.

47. Pollack, M. and Young, L. S. Protective activity of antibodies to exotoxin A and lipopolysaccharide at the onset of *Pseudomonas aeruginosa* septicemia in man, *Journal of Clinical Investigation,* 63, 276-286, 1979.

48. Fomsgaard, A., Dinesen, B., Shand, G. H., Pressler, T., and Hoiby, N. Antilipopolysaccharide antibodies and differential diagnosis of chronic *Pseudomonas aeruginosa* lung infection in cystic fibrosis, *Journal of Clinical Microbiology,* 27, 1222-1229, 1989.

49. Wilkinson, S. G. and Galbraith, L. Studies of lipopolysaccharides from *Pseudomonas aeruginosa, European Journal of Biochemistry,* 52, 331-343, 1975.

50. Darveau, R. P. and Hancock, R. E. W. Procedure for isolation of bacterial lipopolysaccharides from both smooth and rough *Pseudomonas aeruginosa* and *Salmonella typhimurium* strains, *Journal of Bacteriology,* 155, 831-838, 1983.

51. Proctor, R. A., Denlinger, L. C., and Bertics, P. J. Lipopolysaccharide and bacterial virulence, in J. A. Roth, C. A. Bolin, K. A. Brogden, F. C. Minion, and M. J. Wannemuehler (Eds.), *Virulence Mechanisms of Bacterial Pathogens,* Second edition, pp. 173-194. Washington, D.C.: ASM Press, 1995.

52. Masoud, H., Sadovskaya, I., de Kievit, T., Altman, E., Richards, J. C., and Lam, J. S. Structural elucidation of the lipopolysaccharide core region of the O-chain deficient mutant strain A28 from *Pseudomonas aeruginosa* serotype O6 (International Antigenic Typing Scheme), *Journal of Bacteriology*, 177, 6718-6726, 1995.

53. de Kievit, T. R. and Lam, J. S. Monoclonal antibodies that distinguish inner core, outer core, and lipid A regions of *Pseudomonas aeruginosa* lipopolysaccharide, *Journal of Bacteriology*, 176, 7129-7139, 1994.

54. Yokota, S., Terashima, M., Chiba, J., and Noguchi, H. Variable cross-reactivity of *Pseudomonas aeruginosa* lipopolysaccharide-core-specific monoclonal antibodies and its possible relationship with serotype, *Journal of General Microbiology*, 138, 289-296, 1992.

55. Terashima, M., Unezumi, I., Tomio, T., Kato, M., Irie, K., Okuda, T., Yokota, S.-I., and Noguchi, H. A protective human monoclonal antibody directed to the outer core region of *Pseudomonas aeruginosa* lipopolysaccharide, *Infection and Immunity*, 59, 1-6, 1991.

56. Nelson, J. W., Barclay, G. R., Micklem, L. R., Poxton, I. R., and Govan, J. R. W. Production and characterisation of mouse monoclonal antibodies reactive with the lipopolysaccharide core of *Pseudomonas aeruginosa*, *Journal of Medical Microbiology*, 36, 358-365, 1992.

56a. Franklund et al. manuscript in preparation.

57. de Kievit, T. R. and Lam, J. S. Isolation and characterization of two genes, *waaC* (*rfaC*) and *waaF* (*rfaF*), involved in *Pseudomonas aeruginosa* serotype O5 inner-core biosynthesis, *Journal of Bacteriology*, 179, 3451-3457, 1997.

58. Fleiszig, S. M. J., Zaidi, T. S., Fletcher, E. L., Preston, M. J., and Pier, G. B. *Pseudomonas aeruginosa* invades corneal epithelial cells during experimental infection, *Infection and Immunity*, 62, 3485-3493, 1994.

59. Fleiszig, S. M. J., Zaidi, T. S., and Pier, G. B. *Pseudomonas aeruginosa* invasion of and multiplication within corneal epithelial cells *in vitro*, *Infection and Immunity*, 63, 4072-4077, 1995.

60. Zaidi, T. S., Fleiszig, S. M. J., Preston, M. J., Goldberg, J. B., and Pier, G. B. Lipopolysaccharide outer core is a ligand for corneal cell binding and ingestion of *Pseudomonas aeruginosa*, *Investigative Ophthalmology and Visual Sciences*, 37, 976-986, 1996.

61. Pier, G. B., Grout, M., Zaidi, T. S., Olsen, J. C., Johnson, L. G., Yankaskas, J. R., and Goldberg, J. B. Role of mutant CFTR in hypersusceptibility of cystic fibrosis to lung infections, *Science*, 271, 64-67, 1996.

62. Pier, G. B., Grout, M., and Zaidi, T. S. Cystic fibrosis transmembrane conductance regulator is an epithelial cell receptor for clearance of *Pseudomonas aeruginosa* from the lung, *Proceedings of the National Academy of Sciences, U.S.A.*, 94, 12088-12093, 1997.

63. Pier, G. B. and Goldberg, J. B. *Pseudomonas aeruginosa* A-band and B-band lipopolysaccharides: letter to the editor, *Infection and Immunity*, 63, 4964-4965, 1995.

64. Rivera, M. and McGroarty, E. J. Analysis of a common antigen lipopolysaccharide from *Pseudomonas aeruginosa*, *Journal of Bacteriology*, 171, 2244-2248, 1989.

65. Rivera, M., Bryan, L. E., Hancock, R. E. W., and McGroarty, E. J. Heterogeneity of lipopolysaccharide from *Pseudomonas aeruginosa*: analysis of lipopolysaccharide chain length, *Journal of Bacteriology*, 170, 512-521, 1988.

66. Hatano, K., Goldberg, J. B., and Pier, G. B. *Pseudomonas aeruginosa* lipopolysaccharide: evidence that the O side chains and common antigens are on the same molecule, *Journal of Bacteriology,* 175, 5117-5128, 1993.
67. Lightfoot, J. and Lam, J. S. Molecular cloning of genes involved with expression of A-band lipopolysaccharide, an antigenically conserved form, in *Pseudomonas aeruginosa, Journal of Bacteriology,* 173, 5624-5630, 1991.
68. Currie, H. L., Lightfoot, J., and Lam, J. S. Prevalence of *gca,* a gene involved in synthesis of A-band common antigen polysaccharide in *Pseudomonas aeruginosa, Clinical and Diagnostic Laboratory Immunology,* 2, 554-562, 1995.
69. Rocchetta, H. L. and Lam, J. S. Identification and functional characterization of an ABC transport system involved in polysaccharide export of A-band lipopolysaccharide in *Pseudomonas aeruginosa, Journal of Bacteriology,* 179, 4713-4724, 1997.
70. Makin, S. A. and Beveridge, T. J. The influence of A-band and B-band lipopolysaccharide on the surface characteristics and adhesion of *Pseudomonas aeruginosa* to surfaces, *Microbiology,* 142, 299-307, 1996.
71. Dasgupta, T., deKievit, T. R., Masoud, H., Altman, E., Richards, J. C., Sadovskaya, I., Speert, D. P., and Lam, J. S. Characterization of lipopolysaccharide-deficient mutants of *Pseudomonas aeruginosa* derived from serotypes O3, O5, and O6, *Infection and Immunity,* 62, 809-817, 1994.
72. Lam, M. Y. C., McGroarty, E. J., Kropinski, A. M., MacDonald, L. A., Pederson, S. S., Hoiby, N., and Lam, J. S. Occurrence of a common lipopolysaccharide antigen in standard and clinical strains of *Pseudomonas aeruginosa, Journal Clinical Microbiology,* 27, 962-967, 1989.
73. McGroarty, E. J. and Rivera, M. Growth-dependent alterations in production of serotype-specific and common antigen lipopolysaccharides in *Pseudomonas aeruginosa* PAO1, *Infection and Immunity,* 58, 1030-1037, 1990.
74. Makin, S. A. and Beveridge, T. J. *Pseudomonas aeruginosa* PAO1 ceases to express serotype-specific lipopolysaccharide at 45°C, *Journal of Bacteriology,* 178, 3350-3352, 1996.
75. Hatano, K., Goldberg, J. B., and Pier, G. B. Biologic activities of antibodies to the neutral-polysaccharide component of the *Pseudomonas aeruginosa* lipopolysaccharide are blocked by O side chains and mucoid exopolysaccharide (alginate), *Infection and Immunity,* 63, 21-26, 1995.
76. Knirel, Y. A. and Kochetkov, N. K. The structure of lipopolysaccharides of gram-negative bacteria. I. General characterization of the lipopolysaccharides and the structure of lipid A, *Biochemistry (Moscow),* 58, 73-84, 1993.
77. Dotson, G. D., Kaltashov, I. A., Cotter, R. J., and Raetz, C. R. H. Expression cloning of a *Pseudomonas* gene encoding a hydroxydecanolyl-acyl carrier protein-dependent UDP-GlcNAc acyltransferase, *Journal of Bacteriology,* 180, 330-337, 1998.
78. Hyland, S. A., Eveland, S. S., and Anderson, M. S. Cloning, expression and purification of UDP-3-O-Acyl-GlcNAc deacetylase from *Pseudomonas aeruginosa*: a metalloamidase of the lipid A biosynthesis pathway, *Journal of Bacteriology,* 179, 2029-2037, 1997.
79. Liao, X., Charlesbois, I., Ouellet, C., Morency, M.-J., Dewar, K., Lightfoot, J., Foster, J., Siehnel, R., Schweizer, B., Lam, J. S., Hancock, R. E. W., and Leverque, R. C. Physical mapping of 32 genetic markers on the *Pseudomonas aeruginosa* PAO1 chromosome, *Microbiology,* 142, 79-86, 1996.

80. Wyckoff, T. J. O., Raetz, C. R. H., and Jackman, J. E. Antibacterial and anti-inflammatory agents that target endotoxin, *Trends in Microbiology,* 6, 154-159, 1998.
81. Onishi, H. R., Pelak, B. A., Gerckens, L. S., Silver, L. L., Kahan, F. M., Chen, M.-H., Patchett, A. A., Galloway, S. M., Hyland, S. A., Anderson, M. S., and Raetz, C. R. H. Antibacterial agents that inhibit lipid A biosynthesis, *Science,* 274, 980-982, 1996.
82. Goldberg, J. B., Hatano, K., and Pier, G. B. Synthesis of lipopolysaccharide O side chains by *Pseudomonas aeruginosa* PAO1 requires the enzyme phosphomannomutase, *Journal of Bacteriology,* 175, 1605-1611, 1993.
83. Coyne Jr., M. J., Russell, K. S., Coyle, C. L., and Goldberg, J. B. The *Pseudomonas aeruginosa algC* gene encodes phosphoglucomutase, required for the synthesis of a complete lipopolysaccharide core, *Journal of Bacteriology,* 176, 3500-3507, 1994.
84. Zielinski, N., Chakrabarty, A., and Berry, A. Characterization and regulation of the *Pseudomonas aeruginosa algC* gene encoding phosphomannomutase, *Journal of Biological Chemistry,* 266, 9754-9763, 1991.
85. Ye, R. W., Zielinski, N. A., and Chakrabarty, A. M. Purification and characterization of phosphomannomutase/phosphoglucomutase from *Pseudomonas aeruginosa* involved in the biosynthesis of both alginate and lipopolysaccharide, *Journal of Bacteriology,* 176, 4851-4857, 1994.
86. Cryz, S., Pitt, T., Furer, E., and Germanier, R. Role of lipopolysaccharide in virulence of *Pseudomonas aeruginosa, Infection and Immunity,* 44, 508-513, 1984.
87. Holloway, B. W., Romling, U., and Tummler, B. Genomic mapping of *Pseudomonas aeruginosa* PAO, *Microbiology,* 140, 2907-2929, 1994.

chapter two

Molecular genetics
of Yersinia *lipopolysaccharide*

Mikael Skurnik

Contents

Foreword 23
Genus *Yersinia* 24
    Serotypes of *Y. enterocolitica* and related species 24
    Y. pseudotuberculosis serotypes 25
    Y. pestis 25
An overview of LPS biosynthesis 25
Bacteria and bacteriophages used in LPS genetic studies of *Yersinia* 28
Y. enterocolitica serotype O:3 LPS 30
    Serotype O:3 outer core 30
    Serotype O:3 O-antigen 35
Y. enterocolitica serotype O:8 LPS 38
    O-antigen gene cluster of *Y. enterocolitica* O:8 38
O-antigen gene clusters of *Y. pseudotuberculosis* 43
JUMPstart and *ops* sequences 44
The *Yersinia* O-antigen gene clusters are located between
    hemH and *gsk* 45
Yersinia LPS and virulence 46
Future prospects 46
References 47

Foreword

The studies carried out on molecular genetics of *Yersinia* lipopolysaccharide (LPS) serve two major goals: (1) to understand the biology, biochemistry, and genetics of this bacterial surface macromolecule and (2) to provide a basis for

0-8493-0021-5/99/$0.00+$.50
© 1999 by CRC Press LLC

future vaccine development, preventive treatments, and virulence experiments. Most information on molecular genetics has been achieved with *Y. enterocolitica* serotype O:3 LPS. Detailed analysis of the gene clusters directing the biosynthesis of the outer core oligosaccharide and of the O-antigen of that organism is available. The O-antigen gene clusters of *Y. enterocolitica* serotype O:8 and *Y. pseudotuberculosis* serotypes O:2a and O:5a have been cloned and partially characterized. The biosynthesis of LPS in these *Yersinia* species includes examples of the two major variations recognized in the biosynthesis of this macromolecule: (1) the homopolymeric, or O-antigen polymerase independent biosynthesis and (2) the heteropolymeric, or O-antigen polymerase dependent biosynthesis.

The chemical structure of O-antigens of a number of *Yersinia* strains are known, but the lipid A or core structures have been studied in only a couple of cases. The structures were reviewed recently,[1] and here I will only discuss those for which the genetics is worked out.

Genus Yersinia

The genus *Yersinia*, which is a member of family *Enterobacteriaceae* of the γ-purple bacteria,[2] now comprises 11 species, three of which contain human pathogenic strains: *Y. pestis*, *Y. pseudotuberculosis*, and *Y. enterocolitica*. *Y. pestis* is the causative agent of the feared bubonic plague that spreads from infected rodents to humans via a flea vector. Pathogenic types of *Y. pseudotuberculosis* and *Y. enterocolitica* cause yersiniosis, which is usually a mild diarrheal disease, and the infection usually takes place by ingestion of contaminated foodstuffs. *Y. pseudotuberculosis* and *Y. enterocolitica* and related species (*Y. aldovae*, *Y. bercovieri*, *Y. frederikseni*, *Y. intermedia*, *Y. kristenseni*, *Y. mollareti*, *Y. rohdei*) have several bio- and serovariants, whereas *Y. pestis* has three biovariants but is serologically of a single type. Five serotypes have been described for *Y. rückeri* that causes the enteric red mouth disease of fish.[3]

Characteristic for *Yersinia*, expression of many factors is temperature regulated. These factors include O-antigen, flagella, virulence factors, and some biochemical reactions. With respect to the optimal expression temperature, the factors fall into two categories: (1) those that are optimally expressed below 28°C and (2) those that are optimally expressed at 37°C. Expression of the O-antigen falls into the first category.

Serotypes of Y. enterocolitica *and related species*

In *Y. enterocolitica* and related species there are over 70 serotypes, which are mainly determined by the variability of O-antigen.[4] Human and animal pathogenic strains of *Y. enterocolitica* that carry the virulence plasmid, pYV*, belong to certain serotypes: in Scandinavia and Europe, Canada, Japan, and South Africa, to O:3 and O:9; and in the United States, to O:8. Less frequently

* pYV, 70-75 kb virulence plasmid of *Yersinia*

encountered pathogenic serotypes are O:4,32. O:5,27, O:13a,18, and O:21. pYV is not present in certain serotypes, and these are usually called nonpathogenic or environmental serotypes. However, strains belonging to these serotypes are often isolated from stool samples of healthy humans. The virulence of different pathogenic serotypes varies, i.e., serotype O:8 is more virulent than O:3 or O:9. This is most evident as mouse lethality. O:8 strains kill mice, while the others do not unless iron is made available to bacteria by pretreatment of mice with desferroxamine or iron.[5] On the other hand, O:3 strains cause frequently reactive arthritis in humans, while O:8 does that extremely rarely. It is not clear if O-antigen has any role in determining the differences in the virulence between the serotypes or whether that is due to other factors.

Y. pseudotuberculosis *serotypes*

The current *Y. pseudotuberculosis* antigenic scheme lists 14 O-serogroups;[6] the first five contain O-subgroups. The division into different groups is based on different O-factors. Differences in the first seven serogroups are mainly determined by the presence of different DDHs* in the O-antigen. Because of DDHs many *Y. pseudotuberculosis* serotypes share O-antigenic structures with *Salmonella* and *E. coli.*[7,8] *Y. pseudotuberculosis* serotypes O:1a, O:1b, O:3 O-antigens contain Par†, serotypes O:2a, O:2b, O:2c, Abe, serotypes O:4a and O:4b, Tyv, serotype O:5a, Asc and serotypes O:6 and O:7, Col.[9-12] Serotype O:5b contains 6-d-Altf‡.[10,13] Epidemiologically there are some differences in prevalence of the serotypes in human isolates. In Europe, most often the isolates belong to serotypes O:1 to O:3,[14] while in Japan to serotypes O:4b, O:3, O:5a, and O:5b.[15]

Y. pestis

Y. pestis isolates are divided into three biovariants. Serologically, *Y. pestis* strains are very homogeneous because their LPS is rough, i.e., lacking O-antigen.[16] The biovariants are *mediaevalis, orientalis,* and *antiqua,* based on differences in fermentation of glycerol (+, −, +) and ability to reduce nitrate to nitrite (−, +, +), respectively.[17,18] Variety *orientalis* is endemic in India, southwest Asia, South America, and western North America; *antiqua* is present in southeast and northern Asia (China and Russia), and Africa; and *mediaevalis* is limited to Turkey and Iran.[18]

An overview of LPS biosynthesis

The genes directing the biosynthesis of the three structural components of LPS are chromosomally located in most Gram-negative bacteria, and com-

---

* DDH, 3,6-dideoxyhexose
† Par, paratose; Abe, abequose; Tyv, tyvelose; Asc, ascarylose; Col, colitose
‡ 6d-Altf, 6-deoxy-L-altrofuranose

Table 2.1 The Present and Former Names of *Yersinia* LPS Genes

Y. enterocolitica O:3 Outer core cluster Acc.no Z47767		Y. enterocolitica O:3 O-antigen cluster Acc.no. Z18920		Y. enterocolitica O:8 O-antigen cluster Acc.no. U46859		Y. pseudotuberculosis O-antigen clusters Acc.no. L01777, U13685, U29692, U29691	
Present	Old	Present	Old	Present	Old	Present	Old
wzx	trsA	wbbS	rfbA	ddhA	rfbF	ddhD	rfbI
wbcK	trsB	wbbT	rfbB	ddhB	rfbG	ddhA	rfbF
wbcL	trsC	wbbU	rfbC	wbcA	orf4.0	ddhB	rfbG
wbcM	trsD	wzm	rfbD	wbcB	orf4.5	ddhC	rfbH
wbcN	trsE	wzt	rfbE	wbcC	orf5.6	abe	rfbJ
wbcO	trsF	wbbV	rfbF	wzx	rfbX	wbyA	orf8.7
wbcP	trsG	wbbW	rfbG	wbcD	orf7.8	wzz	cld
wbcQ	trsH	wbbX	rfbH	wbcE	orf8.9	prt	rfbS
galE	galE			wbcF	orf9.9	tyv	rfbE
				wbcG	orf10.9	wzx	rfbX
				wbcH	orf11.8		
				wbcI	rfbP		
				gmd	orf13.7		
				fcl	orf14.8, wbcJ		
				manC	rfbM		
				manB	rfbK		
				galE	galE		
				wzy	rfc		

monly the genes for each component form clusters that map to different parts of the chromosome. In a few cases, some LPS genes are located on plasmids or encoded within temperate bacteriophage genomes. In *Yersinia*, all the LPS genes identified thus far are located in the chromosome. A new nomenclature for bacterial surface polysaccharide genes, including those for LPS, was introduced recently.[19] The new nomenclature will be followed here, and the *Yersinia* gene names are listed in Table 2.1. Due to the chemical nature of LPS, most of the LPS genes encode enzymes that act on sugar structures. Based on function, the genes can be classified as shown in Table 2.2.

LPS biosynthesis starts by the activation of sugar-1-P*, such as Glc-1-P or Fru-1-P, by reaction with NTP to form NDP[†]-activated sugars. Different nucleotides are used for activation in pathways leading to different sugars, e.g., GTP for Man and D-Rha, CTP for Abe and Par, UTP for Gal, and dTTP for L-Rha; the reason for the choice of a specific nucleotide for a given sugar is unclear. The activation reactions take place in the cytoplasm, catalyzed by soluble enzymes. Different NDP-sugars are then synthesized in several enzymatic steps, including function of epimerases, reductases, etc. Some NDP-sugars are present in bacteria as intermediates of general metabolism, but

---

* P, phosphate; Glc, D-glucose; Fru, D-fructose; Man, D-mannose; Rha, rhamnose; Gal, D-galactose
† NTP, nucleoside triphosphate; NDP, nucleoside diphosphate

Table 2.2 Classification of LPS Gene Products

Class	Gene products	Function or role in LPS biosynthesis
I	Acyltransferases	Lipid A biosynthesis: Transfer of acyl groups to UDP-GlcNAc-residues
II	Biosynthetic enzymes like epimerases, hydratases, etc.	Involved in the biosynthesis of NDP-activated sugar precursors starting from sugar-1-phosphates
III	Glycosyltransferases	Transfer of sugar residues from NDP-activated sugar precursors to nonreducing end of growing oligosaccharide chain or to Und-P
IV	Kinases, *O*- or *N*-acetylases, deacetylases	Decorating enzymes, adding to the oligosaccharides small prosthetic groups or modifying them
V	Translocases	Proteins that function in the oligosaccharide translocation apparatus
VI	O-antigen polymerase	Transfers O-units into growing O-unit chains to form O-antigen
VII	O-antigen ligase	Transfers full length O-antigen to lipid A core
VIII	Regulatory factors	O-antigen chain length determinator, transcription regulators, etc.

more often they are synthesized specifically for LPS biosynthesis by class II enzymes (Table 2.2).

Biosynthesis of LPS proceeds via two pathways; the lipid A-core and the O-antigen pathways that merge into one pathway after both components are fully completed. In these pathways, the biosynthesis flows so that the product of one step functions as a substrate for the next. Lipid A and core syntheses are coupled; lipid A is assembled first on the cytoplasmic face of IM, and the core oligosaccharide is built on it by sequential transfer of sugar residues from NDP-activated sugar precursors. NDP-sugars are linked to the growing core by specific glycosyltransferases, the class III enzymes that are specific both to the sugar precursor and to the acceptor structure, and generate a specific linkage between the sugar and the acceptor. The completed lipid A core structure is translocated to the periplasmic face of IM to wait for eventual ligation of O-antigen and translocation to OM.

O-antigen biosynthesis similarly takes place in the cytoplasmic face of IM. Depending on the nature of O-antigen, two main synthesis modes are recognized. I call them hetero- and homopolymeric or O-antigen polymerase (Wzy)-dependent and -independent. In the heteropolymeric (Wzy-dependent) pathway, each identical repeat unit is first assembled on a carrier lipid, Und-P,* by dedicated glycosyltransferases that use NDP-activated sugar precursors (class III). The initiation reaction, transfer of the first sugar-1-P to Und-P, is in strains that have GlcNAc in the O-antigen structure, and in many homopolymeric pathways (see below) accomplished by WecA (formerly

---

* Und-P, undecaprenol phosphate, a 55-carbon isoprenoid carrier lipid

known as Rfe), a GlcNAc-1-P transferase, the gene of which is located in the enterobacterial common antigen (ECA) gene cluster. In other strains, a dedicated transferase, such as a Gal-1-P transferase, is used for initiation. Completed O-unit is translocated by a class V protein, "flippase," Wzx,[20] to the periplasmic side of IM where the O-antigen polymerase, Wzy (class VI), polymerizes the O-units into long chains, and the length of the chains is regulated by Wzz, the chain length determinant (class VIII). The full-length chains are translocated from Und-P to the preformed lipid A core structure by O-antigen ligase, WaaL (class VII).

In the homopolymer (Wzy-independent) pathway, the O-antigen polymer is synthesized on Und-P and completely elongated to full length in the cytoplasmic side of IM by sequential transfer of sugar residues to the non-reducing end of the growing polysaccharide chain. The completed O-antigen is translocated to the periplasmic side by an ATP-driven transporter system, composed of Wzt and Wzm, that belongs to the ATP-binding cassette (ABC) transporter family (class V).[21] O-antigen ligase, WaaL, transfers and ligates also the homopolymeric O-antigen onto lipid A core. Finally, the completed LPS molecule is routed in both O-antigen synthesis modes to OM by an unknown mechanism.

Bacteria and bacteriophages used in LPS genetic studies of Yersinia

Bacteriophages have long been utilized as tools in bacterial genetics and systematics. Indeed, the first suspicions that the genus *Yersinia* belongs to the *Enterobacteriaceae* was made on the basis of common sensitivities to phages.[18] Phages have also been used in epidemiological characterization and other studies on *Y. enterocolitica* strains.[22,23] Many phages use different parts of the LPS as receptors, and can be used to select mutants missing respective parts of the molecule. Much of the genetic work in my laboratory was made possible by the use of *Yersinia* LPS-specific bacteriophages isolated by us. Our bacteriophages have all been isolated from raw incoming sewage samples obtained from the Turku city sewage treatment plant.[24-27] *Yersinia*-specific bacteriophages in the sewage were enriched and isolated as described by Baker and Farmer.[22] By using different host strains for enrichment, phages with different specificities were obtained (Table 2.3). Thus, φYeO3-12 was obtained after enrichment using a smooth *Y. enterocolitica* O:3 strain 6471/76-c, and it was shown later to use *Y. enterocolitica* O:3 O-antigen as receptor.[24] φYeO3-12 was used to select rough *Y. enterocolitica* O:3 mutants YeO3-R1 and YeO3-R2;[25] φR1-37, an outer core specific phage, was in turn obtained after enriching the sewage phages using YeO3-R1.[27] Similarly, phages φ80-18 and φWA-1 were obtained after enrichment with *Y. enterocolitica* O:8 strains 8081 and WA, respectively. Bacteriophages φ80-18 and φWA-1 infect several different serotype O:8 strains. Purified O:8 LPS preparation inhibits both phages, indicating that LPS is the phage receptor.

Table 2.3   Bacterial Strains and Bacteriophages Relevant to Genetics of Yersinia LPS Described in this Chapter

Bacterial strain or bacteriophage	Description	LPS and phage phenotype*	Reference
Y. enterocolitica strains			
YeO3 (= 6471/76)	serotype O:3, wild type strain pYV+, patient stool isolate	O-ag+ oc+ φYeO3-12+ φR1-37-	63
YeO3-c (= 6471/76-c)	pYV- derivative of YeO3	O-ag+ oc+ φYeO3-12+ φR1-37-/+	63
YeO3-R1	Rough derivative of YeO3-c selected by phage φYeO3-12	O-ag- oc+ φYeO3-12- φR1-37+	25
YeO3-R2	Rough derivative of YeO3 selected by phage φYeO3-12	O-ag- oc+ φYeO3-12- φR1-37+	25
YeO3-RfbR12	YeO3-c wzx::Tn5-Tc1, tetR	O-ag+ oc+ φYeO3-12+ φR1-37-	27
YeO3-trs11	YeO3, Δwzx-wbcKL::kmR -GenBlock	O-ag+ oc+ φYeO3-12+ φR1-37-	27
IP614	Serotype O:15	O-ag+ oc+ φYeO3-12+ φR1-37+	44
8081	Serotype O:8, pYV+	O-ag+ φWA-1+ φ80-18+	64
8081-c	pYV- derivative of 8081	O-ag+ φWA-1+ φ80-18+	64
8081-res	R-M+ derivative of 8081-c	O-ag+ φWA-1+ φ80-18+	26
8081-R-M+	R-M+ derivative of 8081	O-ag+ φWA-1+ φ80-18+	28
Escherichia coli strains			
C600/PIZ6020	pLZ6020 is a BamHI deletion derivative of pLZ5010. The deletion extends from 3' end of wzy to ushA	O-ag+ φWA-1+ φ80-18+	28
DH1/pLZ5005	pLZ5005, the O:8 gene cluster (Figure 2.3) cloned into cosmid pHC79. The cloned fragment extends ca. 3 kb upstream adk	O-ag+ φWA-1+ φ80-18+	28
DH1/pLZ5010	pLZ5010, the O:8 gene cluster (Figure 2.3) cloned into cosmid pHC79. The cloned fragment extends from adk to rosB	O-ag+ φWA-1+ φ80-18+	28
Bacteriophages			
φYeO3-12	Y. enterocolitica O:3 O-antigen specific phage		24
φR1-37	Y. enterocolitica O:3 outer core specific phage		27
φWA-1	Y. enterocolitica O:8 O-antigen specific phage		Skurnik, M., unpublished
φ80-18	Y. enterocolitica O:8 O-antigen specific phage		26

* O-ag, O-antigen; oc, outer core

φ80-18 was used to isolate a rough O:8 strain 8081-R2 that was used in virulence experiments.[28] Both φ80-18 and φYeO3-12 are able to infect *E. coli* strains expressing O-antigens of cloned O-antigen gene clusters of *Y. enterocolitica* serotypes O:8 and O:3, respectively.[24,28]

Y. enterocolitica *serotype O:3 LPS*

The lipid A structure of *Y. enterocolitica* O:3 has not been studied, but the inner core structure is known.[29] (See also Reference 1 and the preliminary report of the structure of the outer core hexasaccharide composed of FucNAc, 2 × GalNAc, Gal and 2 × Glc.*) (Figure 2.1) (Shashkov et al., Abstract B017, 8th European Carbohydrate Symposium, Sevilla, Spain, 1995.)[27] Our molecular genetic studies indicate that the outer core hexasaccharide of O:3 is an ancestral heteropolymeric O-unit. The structure of the *Y. enterocolitica* O:3 LPS, furthermore, is unique because the present O-antigen which is a homopolymer of 6-deoxy-L-altrose (6d-Alt),[30] is not attached to the outermost tip of the outer core but instead to the inner core.[27]

Serotype O:3 outer core

Biosynthesis of outer core

The biosynthesis of the hexasaccharide outer core (or a single ancestral O-unit) needs (1) enzymes (class II) for the biosynthesis of NDP-sugar precursors, (2) at least six glycosyltransferases (class III), and (3) a flippase (class V). Because there is only a single oligosaccharide unit in the outer core, the O-antigen polymerase Wzy is not needed. The conceivable biosynthetic steps for building the hexasaccharide unit on the cytoplasmic face of IM are depicted in Figure 2.1, which also illustrates the biosynthetic steps of an oligosaccharide unit in general. The biosynthesis starts with the transfer of FucNAc-1-P from UDP-FucNAc to the carrier lipid Und-P (step 1). The rest of the hexasaccharide is then sequentially completed on Und-P-P-FucNAc (steps 2 to 6), where the coming sugar residues are transferred to the non-reducing end of the growing oligosaccharide. In step 7 carbon 4 of the FucNAc residue is decorated with a phosphate group. After completion, the Und-P-P-hexasaccharide unit is flipped to the periplasmic face of the IM, where the hexasaccharide is transferred (ligated) onto a lipid A core structure.

The gene cluster directing the biosynthesis of the outer core was recently characterized (Table 2.4).[27] Bacteriophage φR1-37 (Table 2.3) was used to isolate phage-resistant mutants from a transposon library of O:3, with the original idea to isolate O-antigen temperature-regulation mutants.[27] Some of the obtained mutants, e.g., YeO3-RfbR12 (Table 2.3), turned out to have the phage φR1-37 receptor eliminated and to be defective in the outer core biosynthesis.[27]

---

* GalNAc, N-acetyl-D-galactosamine; FucNAc, N-acetyl-D-fucosamine

Table 2.4   Sequenced LPS Gene Clusters of *Yersinia*

Yersinia species	Cluster	Reference	GenBank acc.no
Y. enterocolitica O:3	Outer core	27	Z47767
Y. enterocolitica O:3	O antigen	39	Z18920
Y. enterocolitica O:8	O antigen	28, 36	U46859
Y. enterocolitica O:8	O antigen, 3' end	32	U43708
Y. pseudotuberculosis O:1a	O antigen, DDH genes	50	U29692
Y. pseudotuberculosis O:2a	O antigen, DDH genes + 3' end	50	L01777, U13685
Y. pseudotuberculosis O:4a	O antigen, DDH genes	50	U29691
Y. pseudotuberculosis O:5a	O antigen, DDH genes	50, 54	S72887
Y. pestis, orientalis	whole genome	http://www.sanger.ac.uk/Projects/Y_pestis/	

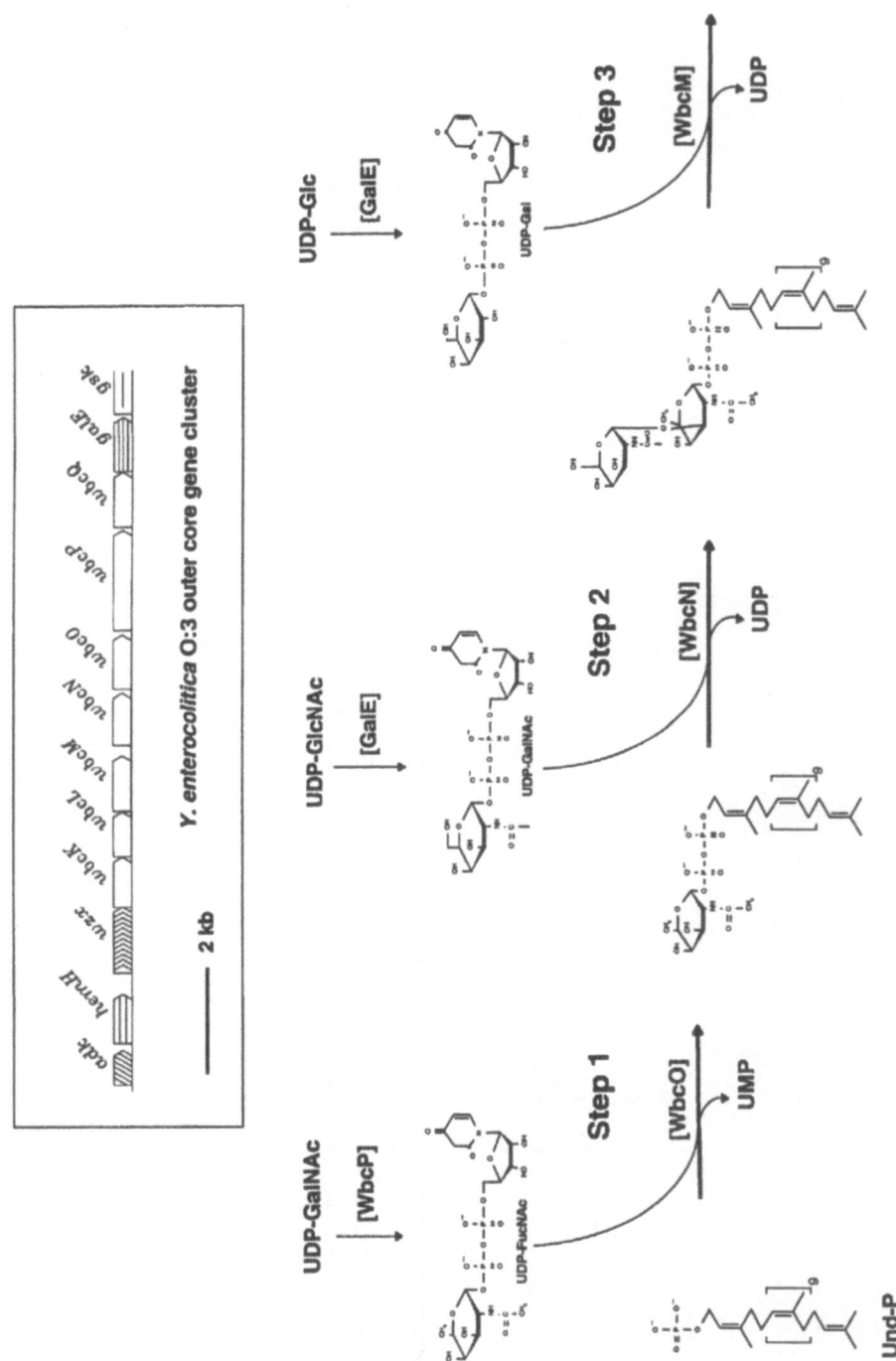

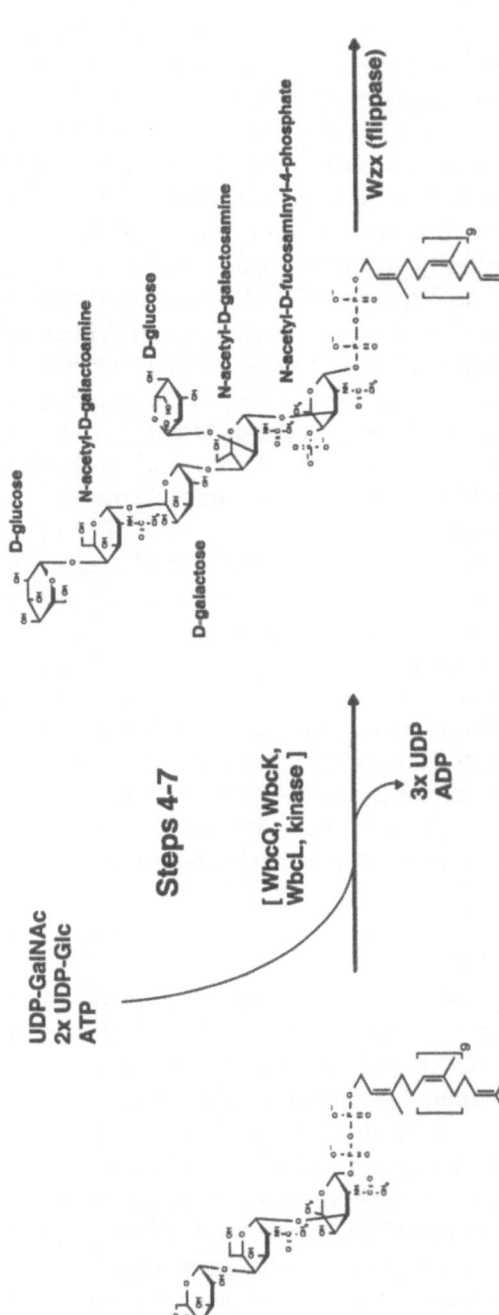

Figure 2.1 Biosynthesis and genetics of the LPS outer core oligosaccharide of *Y. enterocolitica* O:3. The organization of the outer core gene cluster is shown in the box at the top. The genes are drawn to scale, and a total of 13.6 kb of the chromosomal region is shown. The upstream genes *adk* and *hemH* and the downstream gene *gsk* are not outer core genes. The lower part of the figure shows the biosynthetic steps of the outer core pathway and the gene products that catalyze the reactions. Note that these assignments are based on sequence similarity data and that these may be subject to change later. In step 1, in the initiation reaction, it is suggested that FucNAc is transferred by WbcO from UDP-FucNAc to Und-P to generate FucNAc-Und-PP. Biosynthesis of UDP-FucNAc may be catalyzed by WbcP in a dehydration reaction from UDP-GalNAc. In step 2, GalNac is transferred by WbcN from UDP-GalNAc to generate FucNAc-Und-PP. It is suggested that epimerization of UDP-GlcNAc to UDP-GalNAc is catalyzed by the GalE homologue encoded by the gene downstream from the *wbcQ* gene, and that the epimerization of UDP-Glc to UDP-Gal, which is transferred in step 3 by WbcM to the growing oligosaccharide structure, is catalyzed by the "true" GalE, the gene of which is located in the *gal* operon at some unknown location in the chromosome (see also discussion on the O-antigen gene cluster of *Y. enterocolitica* O:8). During steps 4–7 the outer core hexasaccharide is completed to the structure shown. After the structure is flipped to the periplasmic side, it is ligated from the FucNAc residue via a β1→3 glycosidic linkage to the inner core L-Hep-residue (L-*glysero*-D-*mannoheptopyranose*).

The outer core gene cluster

The affected gene cluster was cloned and characterized by deletion and complementation analysis, and about 13.6 kb of DNA was sequenced.[27] In total, 12 genes were identified (Figure 2.1), of which nine are involved in the outer core biosynthesis. The sequence analysis revealed that the transposon is inserted in the first gene, *wzx* (old name *trsA*), of the outer core operon encoding for the flippase. Upstream of *wzx* is a noncoding region of 400 bp, preceded by two genes highly homologous to the *adk* and *hemH* genes of *E. coli* and other bacteria. Downstream of *wzx* are six genes, *wbcK, wbcL, wbcM, wbcN, wbcO,* and *wbcQ* (old names *trsB* to *trsF,* and *trsH,* Table 2.1) the products of which all show similarities to glycosyltransferases. Based on the similarities it is likely that WbcK and WbcL are Glc-transferases, WbcM a Gal-transferase, and WbcN and WbcQ, GalNAc-transferases, and WbcO, an FucNAc-transferase (Figure 2.1). The *wbcP* (*trsG*) gene encodes a product that, based on similarities to enzymes of other bacteria, is apparently involved in UDP-FucNAc biosynthesis. Downstream of *wbcQ,* two genes highly homologous to *galE* and *gsk* were found. In the noncoding 400 bp region between *hemH* and *wzx,* the 39 bp nucleotide sequence characteristic for the JUMPstart sequence was recognized.[31]

Since UDP-Glc and UDP-Gal are products of general metabolism, the biosynthesis of the UDP-GalNAc and UDP-FucNAc residues remains to be taken care of by the outer core genes. Two putative biosynthetic genes were identified, a *galE* homolog and *wbcP.* UDP-glucose-4-epimerase (GalE) catalyzes the epimerization of UDP-Glc to UDP-Gal. Based on our own unpublished experiments and on those of D. Pierson,[32] the *galE* homolog probably encodes the UDP-GlcNAc-4-epimerase, thus producing UDP-GalNAc for the outer core biosynthesis. WbcP is involved in FucNAc biosynthesis, but its function is not known. Since FucNAc is a 6-deoxy derivative of GalNAc WbcP may be catalyzing this conversion. WbcP has local similarities to GalE supporting this hypothesis.[27]

Outer core is an ancestral O-unit

What then is the basis for claiming that the outer core of O:3 is an ancestral O-antigen relic? The outer core operon has a number of analogies with O-antigen gene clusters of other *Enterobacteriaceae.* The flippase gene, *wzx,* is present in all known heteropolymeric O-antigen clusters but nowhere else.[19,20,33-35] In two other members of *Yersinia, Y. pseudotuberculosis* and *Y. enterocolitica* serotype O:8, the O-antigen gene clusters are located at the same location in the chromosome as the outer core gene cluster, i.e., between the *hemH* and *gsk* genes.[36,37] The serotype O:3 homopolymeric O-antigen has a capsule-like structure, and it is strangely linked to the inner core.[27] In the O-antigen cluster of O:8, the O-antigen polymerase gene (*wzy*) is located between the *galE* and *gsk* genes,[28,36] but in the O:3 outer core cluster there are no genes. It is feasible that the O-antigen region of *Y. enterocolitica* O:3 directing the biosynthesis of the homopolymeric capsule-like O-antigen has taken over the O-antigen expressing role from the ancestral outer core operon. The preservation of the outer core biosynthesis in *Y. enterocolitica* O:3 indicates that it is indispensable, and, indeed, we have results that outer core mutants are avirulent, supporting this conclusion.[37a]

Serotype O:3 O-antigen

Cloning of the O-antigen gene cluster of *Y.enterocolitica* O:3 was the beginning of the molecular genetic studies of LPS in my laboratory. A genomic library of strain 6471/76-c (Table 2.3) was constructed in pBR322, and O-antigen expressing *E. coli* clones were screened using Mab A6, specific for the *Y. enterocolitica* O:3 O-antigen.[38] Two O:3 O-antigen expressing clones were obtained carrying 12-13 kb overlapping inserts.[24] The cloned gene cluster has been analyzed by the classical molecular genetic methods such as deletion analysis, transposon mutagenesis, transcomplementation experiments, and finally by determination of the complete nucleotide sequence (Table 2.4).[24,25,39] The genetic organization has further been resolved by identification of the transcriptional start points and functional promoters by primer extension experiments and promoter clonings.[39] The gene cluster is presented in Figure 2.2.

Biosynthesis of the O:3 O-antigen

To direct the synthesis of the homopolymeric *Y. enterocolitica* O:3 O-antigen composed of 6d-Alt, the O-antigen gene cluster must contain genes for (1) the biosynthesis of the NDP-activated sugar precursor, (2) at least two glycosyltransferases (one to initiate the O-antigen synthesis and the other to extend the homopolymer), and (3) a transporter system. Since 6d-Alt is a carbon-3 epimer of L-Rha, the biosynthetic pathways of both sugars are likely to be similar, and dTDP-4-keto-6-deoxy-D-Glc is probably an intermediate also in the dTDP-6d-Alt pathway (Figure 2.2). Biosynthesis of dTDP-L-Rha takes place in four steps, and the same is expected for 6d-Alt. However, details of the dTDP-6d-Alt biosynthesis are not yet known and need to be elucidated in the future.

The genetic organization of the O:3 O-antigen gene cluster is shown in Figure 2.2. Ten ORFs were recognized from the sequence; the first two are not involved in the O-antigen synthesis, as shown by transposon mutagenesis. For these two ORFs, there are no similar sequences in GenBank; thus, the chromosomal location of the O-antigen gene cluster is not known. The remaining eight ORFs form two operons. The first operon contains three genes that are designated *wbbS, wbbT* and *wbbU*.[19] The second operon contains five genes: *wzt, wzm, wbbV, wbbW,* and *wbbX*. Both operons are essential for O-antigen synthesis, as shown by transposon mutagenesis, deletion analysis, and transcomplementation experiments.[24,25,39] In the noncoding 1.5 kb region upstream of *wbbS*, two structures were identified: relics of the *rmlB* gene and a highly conserved JUMPstart sequence. Transposon insertions in the *rmlB* gene relics did not affect the O:3 O-antigen synthesis, indicating

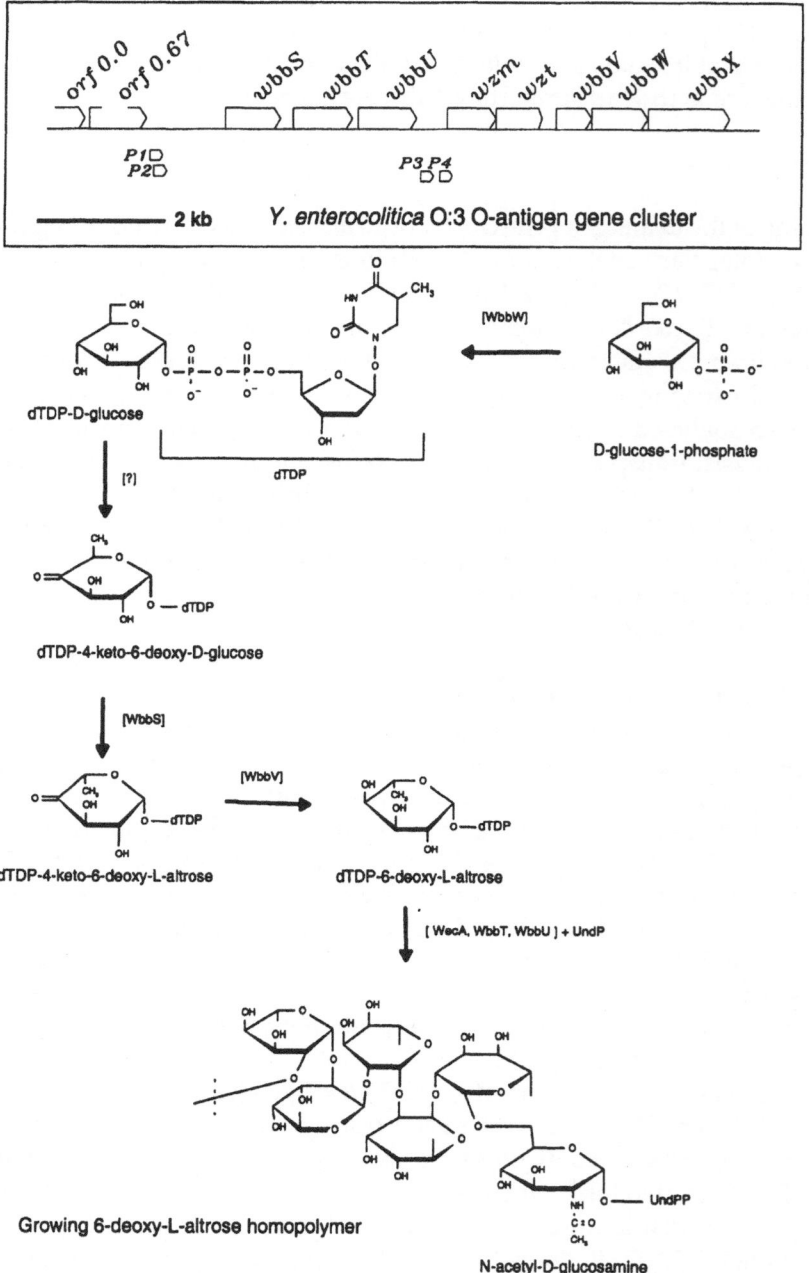

Figure 2.2   Biosynthesis and genetics of the LPS O-antigen of *Y. enterocolitica* O:3. The organization of the O-antigen gene cluster is shown in the box at the top. The genes are drawn to scale and a total of 11.6 kb of the chromosomal region is shown. Putative biosynthetic pathway to dTDP-6d-Alt, transfer of the 6d-Alt residues to the growing homopolymeric O-chain, and the roles of O-antigen gene cluster gene products are shown. The first residue of the chain connected to UndPP is drawn to be GlcNAc because biosynthesis of the O-antigen in *E. coli* is WecA dependent.

that these relics do not have any function.[39] However, they indicate that evolution of the O-antigen cluster included a phase where the biosynthetic genes of dTDP-L-Rha served as a starting point for the biosynthetic genes of dTDP-6d-Alt, and that the *rmlB* gene was not useful for that purpose and could then accumulate mutations, resulting in several frame shifts. On the contrary, similarity searches revealed that WbbS, WbbV, and WbbW show significant similarity to RmlC, RmlD, and RmlA, respectively, the enzymes involved in the dTDP-L-Rha biosynthesis. The steps where WbbS, WbbV, and WbbW function in dTDP-6d-Alt biosynthesis are indicated in Figure 2.2.

The initiation of the O-antigen biosynthesis is WecA (Rfe) dependent
We recently obtained results showing that *Y. enterocolitica* O:3 O-antigen biosynthesis is dependent on GlcNAc transferase[40] (L. Zhang and M. Skurnik, unpublished). Thus O:3 O-antigen biosynthesis most likely utilizes an initiation step where a single GlcNAc residue is transferred to Und-P. After this, a specific transferase transfers the first 6d-Alt residue to GlcNAc-Und-PP, upon which the homopolymer is then sequentially built up (Figure 2.2). Apparently, the transferase that transfers the first 6d-Alt residue to GlcNAc-Und-PP needs to be different from the transferase that uses 6dAlt-GlcNAc-Und-PP as an acceptor. Both WbbT and WbbU share conserved local motifs with a number of glycosyltransferases,[27,41] and fulfill the need for two glycosyltransferases. Here also, the exact roles need further study. It should be noted that transposon insertions into *wbbT* and *wbbU* totally block O-antigen expression, and this is in agreement with their putative roles.

Transport
Wzt and Wzm show significant similarity to a number of ATP-driven polysaccharide transporter systems in a number of bacteria.[21,39] In addition, Wzt has a potential ATP-binding sequence which is characteristic of the ATP-binding components of periplasmic binding protein-dependent transport systems. In line with the transporter role for Wzt and Wzm, transposon insertions into *wzt* and *wzm* genes resulted in accumulation of cytoplasmic O-antigen.[39] *Y. enterocolitica* O:3 was the first organism in which an ATP-driven polysaccharide transport system was recognized in an O-antigen gene cluster; since then, similar transporters have been reported from a number of other organisms.[21]

The O-antigen gene cluster promoters
Both strands of the nucleotide sequence were searched for promoter motifs using the promoter search algorithm of GENEUS.[42,43] The search revealed promoter-like sequence motifs 1.2 kb upstream of *wbbS*, and between *wbbU* and *wzm*. The functional promoters were identified using promoter clonings combined with sequencing and primer extension analysis. Both regions were shown to contain tandem promoters (see Figure 2.2). In addition, the promoter clonings revealed two antisense promoter activities; one activity

located in the first operon, and the other, downstream of the second operon. These activities have not yet been characterized in further detail. The many functional promoters and the two-operon organization indicate that the O-antigen expression has a complex regulation.

Serotype O:3 related serotypes

6d-Alt is the major component of *Y. enterocolitica* O:1,2a,3 and O:2a,2b,3 O-antigens. The genetic organization of the O-antigen gene clusters of these organisms is not known. However, when the O:3 O-antigen cluster DNA is used as a probe, it hybridized strongly with DNA from these organisms, indicating that similar clusters are also present in them.

Kawaoka et al. studied the LPS of *Y. enterocolitica* O:15 strains[44] and found that the O:15 LPS is rough, indicating that the original strain used to characterize serotype O:15 was possibly a spontaneous rough mutant. The serotype O:15 type strain IP614 is a biotype 4 strain (G. Wauters, personal communication), and since biotype 4 includes only serotype O:3 strains, it is likely that IP614 is a rough mutant of serotype O:3. The phage sensitivity of IP614 (obtained from G. Wauters) was studied and was found to be resistant to φYeO3-12 and sensitive to φR1-37 (Table 2.3, M. Skurnik, unpublished). These results and the LPS profile analysis suggest that IP614 is identical to YeO3-R1, i.e., it is a rough strain missing the homopolymeric O-antigen but expressing the complete core.

Y. enterocolitica *serotype O:8 LPS*

O-antigen gene cluster of Y. enterocolitica *O:8*

Structure and biosynthesis of O:8 O-antigen

The structure of the O-antigen repeat unit of *Y. enterocolitica* O:8 is shown in Figure 2.3; it is a branched pentasaccharide of GalNAc, Gal, Man, Fuc, and

---

Figure 2.3 (opposite) Biosynthesis and genetics of the LPS O-antigen of *Y. enterocolitica* O:8. The organization of the O-antigen gene cluster is shown in the box at the top. The genes are drawn to scale and a total of 26.9 kb of the chromosomal region is shown. The upstream genes *adk* and *hemH* and the genes downstream of *gsk* are not O-antigen cluster genes. The nomenclature used by Pierson to the *galE* gene homologue and the *wzy* gene is given below the genes. The lower part of the figure shows the biosynthetic steps of the O-unit synthesis and the gene products that catalyze the reactions. Note that most of these assignments are based on sequence similarity data and that these may be subject to change later. The *manA* and "true" *galE* genes (see text) are not located in the O-antigen cluster. The structure of the O:8 O-unit is shown in the lower middle. The configurations of the glycosidic bonds (α or β) between the residues are not known, except for the Gal–GalNAc bond (α1–3). The dashed vertical lines mark the borders of the O repeat unit and cross the glycosidic bonds, GalNAc-(1–4)-Man.

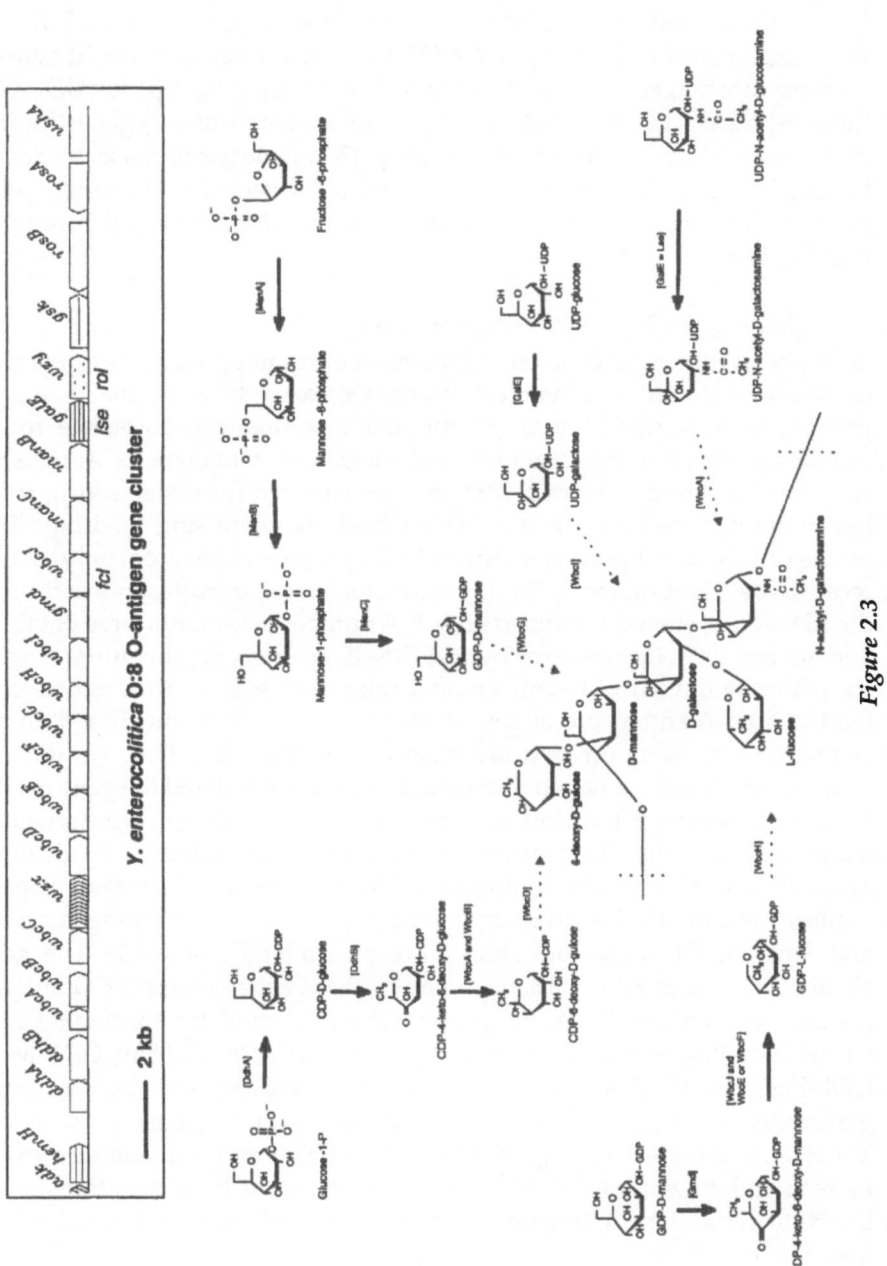

Figure 2.3

6d-Gul*.[45] The adjacent repeat units are joined together by a 1,4 glycosidic bond between GalNAc and Man residues. Interestingly, the O-antigen of serotypes O:7,8 and O:19,8 is almost identical to that of O:8,[46] with the exception that the GalNAc linkage to Man is 1,2, and that the Man residue is substituted with another 6d-Gul with a 1,4 linkage (see Figure 2.3). The gene cluster for biosynthesis of the O:8 O-antigen structure should contain (1) biosynthetic genes for each individual NDP-sugar (except for UDP-Gal, the biosynthesis of which belongs to general metabolism), (2) genes for five glycosyltransferases, one for each residue, (3) a gene for flippase needed in heteropolymeric O-antigen transport, and (4) a gene for O-antigen polymerase, Wzy, and possibly also a gene for the chain length determining protein, Wzz.

Cloning of O:8 O-antigen gene cluster

Transposon mutagenesis is one of the most efficient approaches in identification of gene clusters of interest. When we started to study the O-antigen genetics in *Y. enterocolitica* O:8,[26] our first intention was to isolate rough mutants using an *in vivo* chromosomal transposon mutagenesis. A transposon insertion library in strain 8081-res was screened for phage ϕ80-18 resistant mutants based on the fact that the bacteria expressing wild-type LPS on the cell surface should be sensitive to the phage and be lysed; only phage-receptor deficient mutants, i.e., LPS mutants having transposon insertion in the LPS genes, should be able to grow.[26] A number of mutants were obtained and analyzed for LPS patterns by SDS–PAGE analysis. All the mutants were shown to be devoid of O-antigen, and only expressed LPS molecules containing lipid A and a core oligosaccharide which at that time, based on the estimated size, were erroneously regarded as truncated. Thus we did not continue studying the obtained mutants; instead we isolated the gene cluster by cosmid cloning.[36] The clones expressing the O:8 O-antigen were screened using specific Mab. The isolated cosmids were characterized by deletion analysis and transposon mutagenesis to localize the O-antigen coding region,[36] and finally, the nucleotide sequence of the entire O-antigen region and flanking DNA on both sides were determined (Table 2.4). The gene cluster is shown in Figure 2.3. In summary, the *Y. enterocolitica* O:8 O-antigen gene cluster contains 18 genes, spanning about 19 kb of chromosomal DNA. Genes encoding enzymes for the biosynthesis of UDP-GalNAc, GDP-Man, GDP-Fuc, and CDP-6d-Gul are present in the cluster, and the remaining genes encode for glycosyltransferases, flippase, and O-antigen polymerase. The 3' end of the O:8 gene cluster was also cloned and characterized by D. Pierson and co-workers.[32,47] They identified the locus because a transposon insertion in the *lse* gene (Figure 2.3) increased the Ail-mediated invasion into tissue culture cells.

---

* Fuc, L-fucose; 6d-Gul, 6-deoxy-D-gulose

Y. pseudotuberculosis O:2a O-antigen gene cluster

—— **2 kb**

Figure 2.4 The genetic organization of the O-antigen gene cluster of *Y. pseudotuberculosis* O:2a. The exact size of the cluster in not known because the region between *wbyA* and *wzz* has not been sequenced. Based on the size of a fragment amplified in long-range PCR using primers specific for *hemH* and *gsk*, the size is close to 20 kb (E. Ervelä and M. Skurnik, unpublished).

The O:8 O-antigen biosynthetic genes

At least nine genes are needed for synthesis of NDP-sugar precursors (Figure 2.3). The first four genes of the gene cluster are highly similar to genes in the DDH biosynthetic pathway of *Salmonella* and *Y. pseudotuberculosis*. Transposon mutagenesis and chemical structure analysis indicate that these genes are for CDP-6d-Gul biosynthesis.[28] The genes involved in the synthesis of GDP-Man are *manC* and *manB*. The *gmd* and *fcl(wbcJ)* genes, and perhaps *wbcE* or *wbcF*, are involved in the GDP-Fuc biosynthesis. We also identified a *galE* gene homolog which, however, based on the results of Pierson and Carlson,[32] who named it *lse* (Figure 2.3), is most likely involved in UDP-GalNAc biosynthesis. The authors presented evidence that in *Y. enterocolitica* the "true" *galE* gene is located in the gal operon elsewhere in the chromosome. It is interesting to notice that both O:3 outer core and O:8 O-antigen gene clusters contain the *galE* gene homologue that are 84% identical. The *galE* gene homologue is absent from the *Y. pseudotuberculosis* gene cluster (Figure 2.4). It may be significant for the assignment of the function for the *Y. enterocolitica galE* homologue that *Y. pseudotuberculosis* O:2a O-antigen contains Gal but not GalNAc, while both are present in the outer core of *Y. enterocolitica* O:3 and in the O-antigen of O:8. Thus, in both O:3 and O:8 it might be plausible that the *galE* homolog encodes for UDP-GlcNAc-4-epimerase to yield UDP-GalNAc. This in turn would suggest that there may be a "true" *galE* gene located outside the LPS clusters of these organisms. Future work will answer this question.

The only genes whose products do not show similarities to any known proteins in database are *wbcE* and *wbcF*. These may also be biosynthetic genes as mentioned above.

Glycosyltransferases and transporter

Similarity to protein sequences in databases reveal several candidates for sugar transferases (Figure 2.3). A good candidate for the 6d-Gul transferase, according to similarity searches, would be WbcC that is highly similar (40% identity) to the *Salmonella* WbaV (Abe transferase). However, at the 5′ end of the *wbcC* gene, compared to the *wbaV* gene, the ATG codon has changed to ATA, and there is no apparent ribosomal binding site present where the first ATG codon is encountered; therefore it is uncertain whether WbcC is expressed at all. Based on the local similarity of WbcD to WbaV of *Salmonella*, we assume that WbcD is the 6d-Gul transferase. WbcH is similar to Fuc-transferases, even to eukaryotic ones, and WbcI to Gal-transferases. The most likely candidate for the Man transferase is WbcG, which shows local similarity to a number of bacterial glycosyltransferases.

The O:8 O-antigen cluster also contains the *wzx* gene encoding for the O-unit flippase, Wzx. This assignment was based on the hydropathy profile indicating the presence of about 12 transmembrane segments and on the sequence similarity to other known Wzx proteins,[28] including that of the *Y. enterocolitica* O:3 outer core cluster (Figure 2.1).

The O-antigen polymerase or the chain length determining protein?
Downstream of the *galE* (*lse*) gene homologue is located a gene the function of which is still controversial. Deletion analysis and transcomplementation experiments indicated that this gene is the last gene in the O-antigen gene cluster.[36] The LPS phenotype of and the chemical analysis of the LPS isolated from the *E. coli* clone expressing the O-antigen gene cluster missing this gene indicated clearly that the gene product would be Wzy, the O-antigen poly-merase.[28] The amino acid sequence of the gene product, however, shows significant similarity (36% identity) to a putative Wzz (the old name for Wzz was Rol or Cld[19]) chain length determinor of *Y. pseudotuberculosis* O:2a.[9,37] Furthermore, the role of the gene product as Wzz (Rol) was suggested by Pierson and Carlson.[32] They have a *Y. enterocolitica* O:8 mutant carrying a polar transposon insertion in the *lse* gene such that the downstream gene should not be expressed. LPS of this strain contains only lipid A and core. Complementation of the mutant with an intact *lse* gene allows the strain to produce O-units but results in ladder-like LPS that is lacking the modal distribution of the O-antigen chain lengths, instead a majority of the O-antigens have less than 4 to 5 repeat units.[32] This pattern is closely remi-niscent of that seen in *wzz* (*rol*) mutants,[37,48] especially with that of *Y. pseudotuberculosis* O:2a[9] and therefore the gene was named *rol* (Figure 2.3).[32] In *E. coli* where the chain length determination has been most thoroughly studied, however, the LPS pattern of *wzz* mutants is somewhat different. Instead of the accumulation of very short O chains, the O-antigen chain lengths are evenly distributed to a very broad range from one unit to tens of units per O-antigen.[37,48] There is no reason why a similar phenotype would not be seen in *Yersinia*. One could therefore speculate that the LPS pattern seen in the complemented *Y. enterocolitica* O:8 *lse* mutant[32] could as well indicate very low Wzy activity. The final assignment of the function for the *Y. enterocolitica* O:8 *wzy* (*rol*) gene needs future genetic and biochemical experimentation.

Temperature regulation of O:8 O-antigen biosynthesis

The expression of *Y. enterocolitica* O:8 O-antigen is regulated by temperature such that more abundant O-antigen is expressed on the cell surface at 25°C than at 37°C. The regulation mechanism is unknown, but it involves repression of the O-antigen cluster transcription at 37°C. We noticed that the locus downstream of the O-antigen cluster, between the *gsk* and *ushA* genes (Figure 2.3), mediates temperature regulated expression of O:8 O-antigen in the *E. coli* clones[40, 49] (L. Zhang and M. Skurnik, unpublished). The longest cosmid clone obtained, pLZ5005, carries a 30-kb chromosomal fragment and includes all the genes shown in Figure 2.3. When residing in *E. coli*, pLZ5005 expresses the O-antigen with temperature regulation similar to that of *Y. enterocolitica* O:8. The shorter cosmids pLZ5010 and pLZ6020 carrying only the O-antigen gene cluster did not. Subcloning and trancomplementation experiments located the regulatory locus in the *rosAB* genes (for *r*egulation of *O*-antigen *s*ynthesis)[40] (L. Zhang and M. Skurnik, unpublished).

rosA encodes a polypeptide similar to members involved in a drug efflux system. RosA shares highest similarity (70% identity) with Fsr (fosmidomycin resistance protein) encoded by the *fsr* gene adjacent to *ushA* gene in *E. coli*. *rosB* encodes a polypeptide highly similar (about 73% identity) to the hypothetical protein YbaL of *E. coli*, the gene of which is between the *fsr* and *gsk* genes. The *E. coli* similarities suggest that homologous genes are located at the same chromosomal location in these organisms; however, apart from the *fsr* gene, the role of the *E. coli* genes is not studied.

Interestingly, the cloned *rosAB* operon is able to complement temperature regulation in *E. coli* C600/pLZ5010 but not in C600/pLZ6020. Compared to pLZ5010, pLZ6020 is missing the *gsk* gene and the C-terminus of the Wzy protein is truncated by about 57 aa. This suggests that Wzy and other O-antigen proteins may participate directly or indirectly in the O-antigen regulation, and perhaps even form a membrane-associated complex[40] (L. Zhang and M. Skurnik, unpublished).

RosAB operon has a clear role in temperature regulation of O-antigen expression when the gene cluster is cloned in *E. coli*. However, in *Y. enterocolitica* O:8 a chromosomal *rosAB* knock-out mutation does not affect the temperature regulation of the O-antigen expression[40] (L. Zhang and M. Skurnik, unpublished). This means that there may be backup systems in *Y. enterocolitica* to compensate the *rosAB* defect and that the O-antigen expression belongs to a regulatory network.

O-antigen gene clusters of Y. pseudotuberculosis

The O-antigen structures of a number of *Y. pseudotuberculosis* serotypes are known.[1] O-antigen gene clusters of *Y. pseudotuberculosis* serotypes O:1a, O:2a,

O:4a, and O:5a have been studied in the laboratories of P. Reeves and H. Liu. The known genetic organization of the O-antigen cluster of O:2a is shown in Figure 2.4. The main emphasis in both laboratories has been on the DDH pathway. The DDH biosynthesis genes, named *ddhDABC*, are present at least in serogroups O:1a, O:1b, O:2, O:2b, O:2c, O:3, O:4a, O:4b, O:5a, and O:6, and are highly homologous to those of *Salmonella*.[10,50] In addition to the *ddhDABC* block, the DDH-pathway of each serogroup possesses one or two additional genes that determine the serogroup specificity, e.g., *abe* in serogroups O:2a, O:2b, and O:2c (see Figure 2.4); *prt* and *tyv* in serogroups O:4a and O:4b; *prt* in serogroups O:1a and O:1b; and *ascEF* in serotype O:5a. *ascEF* genes are not present in *Salmonella* and are therefore *Y. pseudotuberculosis* O:5a specific.[51-54] The chromosomal location of the O-antigen gene cluster in *Y. pseudotuberculosis* is the same location as that of *Y. enterocolitica* O:3 outer core and O:8 O-antigen gene clusters, i.e., between the *hemH* and *gsk* genes.[37] In addition to the DDH pathway genes, the O-antigen cluster contains the flippase gene, *wzx*, (Figure 2.4), and must possess genes coding for the biosynthetic enzymes of the other sugar components, for the respective glycosyltransferases, and for the O-antigen polymerase. At present there is very little information available on them. Comparative analysis of a number of *Y. pseudotuberculosis* O-antigen gene clusters downstream of the DDH biosynthetic genes revealed interesting features relating to the evolution of the clusters, including footprints of homologous recombination events and relics of insertion sequence IS630 also observed in *Shigella sonnei*.[50]

DDHs are formed via a complex biosynthetic pathway beginning with CDP-D-hexoses. Biosynthesis of CDP-Asc, one of the naturally occurring DDHs, consists of five enzymatic steps, with CDP-6-deoxy-Δ-3,4-glucoseen reductase (E3), a homogeneous enzyme composed of a single polypeptide with a molecular weight of 39,000, participating as the key enzyme in this catalysis.[55] The corresponding gene (*ascD* = *ddhD*) is located in the ascarylose biosynthetic cluster. E3 is an NADH-dependent* enzyme which catalyzes the key reduction of the C-3 deoxygenation step during the formation of CDP-Asc. One flavine-adenine dinucleotide (FAD) and one plant ferredoxin type iron–sulfur center were found to be associated with each molecule of E3.[56] E3 employs a short electron transport chain composed of both FAD and the iron–sulfur center to shuttle electrons from NADH to its acceptor. Cys-75 and Cys-296 may be important for electron transfer between NADH, FAD, and the iron–sulfur center.[57]

JUMPstart and ops sequences

The gene clusters for *Y. enterocolitica* O:3 outer core and O-antigen, *Y. enterocolitica* O:8 O-antigen, and *Y. pseudotuberculosis* O:2a O-antigen carry a characteristic sequence called the JUMPstart sequence. The sequence is located, at least in the case of O:3, downstream of the functional promoters

---

* NADH, hydrogenated (reduced) nicotinamide-adenine dinucleotide

of the O-antigen gene cluster. The JUMPstart sequence has been identified in all surface polysaccharide biosynthesis gene clusters and also for sex-pilus assembly and hemolysin production.[31,58] Nieto et al.[58] reported that a short 8 bp sequence within the JUMPstart sequence plays a role in gene regulation, and this element was named *ops* (<u>o</u>peron <u>p</u>olarity <u>s</u>uppressor). *ops* was shown to function as a *cis*-acting element that together with RfaH, a transcriptional regulator, allows efficient transcription of long operons, especially of the distal genes.[59] It is postulated that *ops* provides the specificity for RfaH for the subset of operons it is regulating. The role of *ops* and RfaH in regulation of *Yersinia* LPS operons has not yet been studied, but they are very likely to be involved.

The Yersinia *O-antigen gene clusters are located between* hemH *and* gsk

As a common feature, the chromosomal location of the *Yersinia* O-antigen gene clusters lies between the *hemH* and *gsk* genes. The two genes recognized upstream of the sequenced O-antigen gene clusters are the *adk* and *hemH* genes. Downstream of the cluster there is the *gsk* gene followed by the *rosAB* and *ushA** genes (the latter only determined so far in serotype O:8), which are not needed for O-antigen synthesis. This location is different from that found in *E. coli* and *S. enterica* where the O-antigen region is closely linked to the *gnd*[†] locus, which is upstream of the *his* operon at 45 min on the chromosomal map of *E. coli* K-12, and at centisome 45 on the *S. enterica* LT2 map.[60,61] In *E. coli* and *Salmonella*, the *adk*, *hemH*, *gsk*, and *ushA* genes are located at approximately 11 min on the chromosome.[60,61]

It will be interesting to study what is located at this chromosomal location (*hemH–gsk*) in other *Y. enterocolitica* serotypes and in other *Yersinia* in general. We addressed this question indirectly by screening our *Yersinia* collection for phage φR1-37 sensitive strains.[27] The results showed that the phage receptor was present not only in *Y. enterocolitica* serotype O:3 strains, but also in serotype O:1, O:2, O:5, O:5,27, O:6, O:6,31, O:9, O:21, O:25,26,44, O:41(27)43, O:41,43, and O:50 strains, and in *Y. intermedia* O:52,54 strain. Furthermore, we generated phage-resistant variants from some of these strains, and all of them had lost the outer core, as revealed by DOC–PAGE analysis of LPS (E. Ervelä and M. Skurnik, unpublished). Thus, it is plausible that, at least in these strains, the chromosomal location is occupied by a gene cluster closely related to the outer core gene cluster of serotype O:3. Some of these clusters may even be real O-antigen clusters possessing the O-antigen polymerase gene *wzy*. We have also used long-range PCR to amplify the region between the *hemH-gsk* genes from various *Yersinia* strains, including that of *Y. pestis*, and seen that all these strains have 12 to 20 kb of DNA between these genes (E. Ervelä and M. Skurnik, unpublished). It will

---

* *adk*, adenylate kinase gene; *hemH*, ferrochelatase gene; *gsk*, inosine-guanosine kinase gene; *ushA*, UDP-sugar hydrolase

† *gnd*, 6-phosphogluconate dehydrogenase gene, *his*, the histidine biosynthesis operon

be interesting to analyze the *Y. pestis* region because it is very likely that *Y. pestis*, being rough, has an inactivated O-antigen cluster present in this location. Comparison of the *Y. pestis* cluster to those of different *Y. pseudotuberculosis* serotypes should clarify the relationship between these two species. Relevant to this line of research is that the Sanger Centre (Trust Genome Campus, Cambridge, UK) is sequencing the genome of *Y. pestis*, strain CO-92, biovar Orientalis, a fully virulent strain. The sequence was almost completely determined at the time this was written and is available on the Internet at http://www.sanger.ac.uk/Projects/Y_pestis/.

Yersinia *LPS and virulence*

An insertion mutant in the outer core operon of a wild type *Y. enterocolitica* O:3 strain was constructed by site-directed mutagenesis;[27] this mutant, YeO3-trs11 (Table 2.3) expresses no outer core. A rough mutant 8081-R2 was isolated from *Y. enterocolitica* 8081-R⁻-M⁺ by resistance to lysis by bacteriophage φ80-18.[28] A rough mutant of *Y. enterocolitica* O:3, YeO3-R2, was isolated analogously; the phage used was φYeO3-12.[25]

The LD_{50} values for the *Y. enterocolitica* LPS mutants were determined using intragastrically infected DBA/2 mice. All three mutants were less virulent than the wild-type strain; the LD_{50} values were about 50 to 100 times higher for the rough mutants;[25,28] the outer core mutant was completely avirulent.[37a] The outer core seems to be indispensable by giving the bacteria resistance to cationic bactericidal peptides. Apart from the endotoxic activity, the role of *Yersinia* LPS in bacterial pathogenesis has received very little attention, thus its biological role is far from known. There are indications that O-antigen may play a role in resistance to complement-mediated killing,[62] however, LPS may also have roles in adhesion to and invasion of the host tissues, and in the pathogenesis of reactive arthritis.

For *Y. enterocolitica*, we now have information that O-antigen and the outer core play a role in virulence, but there is no clear understanding what that role is. Studies that include characterized LPS mutants tested in relevant experimental animal and *in vitro* models need to be carried out.

Future prospects

During the past 7 to 10 years molecular genetic research of *Yersinia* LPS has produced a wealth of basic information which, when combined with the information obtained from other Gram-negative organisms, has made it possible to write this review. In future, more information is needed on the exact biological roles different parts of LPS have, and in this respect the roles of the specific structures of O-antigens should be studied using substitution mutants, for example, expressing *Y. enterocolitica* O:3 O-antigen in rough O:8 strain. Another important research field will be the regulation of the LPS biosynthesis and its networking with the biosynthesis of the bacterial cell wall and that of OM especially. Finally, for applications such as vaccine and antimicrobial drug development, this new information may open new prospects.

References

1. Skurnik, M. and Zhang, L., Molecular genetics and biochemistry of *Yersinia* lipopolysaccharide, *APMIS*, 104, 1996.
2. Woese, C. R., Bacterial evolution, *Microbiol. Rev.*, 51, 221, 1987.
3. Davies, R. L., O-serotyping of *Yersinia ruckeri* with special emphasis on European isolates, *Vet. Microbiol.*, 22, 299, 1990.
4. Wauters, G., Aleksic, S., Charlier, J., and Schulze, G., Somatic and flagellar antigens of *Yersinia enterocolitica* and related species, *Contrib. Microb. Immunol.*, 12, 239, 1991.
5. Robins-Browne, R. M. and Prpic, J. K., Effects of iron and desferroxamine on infections with *Yersinia enterocolitica*, *Infect. Immun.*, 47, 774, 1985.
6. Tsubokura, M. and Aleksic, S., A simplified antigenic scheme for serotyping of *Yersinia pseudotuberculosis*: Phenotypic characterization of reference strains and preparation of O and H factor sera, *Contrib. Microb. Immunol.*, 13, 99, 1995.
7. Tsubokura, M., Otsuki, K., Kawaoka, Y., Fukushima, H., Ikemura, K., and Kanazawa, Y., Addition of new serogroups and improvement of the antigenic designs of *Yersinia pseudotuberculosis*, *Curr. Microbiol.*, 11, 89, 1984.
8. Tsubokura, M., Aleksic, S., Fukushima, H., Schulze, G., Someya, K., Sanekata, T., Otsuki, K., Nagano, T., Kuratani, Y., Inoue, M., Zheng, X., and Nakajima, H., Characterization of *Yersinia pseudotuberculosis* serogroups O9, O10 and O11 — subdivision of O1 serogroup into O1a, O1b, and O1c subgroups, *Int. J. Med. Microb.*, 278(4), 500, 1993.
9. Kessler, A. C., Brown, P. K., Romana, L. K., and Reeves, P. R., Molecular cloning and genetic characterization of the *rfb* region from *Yersinia pseudotuberculosis* serogroup IIA, which determines the formation of the 3,6-dideoxyhexose abequose, *J. Gen. Microbiol.*, 137, 2689, 1991.
10. Kessler, A. C., Haase, A., and Reeves, P. R., Molecular analysis of the 3,6-dideoxyhexose pathway genes of *Yersinia pseudotuberculosis* serogroup IIA, *J. Bacteriol.*, 175(5), 1412, 1993.
11. Lindqvist, L., Schweda, K. H., Reeves, P. R., and Lindberg, A. A., *In vitro* synthesis of CDP-D-abequose using *Salmonella* enzymes of cloned *rfb* genes — Production of CDP-6-deoxy-D-xylo-4-hexulose, CDP-3,6-dideoxy-D-xylo-4-hexulose and CDP-3,6-dideoxy-D-galactose, and isolation by HPLC, *Eur. J. Biochem.*, 225(3), 863, 1994.
12. Samuelsson, K., Lindberg, B., and Brubaker, R. R., Structure of O-specific side chains of lipopolysaccharides from *Yersinia pseudotuberculosis*, *J. Bacteriol.*, 117, 1010, 1974.
13. Korchagina, N. I., Gorshkova, R. P., and Ovodov, Y. S., Studies on O-specific polysaccharide from *Yersinia pseudotuberculosis* VB serovar, *Bioorg. Khimiya*, 8(12), 1666, 1982.
14. Aleksic, S., Bockemühl, J. and Wuthe, H.-H., Epidemiology of *Y. pseudotuberculosis* in Germany, 1983-1993, *Contrib. Microb. Immunol.*, 13, 55, 1995.

15. Tsubokura, M., Otsuki, K., Sato, K., Tanaka, M., Hongo, T., Fukushima, H., Maruyama, T., and Inoue, M., Special features of distribution of *Yersinia pseudotuberculosis* in Japan, *J. Clin. Microbiol.*, 27, 790, 1989.

16. Brubaker, R. R., Factors promoting acute and chronic diseases caused by *Yersiniae*, *Clin. Microb. Rev.*, 4, 309, 1991.

17. Devignant, R., Variétés de l'éspèce *Pasteurella pestis*. Nouvelle hypothèse, *Bulletin WHO*, 4, 247, 1951.

18. Brubaker, R. R., The genus *Yersinia*: Biochemistry and genetics of virulence, *Curr. Top. Microbiol. Immunol.*, 57, 111, 1972.

19. Reeves, P. R., Hobbs, M., Valvano, M., Skurnik, M., Whitfield, C., Coplin, D., Kido, N., Klena, J., Maskell, D., Raetz, C., and Rick, P., Bacterial polysaccharide synthesis and gene nomenclature, *Trends Microbiol.*, 4, 495, 1996.

20. Liu, D., Cole, R. A., and Reeves, P. R., An O-antigen processing function for Wzx (rfbX): a promising candidate for O-unit flippase, *J. Bacteriol.*, 178, 2102, 1996.

21. Whitfield, C., Biosynthesis of lipopolysaccharide O antigens, *Trends Microbiol.*, 3(5), 178, 1995.

22. Baker, P. M. and Farmer, J. J. I., New bacteriophage typing system for *Yersinia enterocolitica*, *Yersinia kristensenii*, *Yersinia frederiksenii*, and *Yersinia intermedia*: correlation with serotyping, biotyping, and antibiotic susceptibility, *J. Clin. Microbiol.*, 15, 491, 1982.

23. Nicolle, P., Mollaret, H., Hamon, Y., and Vieu, J. F., Étude lysogénique, bactériocinogénique et lysotypique de l'espèce *Yersinia enterocolitica*, *Annales L'Inst. Pasteur*, 112, 86, 1967.

24. Al-Hendy, A., Toivanen, P., and Skurnik, M., Expression cloning of *Yersinia enterocolitica* O:3 *rfb* gene cluster in *Escherichia coli* K12, *Microb. Pathog.*, 10, 47, 1991.

25. Al-Hendy, A., Toivanen, P., and Skurnik, M., Lipopolysaccharide O side chain of *Yersinia enterocolitica* O:3 is an essential virulence factor in an orally infected murine model, *Infect. Immun.*, 60, 870, 1992.

26. Zhang, L. and Skurnik, M., Isolation of an R⁻ M⁺ mutant of *Yersinia enterocolitica* serotype O:8 and its application in construction of rough mutants utilizing mini-Tn5 derivatives and lipopolysaccharide-specific phage, *J. Bacteriol.*, 176(6), 1756, 1994.

27. Skurnik, M., Venho, R., Toivanen, P., and Al-Hendy, A., A novel locus of *Yersinia enterocolitica* serotype O:3 involved in lipopolysaccharide outer core biosynthesis, *Mol. Microbiol.*, 17, 575, 1995.

28. Zhang, L., Radziejewska-Lebrecht, J., Krajewska-Pietrasik, D., Toivanen, P., and Skurnik, M., Molecular and chemical characterization of the lipopolysaccharide O-antigen and its role in the virulence of *Yersinia enterocolitica* serotype O:8., *Mol. Microbiol.*, 23, 63, 1997.

29. Radziejewska-Lebrecht, J., Shashkov, A. S., Stroobant, V., Wartenberg, K., Warth, C., and Mayer, H., The inner core region of *Yersinia enterocolitica* Ye75R (0:3) lipopolysaccharide, *Eur. J. Biochem.*, 221(1), 343, 1994.

30. Hoffman, J., Lindberg, B., and Brubaker, R. R., Structural studies of the O-specific side-chains of the lipopolysaccharide from *Yersinia enterocolitica* Ye 128, *Carbohydr. Res.*, 78, 212, 1980.

31. Hobbs, M. and Reeves, P. R., The JUMPstart sequence: A 39 bp element common to several polysaccharide gene clusters, *Mol. Microbiol.*, 12(5), 855, 1994.

32. Pierson, D. E. and Carlson, S., Identification of the *galE* gene and a *galE* homolog and characterization of their roles in the biosynthesis of lipopolysaccharide in a serotype O:8 strain of *Yersinia enterocolitica, J. Bacteriol.*, 178(20), 5916, 1996.

33. Klena, J. D. and Schnaitman, C. A., Function of the *rfb* gene cluster and the *rfe* gene in the synthesis of O-antigen by *Shigella dysenteriae* 1, *Mol. Microbiol.*, 9, 393, 1993.

34. Schnaitman, C. A. and Klena, J. D., Genetics of lipopolysaccharide biosynthesis in enteric bacteria, *Microbiol. Rev.*, 57(3), 655, 1993.

35. Reeves, P., Biosynthesis and assembly of lipopolysaccharide, in *Bacterial Cell Wall*, Vol. 27, Ghuysen, J.-M. and Hakenbeck, R., Eds., Amsterdam, Elsevier, 1994, 281.

36. Zhang, L., Toivanen, P., and Skurnik, M., The gene cluster directing O-antigen biosynthesis in *Yersinia enterocolitica* serotype O:8: Identification of the genes for mannose and galactose biosynthesis and the gene for the O-antigen polymerase, *Microbiology* — *U.K.*, 142(Part 2), 277, 1996.

37. Stevenson, G., Kessler, A., and Reeves, P. R., A plasmid-borne O-antigen chain length determinant and its relationship to other chain length determinants, *FEMS Microb. Lett.*, 125(1), 23, 1995.

37a. Skurnik, M., Venho, R., Bengoechea, J.A., and Moriyón, I., The lipopolysaccharide outer core of *Yersinia enterocolitica* serotype O:3 is required for virulence and plays a role in outer membrane integrity. *Mol. Microbiol.*, 31, 5, 1999.

38. Pekkola-Heino, K., Viljanen, M. K., Ståhlberg, T. H., Granfors, K., and Toivanen, A., Monoclonal antibodies reacting selectively with core and O-polysaccharide of *Yersinia enterocolitica* O:3 lipopolysaccharide, *APMIS*, 95, 27, 1987.

39. Zhang, L., Al-Hendy, A., Toivanen, P., and Skurnik, M., Genetic organization and sequence of the *rfb* gene cluster of *Yersinia enterocolitica* serotype O:3: Similarities to the dTDP-L-rhamnose biosynthesis pathway of *Salmonella* and to the bacterial polysaccharide transport systems, *Mol. Microbiol.*, 9, 309, 1993.

40. Zhang, L., *Molecular Genetics of the O-Antigen of Yersinia enterocolitica Lipopolysaccharide.* Ph.D. thesis, *Annales Universitatis Turkuensis, Ser. D.*, Vol. 241, University of Turku, Turku, Finland, 1996.

41. Morona, R., Macpherson, D. F., Vandenbosch, L., Carlin, N. I. A., and Manning, P. A., Lipopolysaccharide with an altered O-antigen produced in *Escherichia coli* K-12 harbouring mutated, cloned *Shigella flexneri rfb* genes, *Mol. Microbiol.*, 18(2), 209, 1995.

42. Harr, R., Häggström, M., and Gustafsson, P., Search algorithm for pattern match analysis of nucleic acid sequences, *Nucl. Acids Res.*, 11, 2943, 1983.

43. Harr, R., Fällman, P., Häggström, M., Wahlström, L., and Gustafsson, P., GENEUS, a computer system for DNA and protein sequence analysis containing information retrieval system for EMBL data library, *Nucl. Acids Res.*, 14, 273, 1985.

44. Kawaoka, Y., Wauters, G., Otsuki, K., and Tsubokura, M., Identification of *Yersinia enterocolitica* O15 lipopolysaccharide as a rough antigen, *J. Clin. Microbiol.*, 24, 272, 1986.

45. Tomshich, S. V., Gorshkova, R. P., and Ovodov, Y. S., Structural studies on lipopolysaccharide from *Y. enterocolitica* serovar O:8, *Khimia Prirodnykh Soedinenii*, 657, 1987.

46. Ovodov, Y. S., Gorshkova, R. P., Tomshich, S. V., Komandrova, N. A., Zubkov, V. A., Kalmykova, E. N., and Isakov, V. V., Chemical and immunochemical studies on lipopolysaccharides of some *Yersinia* species — a review of some recent investigations, *J. Carbohydr. Chem.*, 11, 21, 1992.
47. Pierson, D. E., Mutations affecting lipopolysaccharide enhance Ail-mediated entry of *Yersinia enterocolitica* into mammalian cells, *J. Bacteriol.*, 176, 4043, 1994.
48. Bastin, D. A., Stevenson, G., Brown, P. K., Haase, A., and Reeves, P. R., Repeat unit polysaccharides of bacteria — A model for polymerization resembling that of ribosomes and fatty acid synthetase, with a novel mechanism for determining chain length, *Mol. Microbiol.*, 7(5), 725, 1993.
49. Zhang, L. J., Toivanen, P., and Skurnik, M., Genetic characterization of a novel locus of *Yersinia enterocolitica* serotype O:8 for down-regulation of the lipopolysaccharide O side chain at 37 degrees C, in *Yersiniosis: Present and Future*, Ravagnan, G. and Chiesa, C., Eds., Basel, Switzerland, Karger, 1995, 310.
50. Hobbs, M. and Reeves, P., Genetic organization and evolution of *Yersinia pseudotuberculosis* 3,6-dideoxyhexose biosynthetic genes, *Biochim. Biophys. Acta*, 1245, 273, 1995.
51. Thorson, J. S., Lo, S. F., and Liu, H. W., Molecular basis of 3,6-dideoxyhexose biosynthesis — elucidation of CDP-ascarylose biosynthetic genes and their relationship to other 3,6-dideoxyhexose pathways, *J. Am. Chem. Soc.*, 115, 5827, 1993.
52. Thorson, J. S., Lo, S. F., Liu, H. W., and Hutchinson, C. R., Biosynthesis of 3,6-dideoxyhexoses — New mechanistic reflections upon 2,6-dideoxy, 4,6-dideoxy, and amino sugar construction, *J. Am. Chem. Soc.*, 115(15), 6993, 1993.
53. Thorson, J. S. and Liu, H. W., Characterization of the first PMP-dependent iron sulfur-containing enzyme which is essential for the biosynthesis of 3,6-dideoxyhexoses, *J. Am. Chem. Soc.*, 115(16), 7539, 1993.
54. Thorson, J. S., Lo, S. F., Ploux, O., He, X. M., and Liu, H. W., Studies of the biosynthesis of 3,6-dideoxyhexoses: Molecular cloning and characterization of the *asc* (Ascarylose) region from *Yersinia pseudotuberculosis* serogroup VA, *J. Bacteriol.*, 176(17), 5483, 1994.
55. Lo, S. F., Miller, V. P., Lei, Y. Y., Thorson, J. S., Liu, H. W., and Schottel, J. L., CDP-6-deoxy-Δ(3)(4)-glucoseen reductase from *Yersinia pseudotuberculosis* — enzyme purification and characterization of the cloned gene, *J. Bacteriol.*, 176(2), 460, 1994.
56. Miller, V. P., Thorson, J. S., Ploux, O., Lo, S. F., and Liu, H. W., Cofactor characterization and mechanistic studies of CDP-6-deoxy-Δ(3,4)-glucoseen reductase — exploration into a novel enzymatic C-O bond cleavage event, *Biochemistry*, 32(44), 11934, 1993.
57. Ploux, O., Lei, Y. Y., Vatanen, K., and Liu, H. W., Mechanistic studies on CDP-6-deoxy-Delta(3,4)-glucoseen reductase: The role of cysteine residues in catalysis as probed by chemical modification and site-directed mutagenesis, *Biochemistry*, 34(13), 4159, 1995.
58. Nieto, J. M., Bailey, M. J. A., Hughes, C., and Koronakis, V., Suppression of transcription polarity in the *Escherichia coli* haemolysin operon by a short upstream element shared by polysaccharide and DNA transfer determinants, *Mol. Microbiol.*, 19, 705, 1996.
59. Bailey, M. J. A., Hughes, C., and Koronakis, V., Increased distal gene transcription by the elongation factor RfaH, a specialized homologue of NusG, *Mol. Microbiol.*, 22(4), 729, 1996.

60. Berlyn, M. K. B., Low, K. B., Rudd, K. E., and Singer, M., Linkage map of *Escherichia coli* K-12, edition 9., in *Escherichia coli and Salmonella: Cellular and Molecular Biology,* Vol. 2, 2nd ed., Neidhardt, F. C., Curtiss III, R., Ingraham, J. L., Lin, E. C. C., Low, K. B., Magasanik, B., Reznikoff, W. S., Riley, M., Schaechter, M., and Umbarger, H. E., Eds., Washington, D.C., American Society for Microbiology, 1996, 1715.
61. Sanderson, K. E., Hessel, A., Liu, S.-L., and Rudd, K. E., The genetic map of *Salmonella typhimurium,* edition VIII., in *Escherichia coli and Salmonella: Cellular and Molecular Biology,* Vol. 2, 2 ed., Neidhardt, F. C., Curtiss III, R., Ingraham, J. L., Lin, E. C. C., Low, K. B., Magasanik, B., Reznikoff, W. S., Riley, M., Schaechter, M., and Umbarger, H. E., Eds., Washington, D.C., American Society for Microbiology, 1996, 1903.
62. Wachter, E. and Brade, V., Influence of surface modulations by enzymes and monoclonal antibodies on alternative complement pathway activation by *Yersinia enterocolitica, Infect. Immun.,* 57, 1984, 1989.
63. Skurnik, M., Lack of correlation between the presence of plasmids and fimbriae in *Yersinia enterocolitica* and *Yersinia pseudotuberculosis, J. Appl. Bact.,* 56, 355, 1984.
64. Portnoy, D. A., Moseley, S. L., and Falkow, S., Characterization of plasmids and plasmid-associated determinants of *Yersinia enterocolitica* pathogenesis, *Infect. Immun.,* 31, 775, 1981.

chapter three

Rhizobial cell surface carbohydrates: Their structures, biosynthesis, and functions

Russell W. Carlson, Bradley L. Reuhs, L. Scott Forsberg, and Elmar L. Kannenberg

Contents

Introduction 53
Rhizobial extracellular polysaccharides..55
Cyclic glucans...57
Rhizobial lipo-chitin-oligosaccharide (LCO) signal molecules...................58
Rhizobial K-antigens 59
    The structures of the K-antigens from rhizobia and other plant-
       associated bacteria..59
    The genetics of K-antigen expression in rhizobia 60
    The biology of *Rhizobium* K-antigens 61
Rhizobial lipopolysaccharides 63
    The structures of rhizobial lipopolysaccharides.....................................63
    The biosynthesis of *R. etli* and *R. leguminosarum* lipid-A 71
    The genetics of *Rhizobium* LPS synthesis...73
    Rhizobial LPS structural variation during symbiotic
       nodule development..74
Acknowledgments 77
References...77

Introduction

The members of the family *Rhizobiaceae* (this family of bacteria includes *Rhizobium, Azorhizobium, Bradyrhizobium, Mesorhizobium,* and *Sinorhizobium*[201] and will be collectively referred to as rhizobia) are Gram-negative eubacteria.

Rhizobia are soil bacteria which form a symbiosis with legume plants resulting in nitrogen-fixing root nodules. Characterizing the molecular basis for *Rhizobium*–legume symbiosis will lead to a greater understanding (1) of cell–cell recognition between a prokaryotic and eukaryotic cell, (2) of the plant defense mechanism and how it is regulated to permit symbiotic infection, and (3) of both bacterial and plant differentiation processes (bacteroid and nodule development, respectively).

Rhizobial–legume symbiosis involves an exchange of signal molecules which results in a highly coordinated regulation of gene expression on the part of both the rhizobial symbiont and the host legume. The end result of this process is a root nodule which contains nitrogen-fixing bacteria called *bacteroids*. A successful nitrogen-fixing relationship is characterized by several observable phenomena.[74]

1. There is host–symbiont specificity, e.g., *R. leguminosarum* infects peas but not alfalfa, while *S. meliloti* infects alfalfa but not peas.
2. Rhizobia attach to the emerging root hairs of the host legume.
3. Invagination of the root hair membrane occurs, which extends down the root hair to the inner cortex cells of the root. This invagination is called the *infection thread*, and bacteria travel down this infection thread and reach the newly formed meristem in the cortex of the root.
4. The cortex cells are stimulated to divide and de-differentiate, forming a nodule on the root.
5. The infection thread penetrates the cortical cells of the root nodule, and the bacteria, surrounded by infection thread membrane (now called the *peribacteroid membrane*), are released into the cell cytoplasm.
6. The bacteria differentiate into bacteroids, which are altered in size and shape when compared to free-living bacteria, and produce nitrogenase, the enzyme that reduces di-nitrogen to ammonia.

A legume host can form one of two types of nodules, *determinate* or *indeterminate*. The formation of indeterminate nodules (e.g., on pea, alfalfa, vetch, etc.) takes place by the initiation of cortical cell division in the inner cortex and by the formation of an apical meristem which continues to divide in mature nodules. Cell division in indeterminate nodules is mediated by infection threads, and infected cells do not divide. The formation of determinate nodules (e.g., on bean, soybean, etc.) is initiated by cortical cell division in the outer cortex, and by development of a spherical meristem which ceases to divide when nodules mature. Cell invasion in determinate nodules is mediated by infection threads and the dividing meristematic cells (which continue to divide after rhizobial infection).

The outer surface of Gram-negative bacteria, including rhizobia, consists of a complex array of different molecules which include lipopolysaccharides (LPSs), capsular polysaccharides (CPSs), extracellular polysaccharides (EPSs), porins, fimbriae, and flagella. Rhizobia are also known to produce additional molecules, including cyclic glucans and, in response to the host legume, lipo-chitin-oligosaccharides (LCOs). The bacterial surface is the first line of defense against antimicrobial molecules, and against stress caused by changes in the environment surrounding the bacterium. In the case of plant– and animal–microbe interactions, many of these bacterial surface molecules are important virulence determinants. Thus, in order to understand the molecular basis of symbiosis, it is important to characterize the molecular architecture of the rhizobial cell surface and determine how the bacterium modifies this architecture in response to its different environments, including its *in planta* environment.

This chapter will concentrate on rhizobial LPSs and CPSs. Only brief summaries of the work on EPSs, cyclic glucans, and LCOs will be given. There have been a number of relatively recent and extensive reviews on these molecules.

Rhizobial extracellular polysaccharides

The EPSs are released into the growth media and, therefore, can be purified from the cell-free culture supernatant following centrifugation. They consist of polymerized repeating oligosaccharides which differ in size and structure. They have been most extensively examined from strains of *R. leguminosarum*, *R. etli*, *S. meliloti*, *B. japonicum*, and *B. elkanii*. Examples of various EPS structures are given in Table 3.1. Certain mutants of *S. meliloti* that are unable to synthesize the predominant EPS (EPS I), a succinylated β-3-linked glucan, synthesize the second EPS (EPS II).[103,205] For some rhizobia, it has been found that a second EPS can be produced which is not synthesized under normal laboratory culture conditions. Some strains of *B. japonicum* produce a soybean nodule-specific EPS (NPS) which does not have the same structure as that observed for the EPS from cultures grown under laboratory conditions.[2,186]

The precise role of EPSs in symbiotic infection is not known. However, rhizobia that are unable to synthesize EPS are normally defective at infecting indeterminate nodules.[59,70,72,100,101,156,168] These mutants still attach to their legume hosts, form infection threads, and induce nodules. One possibility for the function of the EPS is that it acts as a passive barrier to antimicrobial compounds produced as part of the host legume's defense response. However, other evidence suggests that EPS may have a more active role in infection. When the legume host is inoculated with the mutant together with added amounts of purified EPS, the symbiotic defect is complemented and nitrogen-fixing nodules are formed.[8,56,67] It was found, in the case of *S. meliloti*, that a tetramer of the EPS I octasaccharide repeating unit is the optimum size for this complementation.[8] More recently, it was found that a hexadecamer of the EPS II disaccharide repeating unit is the optimum size which can also complement this symbiotic defect.[67] In both cases, the optimum number of sugar residues for complementation is around 32. Furthermore, only picomolar concentrations, in the case of EPS II, are required,[67] suggesting that, rather than acting as a passive protector from the host defense mechanism, EPS is acting as a type of signal molecule required for infection. It has been hypothesized that this "signal" may act by suppressing the host

Table 3.1 Examples of Rhizobial Extracellular Polysaccharides

Species	Repeat Unit Structures
R. leguminosarum and *R. etli*	→4)-β-D-Glcp*A*-(1→4)-3-O-Ac-β-D-GlcpA-(1→4)-2 or 3-O-Ac-β-D-Glcp-(1→ 4,6-O-Pyr-β-D-Galp-(1→3)-4,6-O-Pyr-β-D-Glcp-(1→4)-β-D-Glcp-(1→4)-β-D-Glcp-(1→6ᴶ
S. meliloti	→4)-β-D-Glcp-(1→4)-6-O-Ac-β-D-Glcp-(1→3)-β-D-Galp-(1→4)-β-D-Glcp-(1→ 4,6-O-Pyr-β-D-Glcp-(1→3)-6-O-Succ-β-D-Glcp-(1→3)-β-D-Glcp-(1→6)-β-D-Glcp-(1→6ᴶ EPS I →3)-6-O-Ac-β-D-Glcp-(1→3)-4,6-O-Pyr-β-D-Galp-(1→ EPS II
R. spp. NGR234	→6)-β-D-Glcp-(1→4)-β-D-Glcp-(1→4)-β-D-Glcp-(1→3)-β-D-Galp-(1→4)-β-D-Glcp-(1→6)-β-D-Glcp-(1→ 2 or 3-O-Ac-4,6-Pyr-α-D-Galp-(1→4)-α-D-GlcpA-(1→3)-β-D-GlcpA-(1→4ᴶ
B. japonicum	→3)-β-D-Glcp-(1→3)-α-D-GalpA-(1→3)-α-D-Manp-(1→3)-α-D-Glcp-(1→ 4-O-Me-α-D-Galp-(1→6ᴶ) EPS →4)-β-L-Rhap-(1→3)-α-D-Galp-(1→3)-β-L-Rhap-(1→4)-β-L-Rhap-(1→ 2-O-Me-β-D-GlcpA-(1→2ᴶ) NPS
B. elkanii	→3)-β-L-Rhap-(1→4)-β-L-Rhap-(1→4)-α-L-Rhap-(1→ 4-O-Me-β-D-GlcpA-(1→3ᴶ)

defense response since examination of symbiotically defective nodules induced by EPS⁻ mutants indicate the induction of a general plant defense response.[124,129] Further work is in progress to characterize the role of EPS in symbiotic infection.

The biosynthetic pathway has been determined for one rhizobial EPS, the EPS I from *S. meliloti*. The biosynthesis of this EPS has been extensively studied in the laboratories of Graham Walker at the Massachusetts Institute of Technology,[65,66,68,105,152,157] Alfred Puhler at the University of Bielefeld,[9,14,36,94,127] and John Leigh at the University of Washington.[3,99,135,203,204] The complete biosynthetic pathway was summarized in a series of articles by Glucksmann et al.,[65,66] and by Reuber and Walker.[157] The general biosynthetic scheme for EPS I, as described in those reports, is as shown in Figure 3.1. The succinoglycan eight-sugar repeat unit is prepared on bactoprenyl phosphate by a series of glycosyl transferases, polymerized, and then exported in what is thought to be an energy-dependent manner. The polymerization and/or transport are thought to be functions of the products encoded by *exoP, exoQ,* and *exoT*.[65,66,157] More recently, a gene, *exsA*, was discovered in the EPS I *exo* cluster, which has homology to ATP-binding cassette (ABC) transport proteins[13] and may be involved in EPS I export. As with other Gram-negative bacteria, the regulation of *S. meliloti* EPS synthesis is complex and involves both positive (e.g., *mucR*[94]) and negative regulatory elements (e.g., *exsB, exoR,* and *exoS*[135]). Both positive and negative regulatory elements have also been reported for EPS synthesis in *R. leguminosarum*.[5,22,23,98,125,126] This general biosynthetic mechanism for *S. meliloti* EPS is analogous to that for the synthesis of other types of bacterial EPSs (for a review see Reference 194), and it is likely that it also applies to the synthesis of other rhizobial EPSs.

Readers are referred to a number of reviews in which the structures and functions of rhizobial EPSs have been described in much more detail.[70,72,99,101]

Cyclic glucans

Cyclic β-linked glucans are cyclic oligosaccharides comprised entirely of glucose. They have been found in all rhizobia examined. These molecules can be excreted from the bacterium, however, they are primarily found in the periplasmic space. In *Rhizobium* and *Sinorhizobium*, as well as in *Agrobacterium* species, these glucans are β-2-linked cyclic molecules,[7,25,52,119,202] while in *Bradyrhizobium* they are cyclic molecules that contain both β-3- and β-6-linkages.[69,117,119,141,169] The cyclic glucans from rhizobia have a minimum size of about 17 glucose units, and have also been found to contain phosphorylated substituents,[26,27,118,121,140] e.g., phosphoglycerol.

The structures, biosynthesis, and function of rhizobial cyclic glucans have been extensively reviewed during the past few years, and for further details readers are referred to References 25, 120, and 122.

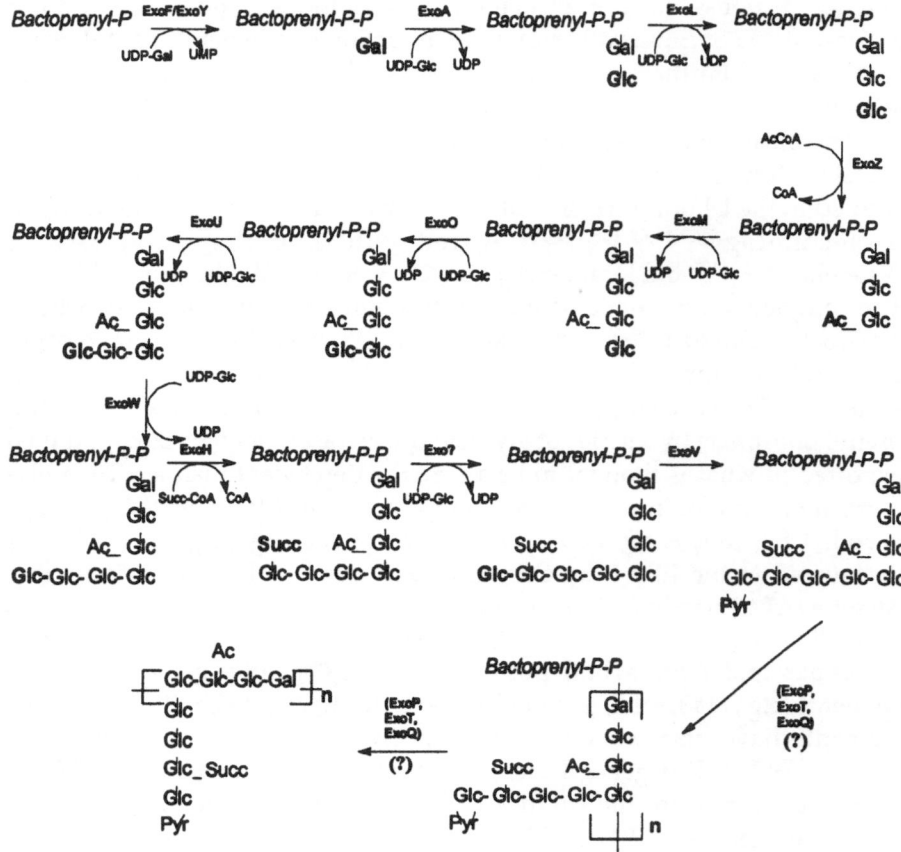

Figure 3.1 The biosynthetic pathway for the synthesis of *S. meliloti* EPS. (From Reuber, T. L. and G. C. Walker. Biosynthesis of succinoglycan, a symbiotically important exopolysaccharide of *Rhizobium meliloti. Cell* 74:269-280, 1993. With permission.)

Rhizobial lipo-chitin-oligosaccharide (LCO) signal molecules

The rhizobial LCO signal molecules have been the subject of numerous recent reviews.[42,54,55,60,97,102,182] Many LCO structures have been published during the past seven years. They all consist of an acylated chitin oligosaccharide backbone which can be modified in numerous ways depending on the rhizobial species from which the LCO is isolated. The LCOs are very active molecules causing root hair curing and inducing division of root cortical cells and nodule formation at nano- to picomolar concentrations. This biological activity can be quite specific in that certain structures are required for activity on certain legume hosts. Thus, LCOs are signal molecules produced by a specific rhizobial species and recognized by a specific legume host. The readers are referred to the recent reviews mentioned above for further information on the structures, biosynthesis, and functions of these very interesting molecules.

Rhizobial K-antigens

Acidic CPSs, including K-antigens, are tightly associated with the bacterial cells, do not impart a mucoid colony morphology, and are common components of many well-studied bacteria.[194] These polysaccharides are often essential for pathogenic virulence and are believed to protect the bacteria from host defense responses. The acidic capsular antigens of *E. coli* were named K-antigens (from "Kapselantigene") because they proved to be bacterial cell surface antigens that were distinct from the O-antigens. During the past several years, CPSs that are analogous to the group II K-antigens of *E. coli* have been found in rhizobia.

The structures of the K-antigens from rhizobia and other plant-associated bacteria

Rhizobial K-antigens were first found in the water layer of hot phenol/water extracts during the purification of LPSs from *S. fredii* strains and were initially identified as being derived from LPS[46] due to the fact that one of their major glycosyl components is D-*manno*-3-deoxy-2-octulosonic acid (Kdo), a component characteristic of LPSs. Subsequently, it was found that these K-antigens were not LPS, but were distinct rhizobial surface antigens.[159] Thus far, the structures of three K-antigens from strains of *S. fredii* have been determined.[62,159] Each is comprised of a disaccharide repeating unit consisting of one Kdo and one hexosyl residue. This general structure is analogous to those of the *E. coli* group II K-antigens.[82,83] The major K-antigen polysaccharide from *S. fredii* USDA205 is $[\rightarrow)$-α-D-Gal$p$-$(1{\rightarrow}5)$-β-D-Kdo$p$ $(2{\rightarrow}]_n$.[159] This strain produces a second minor polysaccharide that consists of a $[\rightarrow2)$-O-methyl-Man-$(1{\rightarrow}?)$-Kdo-$2{\rightarrow}]_n$ disaccharide repeating unit, which has not been completely characterized. *Sinorhizobium fredii* USDA257 also produces two distinct K-antigens[62]: $[\rightarrow)3$-β-D-Man$p$-$(1{\rightarrow}5)$-β-D-Kdo$p$-$(2{\rightarrow}]_n$ being the major polysaccharide, and $[\rightarrow)3$-β-D-2-O-MeMan$p$-$(1{\rightarrow}5)$-β-D-Kdo$p$-$(2{\rightarrow}]_n$ the minor polysaccharide. Similar polysaccharides have also been isolated from an *exoB*⁻ mutant of *S. meliloti* Rm41 (named strain AK631). Since *exoB* encodes for a UDP-Gal 4-epimerase, AK631 does not synthesize either EPS I or II. Present data (unpublished) indicate that the major K-antigen from AK631 has a disaccharide repeating unit consisting of 2-keto-3,5,7,9-tetradeoxy-5,7-diaminononulosonic acid (a variant of Kdo) and an aminohexuronosyl residue. There also may be a second minor K-antigen as was observed for *S. fredii*. Unlike the *S. fredii* polysaccharides, those of *S. meliloti* AK631 are substituted with acetate and β-hydroxybutyrate.[139,159] In summary, the rhizobial K-antigen varies in structure from strain to strain within a single species, and also from species to species. However, within this variability there appears to be a conserved structural motif; namely:

$$R_1 \quad R_2$$
$$\downarrow \quad \downarrow$$
$$[\rightarrow Kdx \rightarrow Sug \rightarrow]_n$$

where Kdx is a 1-carboxy-2-keto-3-deoxy sugar (i.e., Kdo or a variant), Sug is any possible second glycosyl residue, and R_1 and R_2 can be protons or substituents such as acetyl or β-hydroxybutryl groups.

A systematic evaluation of many plant-associated Gram-negative bacteria was undertaken to determine if the presence of these K-antigens is a common feature of these bacteria. This study included examples of many well-studied plant pathogens in addition to rhizobia: *Agrobacterium tumefaciens*, *Erwinia amylovora*, *E. carotovora*, *E. stewartii*, *Pseudomonas solanacearum*, *P. syringae*, and *Xanthomonas campestris*.[161] The results showed that cell-associated, acidic polysaccharides are, in fact, common components of these bacterial cell extracts.[158]

The genetics of K-antigen expression in rhizobia

Genetic investigations have shown that capsule expression in *S. meliloti* AK631 involves at least three separate capsule-specific gene regions[96,139] (Peter Putnoky, personal communication), comprising a minimum of 15 genes, but most likely many more. First, it was shown that expression of the plasmid-borne *rkpZ* gene lowers the size range of the exported polysaccharide.[162] It is the low-molecular-weight form of the K-antigen that is required for infection by the EPS mutants (discussed below). The activity of RkpZ is nonspecific with regard to the K-antigen structure because introduction of *rkpZ* from *S. meliloti* AK631 into *S. fredii* USDA257 resulted in a reduction in the size range of the USDA257 K-antigen, even though it is structurally distinct from that of *S. meliloti* AK631.[162]

The rhizobial K-antigen locus known as the *fix-23* gene region consists of several complementation units.[139,148] One complementation unit of *fix-23*, which comprises six ORFs (termed *rkpABCDEF*), appears to encode for products required for the synthesis of a lipid that may be involved in the biosynthesis and export of the capsule[139,162] since both the gene arrangement in the cistron and the individual ORF sequences were similar to those for the fatty acid biosynthesis genes in rats. The last three complementation units harbor four genes (*rkpGHIJ*) which encode products that may modify the lipid carrier or the polymerization process. Two mutants in *rkpGHIJ* retained sensitivity to a phage which binds K-antigen. These mutants were shown by ELISA and immunoblot assays to partially export K-antigen.[96] However, most mutations in the *fix-23* region result in the intracellular accumulation of incompletely polymerized K polysaccharide.[96] The fact that an *rkpABCDEF*-like cluster of genes has never been reported in *E. coli*, and the fact that there is no evidence of a specific lipid carrier for capsule synthesis in *E. coli* indicates that the mechanism for K-antigen expression in these soil bacteria is fundamentally different from that in *E. coli*. On the other hand, RkpZ shows significant similarity to KpsC[136] of *E. coli*, a protein which affects the size of the *E. coli* K-antigen. Additionally, the predicted amino acid sequence of RkpJ shows significant homology to KpsS of *E. coli*, another protein involved in polymerization and export of the group II K-antigens.[32,137]

The data gathered thus far from studies of *S. meliloti* and *S. fredii* suggest a model for K-antigen expression in these rhizobia:

1. The disaccharide repeating units are produced by gene products encoded by as yet uncharacterized gene regions, and an initial polymerization of the repeating units may take place on a lipid carrier yielding polysaccharide subunits of ~5000 to 7000 daltons (8 to 15 repeating units).
2. Exportable polysaccharides may then be assembled on a capsule-specific lipid carrier, determined by the *fix23* region, resulting in high-molecular-weight polymers formed from the above subunits.
3. The polymerization process would be modified by RkpZ, which promotes the export of the smaller polysaccharides. Interestingly, in enteric bacteria both K-antigen and LPS O-chain sizes are reported to be regulated by a chain length determining gene (known as *cld* or *wzz*),[6,57,63,184] and it may be that RkpZ functions by interacting, in some manner, with a rhizobial "*cld*-like" gene product.
4. Upon termination of the polymerization process, the K-antigens would be exported to the cell surface, and during this transport the lipid carrier may be removed and recycled. In the case of *E. coli* K-antigens, KpsD may be a possible candidate for removal of the lipid during transport to the surface, since this protein has homology to PgpB, a phosphatidylglycerophosphase.[79,80] It is possible that a similar protein may be involved in the export of the rhizobial K-antigen.

A gene region that controls host range in *S. fredii*–soybean interactions has also been shown to affect the expression of the K-antigen.[4,95] Mutations in the *nolWXBTUV* genes of *S. fredii* USDA257 result in extending the host range of this strain to include cultivars of soybeans that are not normally infected by this strain. These extended host-range mutants are significantly altered in their surface chemistry, with clear changes in both the LPS and K-antigens (Reuhs, unpublished). The exact mechanism by which the *nol* gene products produce these changes is unclear at this time. However, the functions of the *nol* gene products appear to be secondary to the more general aspects of K-antigen expression because the *nolWXBTUV* operon is specific to *S. fredii* and has not been reported in *S. meliloti*. Interestingly, several of the genes in the *nolWXBTUV* region are homologous to the *hrp* genes of *Xanthomonas* and *Pseudomonas*.[116]

The biology of Rhizobium K-antigens

Mutants of *S. meliloti* SU47, which are unable to produce either EPS I or EPS II, are defective in invading the root nodule cells of their host, alfalfa. An example is a mutant in the *exoB* gene which encodes UDP-Gal 4-epimerase. Since galactose is a component in both EPS I and II, such a mutant can produce neither of the these polysaccharides. However, strain *S. meliloti*

AK631 (Rm41 *exoB*::Tn5) is a mutant that produces neither EPS I nor EPS II, yet is Fix+ on all hosts tested,[147,148] indicating that another surface component can functionally substitute for EPS I or EPS II. A mutation in either the *rkpZ* gene or the *rkpABCDEFGHIJ* gene region of strain AK631 results in a Fix- phenotype.[147,195,196] Initially, it was assumed that the Fix- phenotypes of the AK631 *rkp* mutants were related to a modified expression of the LPS.[147,195,196] These reports correlated the Fix- phenotypes of these mutants with the altered chromatographic profiles of the extracted polysaccharides. The presence of TBA-positive material (i.e., Kdo) and LPS-specific fatty acids in the eluted material was taken as an indication that it was the LPS that had been affected by the mutation. However, it was subsequently shown that the *S. meliloti* AK631 preparation contained an abundance of K-antigen,[159] and that the *rkp* mutations affected the production of the K-antigen, not the LPS.[139,162] Thus, for *S. meliloti* strains carrying *rkpZ*, it was established that the K-antigen, in the absence of EPS, could promote the infection of alfalfa.

A recent study addressed the effects of purified K-antigens on the expression of certain plant genes involved in the host defense response. Whole cells of *S. meliloti* Rm41, AK631, and three capsule mutants (*rkpA*-, *rkpH*-, and *rkpJ*-), as well as purified K-antigen from strain AK631, were employed in this study.[15] The infusion of the whole cells of AK631 or the purified K-antigen into the leaves of alfalfa seedlings resulted in a significant accumulation of chalcone synthase (CHS) mRNA, an enzyme involved in the synthesis of flavonoids, which are intermediates in phytoalexin (i.e., a plant defense anti-microbial molecule) biosynthesis. This suggested a signal-based response of the host plant to this bacterial product. In contrast, there was no response to whole cells of the *rkp* mutants which lack the K-antigen capsule,[15] indicating that the CHS mRNA induction was due (at least in part) to the K-antigen. Importantly, the kinetics of this CHS induction were different from those observed from the infusion of plant pathogens,[15] indicating that the response to *S. meliloti* AK631 cells is different from the typical defense response to a potential pathogen. Additionally, live cells of *S. meliloti* AK631 elicited a two-phase induction of the CHS mRNA, whereas dead cells or the purified K-antigen elicited only the first phase. This indicated that some component not normally expressed may be responsible for the second phase of induction. It is possible that the host plant, in response to the K-antigen, produces a signal molecule affecting a further modification of, and resulting in, the production of the second-phase elicitor. In fact, the expression of K-antigens has been shown to be affected by host plant-derived compounds: The addition of apigenin, a plant-derived flavonoid which induces the bacterium to produce the LCO signal molecules, to the growth media of *S. fredii* USDA205 was shown to increase the minor to major K-antigen ratio.[160] Also, host root extract was shown to greatly increase the expression of the K-antigen by *S. meliloti* AK631.[162] These effects may be important in the infection process.

Rhizobial lipopolysaccharides

As with other Gram-negative bacteria, the rhizobial LPSs comprise the outer leaflet of the outer membrane and consist of three structural regions: An O-chain polysaccharide that consists of a repeating oligosaccharide, a core oligosaccharide, and a lipid known as the lipid-A. Structurally, the O-chain is linked to the core oligosaccharide which, in turn, is linked to the lipid-A through the mild acid labile bond of a Kdo residue. The lipid-A is an acylated carbohydrate that anchors the LPS in the outer membrane. Early work suggested that LPSs might play an important role in the specific recognition process that occurs between a rhizobial symbiont and its legume host. There was a precedence for such function in the infection of animal host cells by enteric bacterial pathogens, and for the interaction of bacteria with bacteriophages. In addition, LPSs constitute one of the major components of the outer membrane and, therefore, were likely candidates for playing a determining role in such cell–cell recognition processes. These early reports were centered around the idea that there was a specific attachment of the symbiont *Rhizobium* to the legume host root which was thought to be mediated by host plant lectins and the rhizobial LPSs.[91,93,198] Subsequent reports (discussed further below) showed that the LPS is important in later infection events.

While LPSs have been shown to be crucial for symbiotic infection,[40,48,53,133,138,146,183] the function of the LPS during this process is not known. Mutants which lack the O-chain polysaccharide portion of their LPSs are defective in symbiosis in that infection threads abort, or invasion of the host cell is disrupted. The complete loss of the O-chain polysaccharide of the LPS is a rather large change in the rhizobial cell surface. Thus, it was not known whether the symbiotic defect was a direct or indirect result of this rather large alteration in the rhizobial cell wall. However, subsequent work, using monoclonal antibodies specific to the LPS of bacteria or bacteroids, showed that changes in LPS structure occur during symbiosis, and that these structural changes likely reside in the O-chain polysaccharide portion of the LPS.[87,89,90,130,132,180,188,199] These results suggested that the LPS O-chain may have a more direct role during the symbiotic process.

The structures of rhizobial lipopolysaccharides

Rhizobial LPSs, as with enteric bacterial LPSs, are complex molecules and a single LPS preparation contains a heterogeneous mixture of molecules. This mixture usually consists of molecules that contain various lengths of the O-chain polysaccharide, and also of molecules that lack the O-chain polysaccharide. The structures of a number of rhizobial LPSs are quite different from those of enteric bacteria. The structures, biosynthesis, and importance of these molecules in *Rhizobium*–legume interactions are discussed in this section.

Table 3.2 Glycosyl Components of the Lipid-A
from Bacteria Belonging to the *Rhizobiaceae*

Group	Species	GlcN	DAG	Man	GalA	GlcN-onate	Phosphate	Reference
I	S. meliloti	+	–	–	–	–	+	190
	S. fredii	+	–	–	–	–	n.d.	21
	R. leg. bv. trifolii	+	–	–	+	+	–	21
II	R. leg. bv. viciae	+	–	–	+	+	–	21
	R. etli	+	–	–	+	+	–	20,21
III	B. japonicum	–	+	–	–	–	–	113
	B. lupini	–	+	–	–	–	–	113
IV	B. elkanii	–	+	+	–	–	n.d.	unpub.
V	M. loti	–	+	–	–	–	+	174

GlcN, glucosamine; DAG, 2,3-diaminoglucose; Man, mannose; GalA, galacturonic acid; GlcN-onate, 2-aminogluconate; n.d. not determined.

Source: Kannenberg, E.L., Reuhs, G.L., Forsberg, L.S., and Carlson, R.W., Lipopolysaccharides and K-antigens: Their structures, biosynthesis, and functions, in *The Rhizobiaceae, Molecular Biology of Model Plant-Associated Bacteria*, Spaink, H.P., Kondorosi, A., and Hooykaas, P.J.J., Kluwer Academic Publishers, Dordrecht, 1998, 119-154.

The rhizobial lipid-A

The lipid-A regions of rhizobial LPSs vary in structure among the different rhizobial species. The glycosyl residues of the various lipid-A regions are summarized in Table 3.2. *Sinorhizobium meliloti* lipid-A is reported[190] to consist of an acylated bis-phosphorylated glucosamine disaccharide similar to that found in numerous enteric bacteria. *Bradyrhizobium japonicum, B. elkanii, B. lupini,* and *M. loti* lipid-As all contain 2,3-diaminoglucose.[113,174] The one *B. elkanii* lipid-A examined also contains mannose (Carlson, unpublished), and the *M. loti* lipid-A is phosphorylated.[174] The types of fatty acids found in various rhizobial lipid-A molecules can be quite variable among the different species.[21,113,172,174] However, the major fatty acids usually include various β-hydroxy fatty acids and, in some cases, smaller amounts of various saturated and mono-unsaturated fatty acids. Some lipid-As also contain small amounts of oxo-fatty acids.[174]

The lipid-A from *R. leguminosarum* and *R. etli* is unusual compared to that from enteric bacteria since (a) it is totally devoid of phosphate, (b) it does not contain the typical acyloxyacyl substituents found in enteric lipid-As, (c) it contains only hydroxy fatty acids including the very long chain 27-hydroxyoctacosanoic acid (27-OHC28:0), (d) it contains a galacturonosyl (GalA) residue at the 4'-position instead of phosphate, and (e) it contains 2-aminogluconate (GlcNonate) in place of GlcN-1-phosphate. In addition, unlike enteric lipid-A in which the *N*-acyl substituents are exclusively β-hydroxymyristate, this lipid-A contains GlcN which is heterogeneously N-acylated with either β-hydroxymyristate, β-hydroxypalmitate, or β-hydroxystearate.[20,21] The structure of this lipid-A is shown in Figure 3.2. The biosynthesis of this unusual lipid-A is discussed further below.

Structural details of the lipid-A of LPSs have been used to recognize phylogenetic relationships of Gram-negative bacteria.[19,112,114,115,128] The rela-

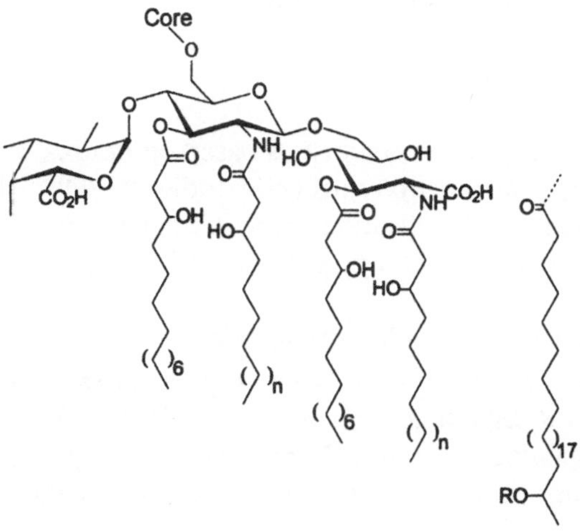

Figure 3.2 The structure of *R. etli* lipid-A. In this structure "n" can vary, being 6, 8, or 10. The "R" designates substitution by 3-hydroxylbutyric acid. In addition, the 27-ORC28:0 fatty acyl moiety is ester linked, but its location is not certain at this time.

tionships determined by 16S or 5S rRNA analyses have been successively correlated with lipid-A structures.[114,128] Therefore, it is of interest to compare LPS structures with the known phylogenetic relationships among the Rhizobiaceae. One of the earliest novelties concerning rhizobial lipid-A was the discovery that *R. leguminosarum* biovar trifolii lipid-A contained, in addition to other fatty acyl residues, the very long chain fatty acid, 27-OHC28:0.[75] This fatty acid was subsequently found in the lipid-A from all members of the Rhizobiaceae examined, with the one exception being *Azorhizobium caulinodans*.[21] (This exception needs to be re-examined.) Since it had been determined from 16S rRNA homology studies[197] that the Rhizobiaceae family belongs to the alpha-2 subgroup of Proteobacteria, the lipid-As from a number of other species in this group, which include phototrophic, nitrifying, nodulating, and intracellular bacteria, including both plant and animal pathogens, were examined. Other lipid-As which contained 27-OHC28:0 were those from *Rhodopseudomonas viridis*, *R. palustris*, *Nitrobacter winogradskyi*, *N. hamburgensis*, *Oligotropha carboxydovorans*, *Brucella abortus*, *Afipia felis*, *Agrobacterium tumefaciens*, *A. radiobacter*, *A. rhizogenes*, and *Thiobacillus* spp.[19] While not all of the alpha-2 subgroup species contain lipid-A with 27-OHC28:0, when this fatty acid was present, the species was found to belong to this phylogenetic group.[19] Examination of the glycosyl residues found in the various rhizobial lipid-As revealed variations in glycosyl components which differed in accordance with known phylogenetic relationships (Table 3.2). Based on these lipid-A compositions, the Rhizobiaceae species examined could be divided into five clusters:

I. *S. meliloti* and *S. fredii*
II. *R. leguminosarum* bv. trifolii, *R. leguminosarum* bv. viciae, *R. leguminosarum* bv. phaseoli, and *R. etli*

III. *B. japonicum* and B. spp. (*Lupinus*)
IV. *B. elkanii*
V. *M. loti.*

Thus, the various designated clusters based on lipid-A compositions were similar to those based on other phylogenetic studies.[84,110,200,201] The one exception is the relatively recent reclassification of the *R. leguminosarum* bv. *phaseoli* strain used in this study (CE3) as belonging to a newly designated species, *R. etli.*[178]

The rhizobial LPS core oligosaccharides

The core regions of rhizobial LPSs have only been examined in some detail for the LPSs from *R. leguminosarum* and *R. etli,*[17,38,39,76,77,206] and *B. elkanii.*[41] In the case of *R. leguminosarum* and *R. etli*, two core region oligosaccharides, isolated by gel filtration after mild acid hydrolysis of the LPSs, have been characterized:

$$\alpha\text{-D-Gal}p\text{-}(1\rightarrow6)\text{-}\alpha\text{-D-Man}p\text{-}(1\rightarrow5)\text{-Kdo}$$
$$4)$$
$$\uparrow$$
$$\alpha\text{-D-Gal}p\text{A-}(1$$

and

$$\alpha\text{-D-Gal}p\text{A-}(1\rightarrow5)\text{-Kdo} .$$
$$4)$$
$$\uparrow$$
$$\alpha\text{-D-Gal}p\text{A-}(1$$

The arrangement of these core oligosaccharides and the point of attachment of the O-chain in the LPS was established by the analysis of LPSs from *R. etli* CE3 mutants, which either completely lacked the O-chain or contained severely truncated forms of the O-chain[43] (Figure 3.3). The position at which the core oligosaccharide was attached to the lipid-A could be determined, since the ketosidic bond involved in this linkage is labile to mild acid. Thus, comparing the lipid-A glucosamine linkages before, with those after, mild acid hydrolysis revealed the point of Kdo attachment as being to O-6 of the lipid-A glucosaminosyl residue. That is, methylation analysis prior to mild acid hydrolysis revealed a glucosaminosyl residue with a component at O-6, while afterwards this was a terminal residue.[43] Similarly, the attachment of the O-chain to the core region was established to be at the O-6 position of the tetrasaccharide galactosyl residue by methylation analysis of a mutant LPS before and after mild acid hydrolysis; before hydrolysis this galactosyl residue contained a component at O-6, and afterwards it was terminal. This result showed that a mild acid labile bond (i.e., the Kdo residue at the reducing end of the O-chain) was attached to this galactosyl residue at O-6.[43] The presence of a third Kdo that is external to the inner Kdo residues is a novel structural feature that has not been reported for any other LPSs.

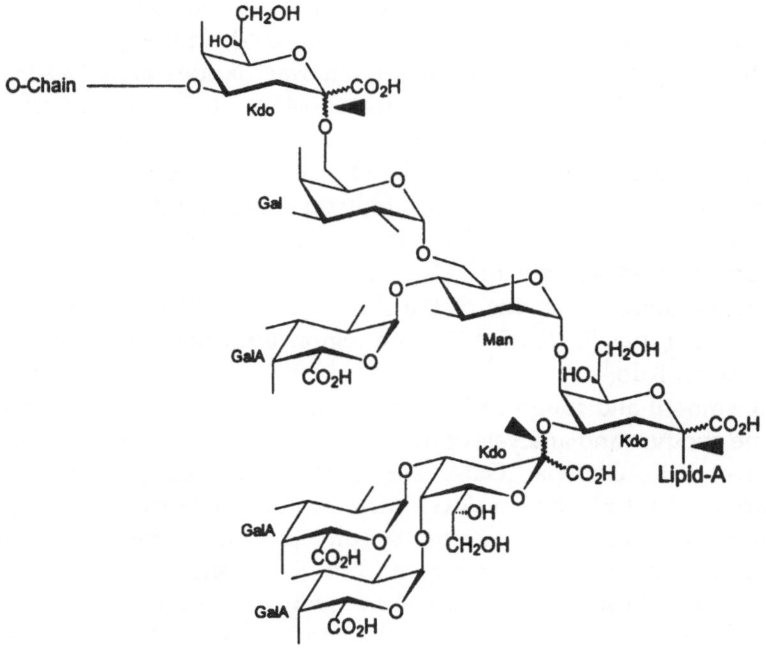

Figure 3.3 The structure of the *R. etli* LPS core region. The arrowheads mark the location of the acid-labile Kdo ketosidic bonds.

Further characterization was accomplished by ESI–MS analysis of the LPSs after removing the ester linked fatty acyl residues (i.e., after de-O-acylation). The removal of the ester linked fatty acyl residues is necessary to facilitate LPS analysis by ESI–MS. The ester linked fatty acyl residues were removed by methanolic sodium methoxide,[20] or by mild hydrazinolysis,[73,111,142] and the resulting de-O-acylated LPS analyzed by ESI–MS. The results showed that the *R. etli* LPS consists of the lipid-A and a core region that is comprised of the core trisaccharide attached to the tetrasaccharide (see above structure).[61] Furthermore, methylation analysis showed that Kdo was present in only two forms: as a terminal residue and as a residue with substituents at both O-4 and O-5. The terminal Kdo is the external residue that is linked to O-6 of the galactosyl tetrasaccharide core component. The remaining Kdo residues containing substituents at both O-4 and O-5 are due to the fact that the Kdo of the trisaccharide is bonded to O-4 of the tetrasaccharide Kdo residue. Thus, the complete structure of the core region is as shown in Figure 3.3. Current evidence indicates that this structure is common among *R. etli* and *R. leguminosarum* strains (discussed further below). One reported exception to this common core structure is that for the LPS from *R. leguminosarum* bv. trifolii 2S, which reportedly lacks the tetrasaccharide portion of the core region.[51] That report also states that the O-chain polysaccharide is attached to a diacylglycerol moiety rather than to the lipid-A.

The structures of the core oligosaccharides released by mild acid hydrolysis of a *B. elkanii* LPS have also been determined. In the case of *B. elkanii* 61A101c, the core region consists of two oligosaccharides: an α-4-O-MeMan-(1→5)-Kdo disaccharide, and an α-Man-(1→4)-α-Glc-(1→4)-Kdo trisaccharide. These oligosaccharides were released by mild acid hydrolysis of LPS isolated from a mutant that lacks the O-chain polysaccharide.[41] Mild acid hydrolysis of the parent LPS released only the trisaccharide and the O-chain, which contained the 4-O-methylmannosyl residue from the core region. Therefore, it is likely that the O-chain polysaccharide is attached to the 4-O-methylmannosyl residue of the core region. Unlike the LPSs from *R. leguminosarum*, the *B. elkanii* LPS core region does not contain any acidic sugars except Kdo.

The smooth and rough LPSs from a strain *S. fredii* have been separated from one another, and analysis of the rough LPS showed that the core region consists of Kdo, Glc, Gal, GalA, and GlcA.[160] The linkage positions and sequence of these glycosyl residues have not been reported. The high performance anion exchange chromatography (HPAEC) profiles of core oligosaccharides from rough and smooth LPSs are identical to one another,[160] indicating that these two LPS forms have identical core structures. Furthermore, HPAEC analyses indicate that the core regions from different strains of *S. fredii* all have a common set of oligosaccharides and a second set of oligosaccharides which varies among the different strains (comparison of the rhizobial LPS core regions is discussed further below).

Sinorhizobium meliloti and *Rhizobium* spp. NGR234 LPSs have core regions that are closely related to those of *S. fredii*. One report also indicates that the LPS from *S. meliloti* has a core region which contains the unusual sugar, 3-deoxy-2-heptulosaric acid (DHA);[171,173] a sugar that is normally part of a plant cell wall component called rhamnogalacturonan II (RG II).[185] It has also been reported that *S. meliloti* LPSs are sulfated.[50] This result may explain why there are several sets of sulfation genes in *S. meliloti*, in addition to those (*nodP*, *nodQ*, and *nodH*) responsible for the sulfation of the lipo-chitin-oligosaccharide Nod factors.[177]

As with the lipid-A composition data, phylogenetic relationships can also be determined by comparison of the various LPS core oligosaccharides. A comparison of the various core regions can be made by high performance anion exchange chromatography (HPAEC) of the LPS mild acid hydrolysate. The HPAEC profile normally shows only the core oligosaccharides, because the O-chain polysaccharide elutes in the void volume, or, if acidic, at much later retention times than the core oligosaccharides. Figure 3.4 compares the HPAEC core profiles of *R. etli* CE3, *R. leguminosarum* bv. *viciae* 3841, and *R. leguminosarum* bv. *trifolii* ANU843, showing that they are essentially identical. For these *R. leguminosarum* profiles the structure of each HPAEC peak has been determined.[43] Thus far, all *R. leguminosarum* and *R. etli* strains examined contain an LPS core region that consists of the tetra- and trisaccharides[17,39,77,78,206] described in the previous section, and most likely arranged as shown in Figure 3.3 for the LPS from *R. etli* CE3. Figure 3.4 also

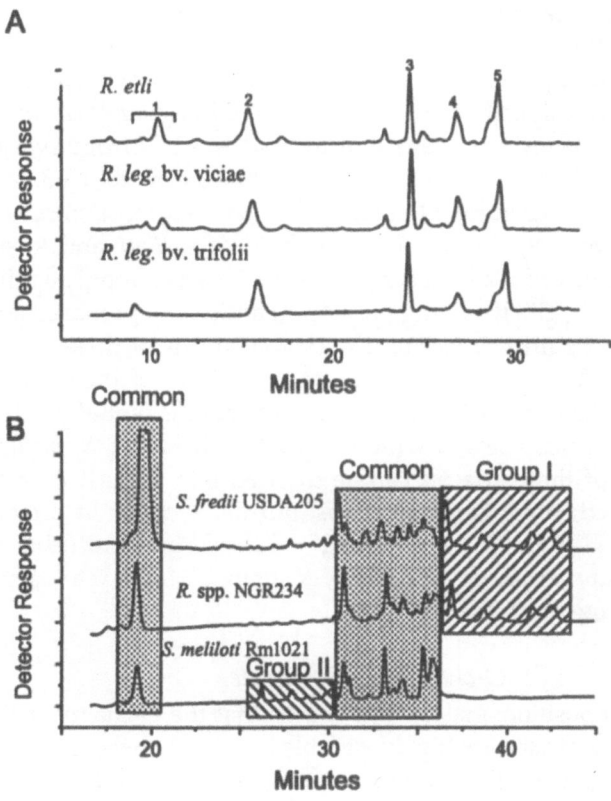

Figure 3.4 The HPAEC core profiles of the LPS from (A) *R. leguminosarum* bv. viciae, *R. leguminosarum* bv. trifolii, and *R. etli*, and (B) *S. fredii, S. meliloti*, and *Rhizobium* sp. NGR234. In panel A, the structures of the numbered peaks have been determined as follows: 1 = Kdo, 2 = GalA, 3 = the Gal-Man[GalA]-Kdo tetrasaccharide, 4 (and the peak just prior to 4) = anhydro-Kdo versions of the tetrasaccharide, and 5 = the GalA-Kdo[GalA] trisaccharide. (From Kannenberg, E.L., Reuhs, G.L., Forsberg, L.S., and Carlson, R.W., Lipopolysaccharides and K-antigens: Their structures, biosynthesis, and functions, in *The Rhizobiaceae, Molecular Biology of Model Plant-Associated Bacteria*, Spaink, H.P., Kondorosi, A., and Hooykaas, P.J.J., Kluwer Academic Publishers, Dordrecht, 1998, 119-154.)

compares the HPAEC profiles of strains *S. fredii* USDA205, *S. meliloti* Rm1021, and *R.* spp. NGR234. The HPAEC profiles are very different from those for *R. leguminosarum* LPSs, but are similar to one another in that there are a number of oligosaccharides that are common to all three species, and other oligosaccharides which vary. The presence of oligosaccharides that are common among all three species supports previous work[58,109,201] which shows the close phylogenetic relatedness of these species. The strains of these three closely related species can be divided into several groups based on the variable oligosaccharides, and based on LPS immunochemical properties (discussed in the next paragraph).

The structural relatedness of the LPS core regions among various rhizobia can also be examined by polyacrylamide gel electrophoresis and immunoblotting. Using this technique, a series of monoclonal antibodies were found that were specific to the core region of *R. leguminosarum* bv. *viciae* LPS. All of these core MAbs were able to bind to LPSs from every strain of *R. leguminosarum* tested, as well as to the LPS from *R. etli* CE3.[85,108] These data confirmed the structural identity of the core region among *R. etli* and *R. leguminosarum* LPSs. In the case of *S. fredii*, *S. meliloti*, and *R.* spp. NGR234, polyclonal antiserum against any one strain reacts strongly against the rough form of its LPS (LPS II). Using polyclonal antiserum prepared against *S. fredii* USDA205 or against *S. meliloti* Rm41, the strains of these species could be divided into four groups based on the reactivity of their rough LPSs with either USDA205 or Rm41 antiserum. These results would predict that future 16S rRNA studies using a wide range of *S. fredii* and *S. meliloti* will show that strains of these species can be divided into at least four different, but closely related, phylogenetic groups. Further work is in progress to determine if the HPAEC "fingerprinting" of the LPSs from other species, e.g., *Bradyrhizobium japonicum*, *B. elkanii*, *R. tropici*, *M. loti*, etc., also reflects the known phylogenetic relationships.

Rhizobial LPS O-chain polysaccharides

Glycosyl composition analysis has shown that the O-chains of *Rhizobium* and *Bradyrhizobium* strains are highly variable in structure even from strain to strain for a single species. They are consistently enriched with various deoxy and methylated deoxyglycosyl residues and can contain uronic acid residues, as well as, in some cases, heptosyl residues.[37,40,44,45,47,149,193] In the case of both *Rhizobium* and *Bradyrhizobium*, the O-chain polysaccharide is the dominant antigenic component since antisera to the bacteria react strongly with those forms of LPS that carry the O-chain.[40,45,47] An interesting feature of *R. leguminosarum* O-chains is that they have a Kdo residue (i.e., the third external Kdo described above) at their reducing ends,[37,40] a feature that is not true of enteric LPSs.

Thus far, the structures of only two *Rhizobium* O-chain repeating units have been published, those from *R. leguminosarum* bv. trifolii 4S,[193] and *R. tropici* CIAT899.[64] Their respective structures are:

α- D-ManNAc

(1

↓

2)

→3)-α-L-Rha-(1→3)-α- L -Rha-(1→4)-β- D -GlcNAc-(1→3)-α- L -Rha-(1→

and

→4)-β- D-Glc-(1→3)-6-deoxy-α- D-Tal-(1→3)-α- L-Fuc-(1→.

2

↑

OAc

The O-chains of *S. fredii*, *S. meliloti*, and *R.* sp. NGR234 appear to be quite different from those of *Rhizobium* or *Bradyrhizobium*. Composition analysis indicates that they can be homopolymers, e.g., a glucan.[160] In addition, in these species the core oligosaccharide is, together with the K-antigen, a dominant antigenic region of the LPS,[139,159,162] i.e., antiserum to these bacterial species reacts strongly with the rough LPS. These results indicate that the O-chains of these species are not very antigenic. Another possible reason for the immunodominance of the LPS core oligosaccharides is that these species produce relatively large amounts of rough LPS compared to smooth LPS (as indicated by PAGE analysis). Thus, the abundance of rough LPS on the cell surface of *S. meliloti*, *S. fredii*, and NGR234 makes it the dominant LPS antigen in these species. *Rhizobium galegae* is another species in which the rough LPS is the abundant form of LPS, making it the dominant antigen.[104] Recently, 16S RNA sequence analysis determined that this species was most closely related to the *Agrobacterium* genus, while 23S sequence analysis places it near the *Rhizobium* genus.[189]

In spite of the large variability of rhizobial O-chain structures (even within a species), the presence of the O-chain polysaccharide is essential for normal interaction with the legume host.

The biosynthesis of R. etli *and* R. leguminosarum *lipid-A*

Biosynthesis of the lipid-A portion of LPS is crucial for viability of *E. coli*.[150,151,181] It is also known that this portion of the LPS from enteric bacteria is responsible for its toxic properties, which result from an overstimulation of the host's immune system, e.g., causing the production of lethal amounts of tumor necrosis factor and other cytokines.[106,107,163,167,175,187] Structural features of the lipid-A that are essential for this toxicity include the presence of a glucosamine disaccharide backbone, phosphate groups, and certain fatty acyl residues.

Due to the unique structure of the *R. leguminosarum* lipid-A and to the requirement of lipid-A for the viability of the Gram-negative bacterial cell, it was of interest to examine its biosynthetic pathway. In *E. coli*, the details of the lipid-A biosynthetic pathway have been worked out by Raetz and co-workers.[150,151] Using the various *E. coli* lipid-A precursors, it was found that *R. leguminosarum* contained the same enzyme activities as those found in *E. coli* that synthesize Kdo_2lipid-IV_A from UDP-GlcNAc,[145] (Figure 3.5). In *E. coli*, the steps leading to the synthesis of Kdo_2lipid-IV_A are crucial for cell viability,[150,151,181] and the presence of the 4'-phosphate is required[16,34] for the transfer of the two Kdo residues to lipid-IV_A from CMP-Kdo, a reaction catalyzed by Kdo transferase (KdtA). After the synthesis of Kdo_2lipid-IV_A, further processing occurs, i.e., the addition of the acyloxyacyl fatty acids, to form the mature *E. coli* lipid-A. These results showed that *R. leguminosarum* likely makes a very close structural analog of Kdo_2lipid-IV_A, and they indicate that the biosynthetic steps leading to Kdo_2lipid-IV_A are probably crucial for the cell viability of a very wide range of Gram-negative bacteria.

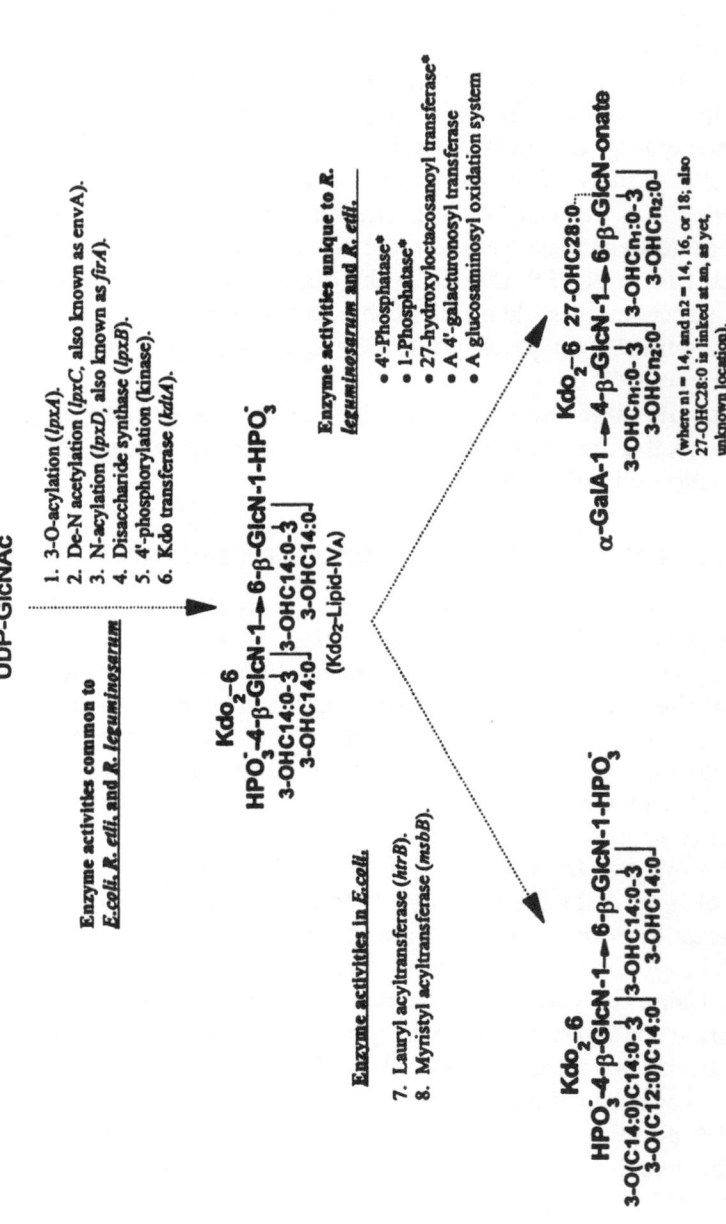

Figure 3.5 A diagram showing the enzyme activities that are common to the biosynthesis of both *E. coli,* *R. etli,* and *R. leguminosarum* lipid-As, and the point of divergence in their biosynthesis. The enzymes that are common to both species are those that are required for cell viability. The unique processing enzymes present in *R. etli* and *R. leguminosarum* are hypothesized not to be required for cell viability. Those marked with "*" have been detected and partially characterized (see text).

The above results also suggested that *R. leguminosarum* possesses unique enzymes which convert its Kdo_2lipid-IV_A precursor analog into the mature rhizobial lipid-A structure. Therefore, these results predict that *R. leguminosarum*:

1. Should not possess usual (i.e., similar to those in *E. coli*) acylating enzymes for the formation of acyloxyacyl substituents
2. Should have 4'- and 1-phosphatases
3. Should have an oxidation system capable of converting the reducing-end glucosamine to 2-aminogluconate
4. Should have a UDP-GalA transferase which transfers a galacturonosyl residue to the 4'-position
5. Should have a unique acyl transferase system for the incorporation of 27-OHC28:0.

Thus far, at least four of these predictions have proven true. First, a 4'-phosphatase activity was reported[144] in *R. leguminosarum*. This enzyme was found in *R. etli* CE3, and in several strains (i.e., found in all strains examined) of *R. leguminosarum* (but not in *E. coli*) it prefers the presence of Kdo in that Kdo_2lipid-IV_A is an efficient substrate (but not lipid-IV_A) and it is a membrane protein. Second, evidence has been reported for the presence of a 1-phosphatase activity.[35] Third, neither *R. leguminosarum* nor *R. etli* has any detectable acylating activity for the formation of the usual acyloxyacyl groups found in *E. coli* lipid-A.[144] Fourth, a recent report described a unique ACP from *R. leguminosarum* that is required for the transfer of 27-OHC28:0 to Kdo_2lipid-IV_A.[33] The sequence of this unique ACP revealed that it is encoded by open reading frame (*orf**), an *lps* gene earlier identified and partially characterized.[179] Further work is in progress regarding the biosynthesis of *R. leguminosarum* lipid-A. The genes which encode for these unique *R. leguminosarum* and *R. etli* lipid-A biosynthetic enzymes have not yet been located.

The genetics of Rhizobium LPS synthesis

The genetics of LPS biosynthesis has been most extensively studied in *R. leguminosarum* and *R. etli*. A number of *lps* regions (e.g., α-, β-, and γ-regions) have been identified in strain *R. etli* CE3.[48,49] One region, the β-region, is on a plasmid (not the symbiotic plasmid),[17,31,48,49] while the other regions are located on the chromosome. The α-region encodes proteins involved in both core oligosaccharide and O-chain polysaccharide synthesis,[43,188] and consists of nine different complementation groups (A to I) on a 17-kb region of DNA.[43,188] Mutations in complementation groups A through F affect O-chain synthesis in that these mutants most frequently contain complete core regions but truncated O-chains, while mutations in G through I affect core oligosaccharide synthesis.[43] Two core mutants were found;[43] one lacks the galacturonosyl residue that is attached to O-4 of the core mannosyl residue, and the second mutant lacks the terminal Kdo residue that is attached to O-6 of the galactosyl residue. Both the β- and γ-regions are likely required for core oligosaccharide synthesis, since mutations in these regions result in an LPS with an altered core region and lacking an O-chain.[17,31,48,49] In enteric bacteria, e.g., *E. coli* or *Salmonella*, the genes for core synthesis are clustered on a region of the chromosome known as the *rfa* region, and those for O-chain polysaccharide synthesis on another chromosomal region called *rfb*.[176] Thus, as stated by Noel et al.,[49] the arrangement of the *lps* genes in *R. etli* CE3 is quite different from that in *Salmonella* or *E. coli*.

Chromosomal regions responsible for the presence of O-chain polysaccharides have also been identified in other strains of *R. leguminosarum*, *R. leguminosarum* bv. viciae,[53,146] and *R. leguminosarum* bv. trifolii.[30] An *R. leguminosarum* bv. trifolii mutant of ANU843 was reported to be complemented by the α-region from *R. etli* CE3, resulting in an LPS which contained the CE3 O-chain and not the ANU843 O-chain.[30] This result supports the above structural data showing that the core regions for *R. etli*, *R. leguminosarum* bv. viciae, and *R. leguminosarum* bv. trifolii, i.e., the acceptors for O-chain transfer, must have identical structures. In addition to the α-, β-, and γ-regions, another region has been identified that is required for smooth LPS synthesis; i.e., mutations in this region result in mutants that have only rough LPS.[1,143] This *lps* region is clustered with genes involved in dicarboxylic acid transport (*dct* genes).[1,143] One such mutant appears to lack the terminal Kdo residue that is attached to O-6 of the core galactosyl residue.[1]

It is apparent from the above studies that the genetics of *R. leguminosarum* LPS synthesis is an area that still needs considerable work. Of particular interest are those genes that convert the Kdo_2lipid-IV_A precursor into the unique *R. leguminosarum* lipid-A. These genes may not be required for cell viability, and, therefore, it should be possible to obtain mutants in which the LPS contains lipid-A that is phosphorylated, that does not contain the 27-OHC28:0 fatty acid (e.g., an *orf*⁺ mutant), or does not contain the galacturonosyl residue at the 4'-position. Given the importance of the bacterial membrane in symbiosis, the effect of such mutations on core biosynthesis, on the presence or absence of an O-chain, and on infection of the host legume would be of great interest.

Rhizobial LPS structural variation during symbiotic nodule development

Detailed biochemical investigations of LPS expression in free-living rhizobia and in nodule bacteria have been employed to investigate whether or not the LPS undergoes structural modifications during symbiosis. Because the plant challenges the bacteria with a series of very different micro-environments, surface adaptations are to be expected. LPS structural modifications and adaptations in rhizobia are, therefore, areas that are currently being investigated in some detail.

Biochemical investigations with monoclonal antibodies have revealed changes in LPS epitope expression during symbiosis. Thus far, these LPS structural changes have been located in the O-chain portion of the LPS. Differential expression of LPS epitopes has been observed in rhizobia from free-living cultures that had been subjected to different growth conditions, and in nodule-derived bacteria. These observations have led to our current understanding that different types of LPSs can be expressed in rhizobia and that this expression depends on their environment.[24,29,86,89,90,134,180,188,191,199] To date, LPS epitope expression has been investigated most thoroughly in strains of *R. leguminosarum* and *R. etli.* LPS epitopes are expressed either constitutively, or their expression is regulated. Among regulated epitopes, expression, or lack of expression, has been observed as a consequence of environmental changes (e.g., lower pH or O_2 concentration). From these investigations, LPS epitopes have been grouped into different classes. One class of regulated epitopes is predominantly expressed on the LPSs from free-living bacteria, while another class is normally found only on the LPS of nodule bacteria.[89,90,180,188,191]

Several lines of evidence have shed some light on the underlying causes of LPS epitope expression and LPS structural adaptation. A number of phys-iological factors have been identified that are important for LPS epitope expression. In *R. leguminosarum* strain 3841, LPS epitope expression has been studied in relation to growth at acidic pH or under low O_2 concentrations. Both conditions induced the expression of certain epitopes and suppressed others.[86,89,90,180] In another investigation, LPSs in bacteroids from strain *R. etli* CE3 reacted weakly with two monoclonal antibodies. The same low reactiv-ity was observed in LPSs isolated from bacteria cultured under a range of regimes: acidic pH, high temperature, low phosphate, or low oxygen con-centrations.[188] In strain *R. leguminosarum* 3841, investigators found that bac-teria cultured at reduced oxygen concentrations and at a near-neutral pH express LPSs that largely resemble those of the corresponding bacteroids.[90]

Recently it has been shown that plant factors play a role in LPS epitope expression. Reuhs et al.[160] have shown that *S. fredii* LPS changes in response to the presence of host root extract in the growth medium. Also, an *R. etli* LPS epitope is suppressed by adding bean seed exudate to the growth medium.[132] The active compound has recently been characterized as an anthocyanin.[131] These findings indicate that LPS expression and molecular adaptation may be generated through mechanisms employing both physio-logical conditions and plant-specific molecular signals.

The structural basis for LPS adaptation in response to physiological factors is not yet adequately characterized. Investigators have achieved only limited insight into the LPS structural alterations that occur during symbiosis. Bhat and Carlson[18] have shown that the O-chain of the LPS from *R. etli* strain CE3 (a derivative of CFN42) grown at pH 4.8 has an altered methylation pattern. These changes correlate with the weak binding of two monoclonal antibodies to the LPS from bacteria cultured at low pH (4.8) compared with the strong binding when bacteria are cultured at neutral pH. This switch in the methylation pattern of LPS was also observed in the LPS from nodule bacteria.[18] The LPSs from free-living and pea nodule bacteria of strain *R. leguminosarum* 3841 have also been isolated by hot phenol–water extraction and partially characterized. The LPSs from free-living bacteria extracted predominantly into the water phase, while those from nodule bacteria were more evenly distributed between the two phases. These findings suggest that LPSs expressed in pea nodules are considerably more hydrophobic in character than those obtained from free-living bacteria. Partial chemical characterization indicated that this increase in hydrophobicity in nodule bacterial LPS may be due to changes in the carbohydrate part of the molecule, and to a higher proportion of long-chain fatty acids.[88]

Three classes of LPS mutants complete the current picture of what is known about LPS structures necessary for symbiotic functioning. One class comprises LPS mutant derivatives with severely truncated O-chains but which still retain a certain amount of O-chain. In nodulation assays with bean and pea, these mutants had the same phenotypes as those which completely lack the O-chain.[48,53,90,138] This class of mutants shows that a single, or even a very few, O-chain repeating units are not sufficient to restore full symbiotic competence. A second class of mutants has been described that produces an LPS-I structure but at approximately one third the normal level.[48] This mutant is as deficient in infection ability as mutants that entirely lack an LPS-I. A third class of mutants has been described that expresses distinctly different or modified O-chain polysaccharides than does the parent strain.[30,90] This class of mutants was as fully effective in symbiosis as was the parent, suggesting that the specific structure of the O-chain is not essential for LPS function. However, it is possible that the LPS O-chain serves as a rather nonspecific scaffold for more specific structural decorations; e.g., acetyl or methyl groups, or a branching sugar; that are important in symbiotic infection.

During tissue invasion the infection threads induced by class I or II LPS mutants in hosts which form either determinate or indeterminate nodules showed severe distortions.[53,133,138,146,183] These distortions have been claimed to constitute a host defense reaction,[138] indicating that a complete LPS is important for repressing such a reaction. Generally, in hosts forming determinate nodules, the tissue invasion process by LPS mutants seems more severely affected compared with that in hosts forming indeterminate nodules, and observations of necrosis and cell death indicate that host defenses are elicited in these nodules.[133,183] The reasons for the differences in the phenotypes of LPS mutants between determinate and indeterminate nodule-forming hosts are unclear. Brewin[28] has argued that the difference in phenotypes may reflect their different modes of infection.

Infection of nodule cells occurs when bacteria are released from the tips of the infection threads when they meet a newly formed meristematic host cell. At this point, the tip of the infection thread is no longer sheeted by a cell wall (in some legume hosts the tip expands at this point to form a ballooning structure, known as an infection droplet), and the rhizobia are engulfed by the "naked" cytoplasmic membrane to form symbiosomes.[28,170] The mechanism seems to resemble endocytosis. However, mechanistically it is not clear if endocytosis or phagocytosis is taking place and if it is receptor- or integrin-mediated.[192] In any case, close contact of plant cytoplasmic membrane with the rhizobial surface may be critical; however, the molecules involved in that step are unknown. A model in which LPSs play a role in attachment to the cytoplasmic membrane during infection has some precedence in animal systems where the involvement of LPSs in cell adherence and invasion is under discussion (for a recent review see Reference 81). The fact that the LPS from nodule bacteria, or from bacteria grown under nodule-like conditions, is more hydrophobic may imply that it has a role in a membrane–membrane contact between the bacterium and the host cell that may be essential for symbiotic infection.

Acknowledgments

This work has been supported by grants from the NIH (GM895832, RWC), the NSF (IBN9305022, RWC and BLR), from the Center for Plant and Microbial Complex Carbohydrate Research funded by the DOE (DE-FG09-87ER13810, to the CCRC), and from the Deutsche Forschungsgemeinschaft (PO 117/16-1) and the University of Tübingen (to ELK). The authors also thank Malcolm O'Neill and A. Perlick for helpful discussion, and Rosmary Nuri and Ann Dunn for secretarial help. In addition, many individuals in the laboratory of RWC have made contributions to the research described in this chapter. These include U. Ramadas Bhat, Tong Bin-Chen, Daniel Geller, Benjamin Jeyaretnam, John Kim, Bhagya Krishnaiah, Neil Price, and Brent Ridley.

References

1. Allaway, D., B. Jeyaretnam, R. W. Carlson, and P. S. Poole. Genetic and chemical characterization of a mutant that disrupts synthesis of the lipopolysaccharide core tetrasaccharide in *Rhizobium leguminosarum*. *J. Bacteriol.* 178:6403-6406, 1996.
2. An, J., R. W. Carlson, J. Glushka, and J. G. Streeter. The structure of a novel polysaccharide produced by *Bradyrhizobium* species within soybean nodules. *Carbohydr. Res.* 269:303-317, 1995.
3. Astete, S. G. and J. A. Leigh. *mucS*, a gene involved in activation of galacto-glucan (EPS II) synthesis gene expression in *Rhizobium meliloti*. *Mol. Plant Microbe Interact.* 9:395-400, 1996.
4. Balatti, P. A. and S. G. Pueppke. Nodulation of soybean by a transposon-mutant of *Rhizobium fredii* USDA257 is subject to competitive nodulation blocking by other rhizobia. *Plant Physiol.* 94:1276-1281, 1990.

5. Barthakur, D., C. E. Barber, J. W. Lamb, M. J. Daniels, J. A. Downie, and A. W. B. Johnston. A mutation that blocks exopolysaccharide synthesis prevents nodulation of peas by *Rhizobium* leguminosarum but not of beans by Rhizobium phaseoli and is corrected by dried DNA from *Rhizobium* or the phytopathogen xanthomonas. *Mol. Gen. Genet.* 203:320-323, 1986.

6. Batchelor, R. A., P. Alifano, E. Biffali, S. I. Hull, and R. A. Hull. Nucleotide sequences of the genes regulating O-polysaccharide antigen chain length (*rol*) from *Escherichia coli* and *Salmonella typhimurium*: Protein homology and functional complementation. *J. Bacteriol.* 174:5228-5236, 1992.

7. Batley, M., J. W. Redmond, S. P. Djordjevic, and B. G. Rolfe. Characterisation of glycerophosphorylated cyclic β-1,2-glucans from a fast-growing *Rhizobium* species. *Biochim. Biophys. Acta* 901:119-126, 1987.

8. Battisti, L., J. C. Lara, and J. A. Leigh. Specific oligosaccharide form of the *Rhizobium meliloti* exopolysaccharide promotes nodule invasion in alfalfa. *Proc. Natl. Acad. Sci. U.S.A.* 89:5625-5629, 1992.

9. Becker, A., A. Kleickmann, W. Arnold, and A. Pühler. Analysis of the *Rhizobium meliloti exoH/exoK/exoL* fragment: ExoK shows homology to excreted endo-β-1,3-1,4-glucanases and ExoH resembles membrane proteins. *Mol. Gen. Genet.* 238:145-154, 1993.

10. Becker, A., A. Kleickmann, M. Keller, W. Arnold, and A. Pühler. Identification and analysis of the *Rhizobium meliloti exoAMONP* genes involved in exopolysaccharide biosynthesis and mapping of promoters located on the *exo*-HKLAMONP fragment. *Mol. Gen. Genet.* 241:367-379, 1993.

11. Becker, A., A. Kleickmann, H. Küster, M. Keller, W. Arnold, and A. Pühler. Analysis of the *Rhizobium meliloti* genes *exoU*, *exoV*, *exoW*, *exoT*, and *exoI* involved in exopolysaccharide biosynthesis and nodule invasion: *exoU*, *exoW* probably encode glucosyltransferases. *The American Phytopathological Society* 6:735-744, 1993.

12. Becker, A., A. Kleickmann, H. Küster, M. Keller, W. Arnold, and A. Pühler. Analysis of the *Rhizobium meliloti* genes *exoU*, *exoV*, *exoW*, *exoT*, and *exoI* involved in exopolysaccharide biosynthesis and nodule invasion: *exoU* and *exoW* probably encode glucosyltransferases. *Mol. Plant Microbe Interact.* 6:735-744, 1993.

13. Becker, A., H. Küster, K. Niehaus, and A. Pühler. Extension of the *Rhizobium meliloti* succinoglycan biosynthesis gene cluster: identification of the *exsA* gene encoding an ABC transporter protein, and the *exsB* gene which probably codes for a regulator of succinoglycan biosynthesis. *Mol. Gen. Genet.* 249:487-497, 1995.

14. Becker, A., K. Niehaus, and A. Pühler. Low-molecular-weight succinoglycan is predominantly produced by *Rhizobium meliloti* strains carrying a mutated ExoP protein characterized by a periplasmic N-terminal domain and a missing C-terminal domain. *Mol. Microbiol.* 16:191-203, 1995.

15. Becquart-de Kozak, I., B. L. Reuhs, D. Buffard, C. Breda, J. S. Kim, R. Esnault, and A. Kondorosi. Role of the K-antigen subgroup of capsular polysaccharides in the early recognition process between *Rhizobium meliloti* and alfalfa leaves. *Mol. Plant Microbe Interact.* 10:114-123, 1997.

16. Belunis, C. J. and C. R. H. Raetz. Biosynthesis of endotoxins. Purification and catalytic properties of 3-deoxy-D-*manno*-octulosonic acid transferase from *Escherichia coli*. *J. Biol. Chem.* 267:9988-9997, 1992.

17. Bhat, U. R., S. K. Bhagyalakshmi, and R. W. Carlson. Re-examination of the structures of the lipopolysaccharide core oligosaccharides from *Rhizobium leguminosarum* biovar phaseoli. *Carbohydr. Res.* 220:219-227, 1991.

18. Bhat, U. R. and R. W. Carlson. Chemical characterization of pH-dependent structural epitopes of lipopolysaccharides from *Rhizobium leguminosarum* biovar phaseoli. *J. Bacteriol.* 174:2230-2235, 1992.

19. Bhat, U. R., R. W. Carlson, M. Busch, and H. Mayer. Distribution and phylogenetic significance of 27-hydroxy-octacosanoic acid in lipopolysaccharides from bacteria belonging to the alpha-2 subgroup of *Proteobacteria. Int. J. Syst. Bacteriol.* 41:213-217, 1991.

20. Bhat, U. R., L. S. Forsberg, and R. W. Carlson. The structure of the lipid A component of *Rhizobium leguminosarum* bv. phaseoli lipopolysaccharide. A unique non-phosphorylated lipid A containing 2-amino-2-deoxy-gluconate, galacturonate, and glucosamine. *J. Biol. Chem.* 269:14402-14410, 1994.

21. Bhat, U. R., H. Mayer, A. Yokota, R. I. Hollingsworth, and R. W. Carlson. Occurrence of lipid A variants with 27-hydroxyoctacosanoic acid in lipopolysaccharides from the *Rhizobiaceae* group. *J. Bacteriol.* 173:2155-2159, 1991.

22. Borthakur, D., J. A. Downie, A. W. B. Johnston, and J. W. Lamb. *psi*, a plasmid-linked *Rhizobium phaseoli* gene that inhibits exopolysaccharide production and which is required for symbiotic nitrogen fixation. *Mol. Gen. Genet.* 200:278-282, 1985.

23. Borthakur, D. and A. W. B. Johnston. Sequence of *psi*, a gene on the symbiotic plasmid of *Rhizobium phaseoli* which inhibits exopolysaccharide synthesis and nodulation and demonstration that its transcription is inhibited by *psr*, another gene on the symbiotic plasmid. *Mol. Gen. Genet.* 207:149-154, 1987.

24. Bradley, D. J., E. A. Wood, A. P. Larkins, G. Galfre, G. W. Butcher, and N. J. Brewin. Isolation of monoclonal antibodies reacting with peribacteroid membranes and other components of pea root nodules containing *Rhizobium*-leguminosarum. *Planta* 173:149-160, 1988.

25. Breedveld, M. W. and K. J. Miller. Cyclic β-glucans of members of the family *Rhizobiaceae. Microbiol. Rev.* 58:145-161, 1994.

26. Breedveld, M. W. and K. J. Miller. Synthesis of glycerophosphorylated cyclic (1,2)-β-glucans in *Rhizobium meliloti* strain 1021 after osmotic shock. *Microbiology* 141:583-588, 1995.

27. Breedveld, M. W., J. S. Yoo, V. N. Reinhold, and K. J. Miller. Synthesis of glycerophosphorylated cyclic β-(1,2)-glucans by *Rhizobium meliloti ndv* mutants. *J. Bacteriol.* 176:1047-1051, 1994.

28. Brewin, N. J. Development of the legume root nodule. *Annu. Rev. Cell Biol.* 7:191-226, 1991.

29. Brewin, N. J., E. A. Wood, A. P. Larkins, G. Galfre, and G. W. Butcher. Analysis of lipopolysaccharide from root nodule bacteroids of *Rhizobium leguminosarum* using monoclonal antibodies. *J. Gen. Microbiol.* 132:1959-1968, 1986.

30. Brink, B. A., J. Miller, R. W. Carlson, and K. D. Noel. Expression of *Rhizobium leguminosarum* CFN42 genes for lipopolysaccharide in strains derived from different *R. leguminosarum* soil isolates. *J. Bacteriol.* 172:548-555, 1990.

31. Brom, S., A. García de los Santos, T. Stepkowsky, M. Flores, G. Dávila, D. Romero, and R. Palacios. Different plasmids of *Rhizobium leguminosarum* bv. phaseoli are required for optimal symbiotic performance. *J. Bacteriol.* 174:5183-5189, 1992.

32. Bronner, D., V. Sieberth, C. Pazzani, I. S. Roberts, G. J. Boulnois, B. Jann, and K. Jann. Expression of the capsular K5 polysaccharide of *Escherichia coli*: Biochemical and electron microscopic analyses of mutants with defects in region 1 of the K5 gene cluster. *J. Bacteriol.* 175:5984-5992, 1994.

33. Brozek, K. A., R. W. Carlson, and C. R. H. Raetz. A special acyl carrier protein for transferring long hydroxylated fatty acids to lipid a in *Rhizobium. J. Biol. Chem.* 271:32126-32136, 1996.

34. Brozek, K. A., K. Hosaka, A. D. Robertson, and C. R. H. Raetz. Biosynthesis of Lipopolysaccharide in *Escherichia coli* Cytoplasmic Enzymes that Attach 3-Deoxy-D-Manno-Octulosonic Acid to Lipid A. *J. Biol. Chem.* 264(12):6956-6966, 1989.

35. Brozek, K. A., J. L. Kadrmas, and C. R. H. Raetz. Lipopolysaccharide biosynthesis in *Rhizobium leguminosarum* — Novel enzymes that process precursors containing 3-deoxy-D-manno-octulosonic acid. *J. Biol. Chem.* 271:32112-32118, 1996.

36. Buendia, A. M., B. Enenkel, R. Köplin, K. Niehaus, W. Arnold, and A. Pühler. The *Rhizobium meliloti exoZ/exoB* fragment of megaplasmid 2: ExoB functions as a UDP-glucose 4-epimerase and ExoZ shows homology to NodX of *Rhizobium leguminosarum* biovar *viciae* strain TOM. *Mol. Microbiol.* 5:1519-1530, 1991.

37. Carlson, R. W. The heterogeneity of *Rhizobium* lipopolysaccharides. *J. Bacteriol.* 158:1012-1017, 1984.

38. Carlson, R. W., F. Garcia, K. D. Noel, and R. I. Hollingsworth. The structures of the lipopolysaccharide core components from *Rhizobium leguminosarum* biovar *phaseoli* CE3 and two of its symbiotic mutants, CE109 and CE309. *Carbohydr. Res.* 195:101-110, 1990.

39. Carlson, R. W., R. L. Hollingsworth, and F. B. Dazzo. A core oligosaccharide component from the lipopolysaccharide of *Rhizobium trifolii* ANU843. *Carbohydr. Res.* 176:127-135, 1988.

40. Carlson, R. W., S. Kalembasa, D. Turowski, P. Pachori, and K. D. Noel. Characterization of the lipopolysaccharide from a *Rhizobium phaseoli* mutant that is defective in infection thread development. *J. Bacteriol.* 169:4923-4928, 1987.

41. Carlson, R. W. and B. S. Krishnaiah. Structures of the oligosaccharides obtained from the core regions of the lipopolysaccharides of *Bradyrhizobium japonicum* 61A101c and its symbiotically defective lipopolysaccharide mutant, JS314. *Carbohydr. Res.* 231:205-219, 1992.

42. Carlson, R. W., N. P. J. Price, and G. Stacey. The biosynthesis of rhizobial lipo-oligosaccharide nodulation signal molecules. *Mol. Plant-Microbe Interact.* 7:684-695, 1994.

43. Carlson, R. W., B. Reuhs, T.-B. Chen, U. R. Bhat, and K. D. Noel. Lipopolysaccharide core structures in *Rhizobium etli* and mutants deficient in O-antigen. *J. Biol. Chem.* 270:11783-11788, 1995.

44. Carlson, R. W., R. E. Sanders, C. Napoli, and P. Albersheim. Host-symbiont interactions III. Purification and characterization of *Rhizobium* lipopolysaccharides. *Plant Physiol.* 62:912-917, 1978.

45. Carlson, R. W., R. Shatters, J.-L. Duh, E. Turnbull, B. Hanley, B. G. Rolfe, and M. A. Djordjevic. The isolation and partial characterization of the lipopolysaccharides from several *Rhizobium trifolii* mutants affected in root hair infection. *Plant Physiol.* 84:421-427, 1987.

46. Carlson, R. W. and M. Yadav. Isolation and partial characterization of the extracellular polysaccharides and lipopolysaccharides from fast-growing *Rhizobium japonicum* USDA205 and its NOD⁻ mutant, HC205, which lacks the symbiotic plasmid. *Appl. Environ. Microbiol.* 50:1219-1224, 1985.

47. Carrion, M., U. R. Bhat, B. Reuhs, and R. W. Carlson. Isolation and characterization of the lipopolysaccharides from *Bradyrhizobium japonicum*. *J. Bacteriol.* 172:1725-1731, 1990.

48. Cava, J. R., P. M. Elias, D. A. Turowski, and K. D. Noel. *Rhizobium leguminosarum* CFN42 genetic regions encoding lipopolysaccharide structures essential for complete nodule development on bean plants. *J. Bacteriol.* 171:8-15, 1989.

49. Cava, J. R., H. Tao, and K. D. Noel. Mapping of complementation groups within a *Rhizobium leguminosarum* CFN42 chromosomal region required for lipopolysaccharide synthesis. *Mol. Gen. Genet.*, 221:125-128, 1990.

50. Cedergren, R. A., J. Lee, K. L. Ross, and R. I. Hollingsworth. Common links in the structure and cellular localization of *Rhizobium* chitolipooligosaccharides and general *Rhizobium* membrane phospholipid and glycolipid components. *Biochemistry* 34:4467-4477, 1995.

51. Cedergren, R. A., Y. Wang, and R. I. Hollingsworth. The "missing" typical *Rhizobium leguminosarum* O antigen is attached to a fatty acylated glycerol in *R-leguminosarum* bv trifolii 4S, a strain that also lacks the usual tetrasaccharide "core" component. *J. Bacteriol.* 178:5529-5532, 1996.

52. Dell, A., W. S. York, M. McNeil, A. G. Darvill, and P. Albersheim. The cyclic structure of β-D-(1\2)-linked D-glucans secreted by *Rhizobia* and *Agrobacteria*. *Carbohydr. Res.* 117:185-200, 1983.

53. deMaagd, R. A., A. S. Rao, I. H. M. Mulders, L. G. Roo, M. C. M. van Loosdrecht, C. A. Wijffelman, and B. J. J. Lugtenberg. Isolation and characterization of mutants of *Rhizobium leguminosarum* bv. *viciae* 248 with altered lipopolysaccharides: possible role of surface charge or hydrophobicity in bacterial release from the infection thread. *J. Bacteriol.* 171:1143-1150, 1989.

54. Denarie, J., F. Debellé, and J.-C. Promé. *Rhizobium* lipo-chitooligosaccharide nodulation factors: Signaling molecules mediating recognition and morphogenesis. 503 (Abstract), 1996.

55. Denarie, J. and P. Roche. *Rhizobium* nodulation signals, *in Molecular Signals in Plant-Microbe Communications*. D.P.S. Verma (Ed.), CRC Press, Boca Raton, 1992, 295-324.

56. Djordjevic, S. P., H. Chen, M. Batley, J. W. Redmond, and B. G. Rolfe. Nitrogen fixation ability of exopolysaccharide synthesis mutants of *Rhizobium* sp. strain NGR234 and *Rhizobium trifolii* is restored by the addition of homologous exopolysaccharides. *J. Bacteriol.* 169:53-60, 1987.

57. Dodgson, C., P. Amor, and C. Whitfield. Distribution of the *rol* gene encoding the regulator of lipopolysaccharide O-chain length in *Escherichia coli* and its influence on the expression of group I capsular K antigens. *J. Bacteriol.* 178:1895-1902, 1996.

58. Elkan, G. H. Taxonomy of the rhizobia. *Can. J. Microbiol.* 38:446-450, 1992.

59. Finan, T. M., A. M. Hirsch, J. A. Leigh, E. Johnsen, G. A. Kuldau, S. Deegan, G. C. Walker, and E. R. Signer. Symbiotic mutants of *Rhizobium meliloti* that uncouple plant from bacterial differentiation. *Cell* 40:869-877, 1985.

60. Fisher, R. F. and S. R. Long. *Rhizobium*-plant signal exchange. *Nature* 357:655-660, 1992.

61. Forsberg, L. S. and R. W. Carlson. The structures of the lipopolysaccharides from *Rhizobium etli* strains CE358 and CE359 — The complete structure of the core region of *R-etli* lipopolysaccharides. *J. Biol. Chem.* 273:2747-2757, 1998.
62. Forsberg, L. S. and B. L. Reuhs. Structural characterization of the K antigens from *Rhizobium fredii* USDA257: Evidence for a common structural motif, with strain-specific variation, in the capsular polysaccharides of *Rhizobium* sp. *J. Bacteriol.* 179:5366-5371, 1997.
63. Franco, A. V., D. Liu, and P. R. Reeves. A Wzz (Cld) protein determines the chain length of K lipopolysaccharide in *Escherichia coli* O8 and O9 strains. *J. Bacteriol.* 178:1903-1907, 1996.
64. Gil-Serrano, A. M., I. González-Jiménez, P. T. Mateo, M. Bernabé, J. Jiménez-Barbero, M. Megías, and M. J. Romero-Vázquez. Structural analysis of the *O*-antigen of the lipopolysaccharide of *Rhizobium tropici* CIAT899. *Carbohydr. Res.* 275:285-294, 1995.
65. Glucksmann, M. A., T. L. Reuber, and G. C. Walker. Family of glycosyl transferases needed for the synthesis of succinoglycan by *Rhizobium meliloti*. *J. Bacteriol.* 175:7033-7044, 1993.
66. Glucksmann, M. A., T. L. Reuber, and G. C. Walker. Genes needed for the modification, polymerization, export, and processing of succinoglycan by *Rhizobium meliloti*: A model for succinoglycan biosynthesis. *J. Bacteriol.* 175:7045-7055, 1993.
67. Gonzáles, J. E., B. L. Reuhs, and G. C. Walker. Low molecular weight EPS II of *Rhizobium meliloti* allows nodule invasion in *Medicago sativa*. *Proc. Natl. Acad. Sci. U.S.A.* 93:8636-8641, 1996.
68. González, J. E., G. M. York, and G. C. Walker. *Rhizobium meliloti* exopolysaccharides: Synthesis and symbiotic function. *Gene* 179:141-146, 1996.
69. Gore, R. S. and K. J. Miller. Cyclic β-1,6-1,3 glucans are synthesized by *Bradyrhizobium japonicum* bacteroids within soybean (*Glycine max*) root nodules. *Plant Physiol.* 102:191-194, 1993.
70. Gray, J. X., R. A. de Maagd, B. G. Rolfe, A. W. B. Johnston, and B. J. J. Lugtenberg. The role of *Rhizobium* cell surface during symbiosis, in *Molecular Signals in Plant-Microbe Communications*. D.P.S. Verma (Ed.), CRC Press, Boca Raton, 1992.
71. Gray, J. X., M. A. Djordjevic, and B. G. Rolfe. Two genes that regulate exopolysaccharide production in *Rhizobium* sp. strain NGR234: DNA sequences and resultant phenotypes. *J. Bacteriol.* 172:193-203, 1990.
72. Gray, J. X. and B. G. Rolfe. Exopolysaccharide production in *Rhizobium* and its role in invasion. *Mol. Microbiol.* 4:1425-1431, 1990.
73. Haishima, Y., O. Holst, and H. Brade. Structural investigation on the lipopolysaccharide of *Escherichia coli* rough mutant F653 representing the R3 core type. *Eur. J. Biochem.* 203:127-134, 1992.
74. Hirsch, A. M. Tansley review no. 40. Developmental biology of legume nodulation. *New Phytol.* 122:211-237, 1992.
75. Hollingsworth, R. I. and R. W. Carlson. 27-Hydroxyoctacosanoic acid is a major structural fatty acyl component of the lipopolysaccharide of *Rhizobium trifolii* ANU 843. *J. Biol. Chem.* 264:9300-9303, 1989.
76. Hollingsworth, R. I., R. W. Carlson, F. Garcia, and D. A. Gage. A new core tetrasaccharide component from the lipopolysaccharide of *Rhizobium trifolii* ANU 843. *J. Biol. Chem.* 264:9294-9299, 1989.

77. Hollingsworth, R. I., R. W. Carlson, F. Garcia, and D. A. Gage. A new core tetrasaccharide component from the lipopolysaccharide of *Rhizobium trifolii* ANU 843 (Errata). *J. Biol. Chem.* 265:12752, 1990.

78. Hollingsworth, R. I., Y. Zhang, and U. B. Priefer. Structure of the unusual trisaccharide lipopolysaccharide component produced by a symbiotically defective mutant of *Rhizobium leguminosarum* biovar *viciae. Carbohydr. Res.* 264:271-280, 1994.

79. Icho, T. Membrane bound phosphatases in *Escherichia coli*: Sequence of the *pgpA* gene. *J. Bacteriol.* 170:5110-5116, 1988.

80. Icho, T. Membrane-bound phosphatases in *Escherichia coli*: Sequence of the *pgpB* gene and dual subcellular localization of the *pgbB* product. *J. Bacteriol.* 170:5117-5124, 1988.

81. Jacques, M. Role of lipo-oligosaccharides and lipopolysaccharides in bacterial adherence. *Trends Microbiol.* 4:408-410, 1996.

82. Jann, B. and K. Jann. Structure and biosynthesis of the capsular antigens of *Escherichia coli. Curr. Top. Microbiol. Immunol.* 150:19-42, 1990.

83. Jann, K. and B. Jann. Biochemistry and expression of bacterial capsules. *Biochem. Soc. Trans.* 19:623-628, 1991.

84. Jarvis, B. D. W., H. L. Downer, and J. P. W. Young. Phylogeny of fast-growing soybean-nodulating rhizobia supports synonymy of *Sinorhizobium* and *Rhizobium* and assignment to *Rhizobium fredii. Int. J. Syst. Bacteriol.* 42:93-96, 1992.

85. Kannenberg, E., L. S. Forsberg, and R. W. Carlson. Lipopolysaccharide core components of *Rhizobium etli* reacting with a panel of monoclonal antibodies. *Plant and Soil* 186:161-166, 1996.

86. Kannenberg, E. L. and N. J. Brewin. Expression of a cell surface antigen from *Rhizobium leguminosarum* 3841 is regulated by oxygen and pH. *J. Bacteriol.* 171:4543-4548, 1989.

87. Kannenberg, E. L. and N. J. Brewin. Host-plant invasion by *Rhizobium*: the role of cell-surface components. *Trends Microbiol.* 2:277-283, 1994.

88. Kannenberg, E. L., N. J. Brewin, and R. W. Carlson. Biochemical separation of lipopolysaccharides expressed in *Rhizobium* nodule bacteria, in *Proceedings of the 1st European Nitrogen Fixation Conference.* G.B. Kiss and G. Endre (Eds.), Officina Press, Szeged, Hungary, 1994.

89. Kannenberg, E. L., S. Perotto, V. Bianciotto, E. A. Rathburn, and N. J. Brewin. Lipopolysaccharide epitope expression of *Rhizobium* bacteroids as revealed by *in situ* immunolabelling of pea root nodule sections. *J. Bacteriol.* 176:2021-2032, 1994.

90. Kannenberg, E. L., E. A. Rathbun, and N. J. Brewin. Molecular dissection of structure and function in the lipopolysaccharide of *Rhizobium leguminosarum* strain 3841 using monoclonal antibodies and genetic analysis. *Mol. Microbiol.* 6(17):2477-2487, 1992.

91. Kato, G., Y. Maruyama, and M. Nakamura. Role of lectins and lipopolysaccharides in the recognition process of specific *Rhizobium*-legume symbiosis. *Agric. Biol. Chem.* 43:1085-1092, 1979.

92. Kato, G., Y. Maruyama, and M. Nakamura. Role of bacterial polysaccharides in the adsorption process of the *Rhizobium*-pea symbiosis. *Agric. Biol. Chem.* 44:2843-2855, 1980.

93. Kato, G., Y. Maruyama, and M. Nakamura. Involvement of lectins in *Rhizobium*-pea recognition. *Plant Cell Physiol.* 22:759-771, 1981.

94. Keller, M., A. Roxlau, W. M. Weng, M. Schmidt, J. Quandt, K. Niehaus, D. Jording, W. Arnold, and A. Pühler. Molecular analysis of the *Rhizobium meliloti* *mucR* gene regulating the biosynthesis of the exopolysaccharides succinogly-can and galactoglucan. *Mol. Plant Microbe Interact.* 8:267-277, 1995.

95. Kim, J. S. and B. L. Reuhs. Extended host range mutants of *Rhizobium fredii* USDA257 show modified expression of the K antigens and lipopolysaccha-rides. *Mol. Plant-Microbe Interact. 8th Int. MPMI Congr.* (Abstract), 1996.

96. Kereszt, A., E. Kiss, B. L. Reuhs, R. W. Carlson, A. Kondorosi, and P. Putnoky, Novel *rkp* gene clusters of *Sinorhizobium meliloti* involved in capsular polysac-charide production and invasion of the symbiotic nodule: the *rkpK* gene en-codes a UDP-glucose dehydrogenase. *J. Bacteriol.*, 180, 5426-5431, 1998.

97. Kondorosi, A. Regulation of nodulation genes in rhizobia, in *Molecular Signals in Plant-Microbe Communications*. D.P.S. Verma (Ed.), CRC Press, Boca Raton, 1992.

98. Latchford, J. W., D. Borthakur, and A. W. B. Johnston. The products of *Rhizo-bium* genes, *psi* and *pss*, which affect exopolysaccharide production, are asso-ciated with the bacterial cell surface. *Mol. Microbiol.* 5:2107-2114, 1991.

99. Leigh, J. A. and D. L. Coplin. Exopolysaccharides in plant-bacterial interac-tions. *Annu. Rev. Microbiol.* 46:307-346, 1992.

100. Leigh, J. A., E. R. Signer, and G. C. Walker. Exopolysaccharide-deficient mu-tants of *Rhizobium meliloti* that form ineffective nodules. *Proc. Natl. Acad. Sci. U.S.A.* 82:6231-6235, 1985.

101. Leigh, J. A. and G. C. Walker. Exopolysaccharides of *Rhizobium*: Synthesis, regulation and symbiotic function. *Trends Genet.* 10:63-67, 1994.

102. Lerouge, P. Symbiotic host specificity between leguminous plants and rhizo-bia is determined by substituted and acylated glucosamine oligosaccharide signals. *Glycobiology* 4:127-134, 1994.

103. Levery, S. B., H. Zahn, C. C. Lee, J. A. Leigh, and S. Hakomori. Structural analyses of a second acidic exopolysaccharide of *Rhizobium meliloti* that can function in alfalfa root nodule invasion. *Carbohydr. Res.* 210:339-348, 1991.

104. Lipsanen, P. and K. Lindstrom. Lipopolysaccharide and protein patterns of *Rhizobium* sp. (*Galega*). *FEMS Microbiol. Lett.* 58:323-328, 1989.

105. Long, S., J. W. Reed, J. Himawan, and G. C. Walker. Genetic analysis of a cluster of genes required for synthesis of the calcofluor-binding exopolysaccharide of *Rhizobium meliloti*. *J. Bacteriol.* 170:4239-4248, 1988.

106. Loppnow, H., I. Dürrbaum, H. Brade, C. A. Dinarello, S. Kusumoto, E. T. Rietschel, and H.-D. Flad. Lipid A, the immunostimulatory principle of li-popolysaccharides. *Adv. Exp. Med. Biol.* 256:561-566, 1990.

107. Loppnow, H., I. Dürrbaum, H. Brade, C. A. Dinarello, S. Kusumoto, E. T. Rietschel, and H.-D. Flad. Lipid A, the immunostimulatory principle of li-popolysaccharides? in *Endotoxin*. H. Friedman, T.W. Klein, M. Nakano, and A. Nowotny (Eds.), Plenum Publishing Corp, 1990.

108. Lucas, M. M., J. L. Peart, N. J. Brewin, and E. L. Kannenberg. Isolation of monoclonal antibodies reacting with the core component of lipopolysaccha-ride from *Rhizobium leguminosarum* strain 3841 and mutant derivatives. *J. Bacteriol.* 178:2727-2733, 1996.

109. Martínez-Romero, E. Recent development in *Rhizobium* taxonomy. *Plant and Soil* 161:11-20, 1994.

110. Martínez-Romero, E. and J. Caballero-Mellado. *Rhizobium* phylogenies and bacterial genetic diversity. *Crit. Rev. Plant Sci.* 15:113-140, 1996.

111. Masoud, H., E. Altman, J. C. Richards, and J. S. Lam. General strategy for structural analysis of the oligosaccharide region of lipooligosaccharides. Structure of the oligosaccharide component of *Pseudomonas aeruginosa* IATS serotype 06 mutant R5 rough-type lipopolysaccharide. *Biochemistry* 33:10568-10578, 1994.
112. Mayer, H., U. R. Bhat, H. Masoud, J. Radziejewska-Lebrecht, C. Widemann, and J. H. Krauss. Bacterial lipopolysaccharides. *Pure Appl. Chem.* 61:1271-1282, 1989.
113. Mayer, H., J. H. Krauss, T. Urbanik-Sypniewska, V. Puvanesarajah, G. Stacey, and G. Auling. Lipid A with 2,3-diamino-2,3-dideoxyglucose in lipopolysaccharides from slow-growing members of *Rhizobiaceae* and from *Pseudomonas carboxydovorans*. *Arch. Microbiol.* 151:111-116, 1989.
114. Mayer, H., E. Moreno, E. Stackebrandt, M. Dorsch, J. Wolters, and M. Busch. *Brucella abortus* 16S rRNA and lipid a reveal a phylogenetic relationship with members of the alpha-2 subdivision of the class *Proteobacteria*. *J. Bacteriol.* 172(7):3569-3576, 1990.
115. Mayer, H. and J. Weckesser. Unusual lipid A's: structures, taxonomical relevance and potential value for endotoxin research, in *Handbook of Endotoxins*, Vol. 1, *Chemistry of Endotoxins*. E.T. Reitschel (Ed.), Elsevier, Amsterdam, 1984.
116. Meinhardt, L. W., H. B. Krishnan, P. A. Balatti, and S. G. Pueppke. Molecular cloning and characterization of a sym plasmid locus that regulates cultivar-specific nodulation of soybean by *Rhizobium fredii* USDA257. *Mol. Microbiol.* 9:17-29, 1993.
117. Miller, K. J. and R. S. Gore. Cyclic beta-1,6 -1,3 glucans of *Bradyrhizobium*: Functional analogs of the cyclic beta-1,2-glucans of *Rhizobium*. *Curr. Microbiol.* 24:101-104, 1992.
118. Miller, K. J., R. S. Gore, and A. J. Benesi. Phosphoglycerol substituents present on the cyclic beta-1,2-glucans of *Rhizobium-meliloti* 1021 are derived from phosphatidylglycerol. *J. Bacteriol.* 170:4569-4575, 1988.
119. Miller, K. J., R. S. Gore, R. Johnson, A. J. Benesi, and V. N. Reinhold. Cell-associated oligosaccharides of *Bradyrhizobium* spp. *J. Bacteriol.* 172:136-142, 1990.
120. Miller, K. J., E. P. Kennedy, and V. N. Reinhold. Osmotic adaptation by gram-negative bacteria: possible role for periplasmic oligosaccharides. *Science* 231:48-51, 1986.
121. Miller, K. J., V. N. Reinhold, A. C. Weissborn, and E. P. Kennedy. Cyclic glucans produced by *Agrobacterium tumefaciens* are substituted with *sn*-1-phosphoglycerol residues. *Biochim. Biophys. Acta* 901:112-118, 1987.
122. Miller, K. J. and J. M. Wood. Osmoadaptation by rhizosphere bacteria. *Annu. Rev. Microbiol.* 50:101-136, 1996.
123. Reference omitted.
124. Milner, J. L., R. S. Araujo, and J. Handelsman. Molecular and symbiotic characterization of exopolysaccharide-deficient mutants of *Rhizobium tropici* strain CIAT899. *Mol. Microbiol.* 6:3137-3147, 1992.
125. Mimmack, M. L., D. Borthakur, M. A. Jones, J. A. Downie, and A. W. B. Johnston. The *psi* operon of *Rhizobium leguminosarum* biovar *phaseoli*: Identification of two genes whose products are located at the bacterial cell surface. *J. Gen. Microbiol.* 140:1223-1229, 1994.
126. Mimmack, M. L., G. F. Hong, and A. W. B. Johnston. Sequence and regulation of *psrA*, a gene on the Sym plasmid of *Rhizobium leguminosarum* biover *phaseoli* which inhibits transcription of the *psi* genes. *J. Gen. Microbiol.* 140:455-461, 1994.

127. Muller, P., M. Keller, W. M. Weng, J. Quandt, W. Arnold, and A. Pühler. Genetic analysis of the *Rhizobium meliloti exo*YFQ operon: ExoY is homologous to sugar transferases and ExoQ represents a transmembrane protein.*Mol. Plant-Microbe Interact.* 6:55-65, 1993.

128. Neumann, U., H. Mayer, E. Schiltz, R. Benz, and J. Weckesser. Lipopolysaccharide and porin of *Roseobacter denitrificans*, confirming its phylogenetic relationship to the α-3 subgroup of Proteobacteria. *Microbiology* 141:2013-2017, 1995.

129. Niehaus, K., D. Kapp, and A. Pühler. Plant defence and delayed infection of alfalfa pseudonodules induced by an exopolysaccharide (EPS I)-deficient *Rhizobium meliloti* mutant. *Planta* 190:415-425, 1993.

130. Noel, K. D. Rhizobial polysaccharides required in symbioses with legumes, in *Molecular Signals in Plant-Microbe Communications*. D.P.S. Verma (Ed.), CRC Press, Boca Raton, 1992.

131. Noel, K. D., D. M. Duelli, and V. J. Neumann.*Rhizobium etli* lipopolysaccharide alterations triggered by host exudate compounds, in *Biology of Plant-Microbe Interactions*. G. Stacey, B. Mullin and P.M. Gresshoff (Eds.), International Society for Molecular Plant-Microbe Interactions, St. Paul, 1996.

132. Noel, K. D., D. M. Duelli, H. Tao, and N. J. Brewin. Antigenic change in the lipopolysaccharide of *Rhizobium etli* CFN42 induced by exudates of *Phaseolus vulgaris*. *Mol. Plant-Microbe Interact.* 9:180-186, 1996.

133. Noel, K. D., K. A. VandenBosch, and B. Kulpaca. Mutations in *Rhizobium phaseoli* that lead to arrested development of infection threads. *J. Bacteriol.* 168:1392-1401, 1986.

134. Olsen, P., S. Wright, M. Collins, and W. Rice. Patterns of reactivity between a panel of monoclonal antibodies and forage *Rhizobium* strains. *Appl. Environ. Microbiol.* 60:654-661, 1994.

135. Ozga, D. A., J. C. Lara, and J. A. Leig. The regulation of exopolysaccharide production is important at two levels of nodule development in *Rhizobium meliloti*. *Mol. Plant. Microbe Interact.* 7:758-765, 1994.

136. Pazzani, C., C. Rosenow, G. J. Boulnois, D. Bronner, K. Jann, and I. S. Roberts. Molecular analysis of region 1 of the *Escherichia coli* K5 antigen gene cluster: A region encoding proteins involved in cell surface expression of capsular polysaccharide. *J. Bacteriol.* 175:5978-5983, 1993.

137. Pazzani, C., C. Rosenow, G. J. Boulnois, D. Bronner, K. Jann, and I. S. Roberts. Molecular analysis of region 1 of the *Escherichia coli* K5 antigen gene cluster: a region encoding proteins involved in cell surface expression of capsular polysaccharide. *J. Bacteriol.* 175:5978-5983, 1994.

138. Perotto, S., N. J. Brewin, and E. L. Kannenberg. Cytological evidence for a host defense response that reduces cell and tissue invasion in pea nodules by lipopolysaccharide-defective mutants of *Rhizobium leguminosarum* strain 3841. *Mol. Plant. Microbe Interact.* 7:99-112, 1994.

139. Petrovics, G., P. Putnoky, B. Reuhs, J. Kim, T. A. Thorp, D. Noel, R. W. Carlson, and A. Kondorosi. The presence of a novel type of surface polysaccharide in *Rhizobium meliloti* requires a new fatty acid synthase-like gene cluster involved in symbiotic nodule development. *Mol. Microbiol.* 8:1083-1094, 1993.

140. Pfeffer, P. E., G. Bécard, D. B. Rolin, J. Uknalis, P. Cooke, and S.-I. Tu. *In vivo* nuclear magnetic resonance study of the osmoregulation of phosphocholine-substituted β-1,3;1,6 cyclic glucan and its associated carbon metabolism in *Bradyrhizobium japonicum* USDA 110. *Appl. Environ. Microbiol.* 60:2137-2146, 1994.

141. Pfeffer, P. E., G. Bécard, D. B. Rolin, J. Uknalis, P. Cooke, and S.-I. Tu. *In vivo* nuclear magnetic resonance study of the osmoregulation of phosphocholine-substituted β-1,3;1,6 cyclic glucan and its associated carbon metabolism in *Bradyrhizobium japonicum* USDA 110. *Appl. Environ. Microbiol.* 60:2137-2146, 1994.

142. Phillips, N. J., M. A. Apicella, J. M. Griffiss, and B. W. Gibson. Structural characterization of the cell surface lipooligosaccharides from a nontypable strain of *Haemophilus influenzae. Biochemistry* 31:4515-4526, 1992.

143. Poole, P. S., N. A. Schofield, C. J. Reid, E. M. Drew, and D. L. Walshaw. Identification of chromosomal genes located downstream of *dctD* that affect the requirement for calcium and the lipopolysaccharide layer of *Rhizobium leguminosarum. J. Gen. Microbiol.* 140:2797-2809, 1994.

144. Price, N. P. J., B. Jeyaretnam, R. W. Carlson, J. L. Kadrmas, C. R. H. Raetz, and K. A. Brozek. Lipid A biosynthesis in *Rhizobium leguminosarum*: Role of a 2-keto-3-deoxyoctulosonate-activated 4′ phosphatase. *Proc. Natl. Acad. Sci. U.S.A.* 92:7352-7356, 1995.

145. Price, N. P. J., T. M. Kelly, C. R. H. Raetz, and R. W. Carlson. Biosynthesis of a structurally novel lipid A in *Rhizobium leguminosarum*: Identification and characterization of six metabolic steps leading from UDP-GlcNAc to Kdo$_2$-lipid IV$_A$. *J. Bacteriol.* 176:4646-4655, 1994.

146. Priefer, U. B. Genes involved in lipopolysaccharide production and symbiosis are clustered on the chromosome of *Rhizobium leguminosarum* biovar *viciae* VF39. *J. Bacteriol.* 171:6161-6168, 1989.

147. Putnoky, P., E. Grosskopf, D. T. C. Ha, G. B. Kiss, and A. Kondorosi. *Rhizobium fix* genes mediate at least two communication steps in symbiotic nodule development. *J. Cell Biol.* 106:597-607, 1988.

148. Putnoky, P., G. Petrovics, A. Kereszt, E. Grosskopf, D. T. C. Ha, Z. Bánfalvi, and A. Kondorosi. *Rhizobium meliloti* lipopolysaccharide and exopolysaccharide can have the same function in the plant–bacterium interaction. *J. Bacteriol.* 172:5450-5458, 1990.

149. Puvanesarajah, V., F. M. Schell, D. Gerhold, and G. Stacey. Cell surface polysaccharides from *Bradyrhizobium japonicum* and a nonnodulating mutant. *J. Bacteriol.* 169:137-141, 1987.

150. Raetz, C. R. H. Biochemistry of endotoxins. *Annu. Rev. Biochem.* 59:129-170, 1990.

151. Raetz, C. R. H. Bacterial endotoxins: extraordinary lipids that activate eucaryotic signal transduction. *J. Bacteriol.* 175:5745-5753, 1993.

152. Reed, J. W., M. Capage, and G. C. Walker. *Rhizobium meliloti exoG* and *exoJ* mutations affect the exoX-exoY system for modulation of exopolysaccharide production. *J. Bacteriol.* 173:3776-3788, 1991.

153. Reed, J. W., J. Glazebrook, and G. C. Walker. The *exoR* gene of *Rhizobium meliloti* affects RNA levels of other *exo* genes but lacks homology to known transcriptional regulators. *J. Bacteriol.* 173:3789-3794, 1991.

154. Reed, J. W. and G. C. Walker. The *exoD* gene of *Rhizobium meliloti* encodes a novel function needed for alfalfa nodule invasion. *J. Bacteriol.* 173:664-677, 1991.

155. Reuber, T. L., S. Long, and G. C. Walker. Regulation of *Rhizobium meliloti exo* genes in free-living cells and in planta examined by using Tn*phoA* fusions. *J. Bacteriol.* 173:426-434, 1991.

156. Reuber, T. L., J. Reed, J. Glazebrook, M. A. Glucksmann, D. Ahmann, A. Marra, and G. C. Walker. *Rhizobium meliloti* exopolysaccharides: Genetic analyses and symbiotic importance. *Biochem. Soc. Trans.* 19:636-641, 1991.

157. Reuber, T. L. and G. C. Walker. Biosynthesis of succinoglycan, a symbiotically important exopolysaccharide of *Rhizobium meliloti*. *Cell* 74:269-280, 1993.
158. Reuhs, B. L. Acidic capsular polysaccharides (K antigens) of *Rhizobium*, in *Biology of Plant-Microbe Interactions*. G. Stacey, B. Mullin and P.M. Gresshoff (Eds.), IS-MPMI, St. Paul, 1996.
159. Reuhs, B. L., R. W. Carlson, and J. S. Kim. *Rhizobium fredii* and *Rhizobium meliloti* produce 3-deoxy-D-*manno*-2-octulosonic acid-containing polysaccharides that are structurally analogous to group K antigens (capsular polysaccharides) found in *Escherichia coli*. *J. Bacteriol.* 175:3570-3580, 1993.
160. Reuhs, B. L., J. S. Kim, A. Badgett, and R. W. Carlson. Production of cell-associated polysaccharides of *Rhizobium fredii* USDA205 is modulated by apigenin and host root extract. *Mol. Plant. Microbe Interact.* 7:240-247, 1994.
161. Reuhs, B. L., J. S. Kim, D. A. Geller, R. W. Carlson, M. N. V. Williams, and S. G. Pueppke. Cell-associated, acidic polysaccharides, which are structurally analogous to the capsular polysaccharides (K-antigens) of *E. coli*, are widespread in mutualistic and pathogenic species of plant symbionts. *Mol. Plant-Microbe Interact. Edinburgh*:71 (Abstract), 1994.
162. Reuhs, B. L., M. N. V. Williams, J. S. Kim, R. W. Carlson, and F. Côté. Suppression of the Fix⁻ phenotype of *Rhizobium meliloti exoB* mutants by *lpsZ* is correlated to a modified expression of the K polysaccharide. *J. Bacteriol.* 177:4289-4296, 1995.
163. Rietschel, E. T. and H. Brade. Bacterial endotoxins. *Sci. Am.* 267:54-61, 1992.
164. Rietschel, E. T., L. Brade, U. Schade, U. Seydel, U. Zähringer, K. Brandenburg, I. Helander, O. Holst, S. Kondo, H. M. Kuhn, B. Lindner, E. Röhrscheidt, R. Russa, H. Labischinski, D. Naumann, and H. Brade. Bacterial lipopolysaccharides: Relationship of structure and conformation to endotoxic activity, serological specificity and biological function. *Adv. Exp. Med. Biol.* 256:81-100, 1990.
165. Rietschel, E. T., T. Kirikae, F. U. Schade, U. Mamat, G. Schmidt, H. Loppnow, A. J. Ulmer, U. Zähringer, U. Seydel, F. Di Padova, M. Schreier, and H. Brade. Bacterial endotoxin: Molecular relationships of structure to activity and function. *FASEB J.* 8:217-225, 1994.
166. Rietschel, E. T., T. Kirikae, F. U. Schade, A. J. Ulmer, O. Holst, H. Brade, G. Schmidt, U. Mamat, H.-D. Grimmecke, S. Kusumoto, and U. Zähringer. The chemical structure of bacterial endotoxin in relation to bioactivity. *Immunobiology* 187:169-190, 1993.
167. Rietschel, E. T., L. Brade, B. Lindner, and U. Zähringer. Biochemistry of lipopolysaccharides, in *Bacterial Endotoxic Lipopolysaccharides*, Vol. I, *Molecular Biochemistry and Cellular Biology*. D.C. Morrison and J.L. Ryan (Eds.), CRC Press, Boca Raton, 1992.
168. Rolfe, B. G., R. W. Carlson, R. W. Ridge, F. B. Dazzo, P. F. Mateos, and C. E. Pankhurst. Defective infection and nodulation of clovers by exopolysaccharide mutants of *Rhizobium leguminosarum* bv *trifolii*. *Aust. J. Plant Physiol.* 23:285-303, 1996.
169. Rolin, D. B., P. E. Pfeffer, S. F. Osman, B. S. Szwergold, F. Kappler, and A. J. Benesi. Structural studies of a phosphocholine substituted β-(1, 3);(1,6) macrocyclic glucan from *Bradyrhizobium japonicum* USDA 110. *Biochim. Biophys. Acta Gen. Subj.* 1116:215-225, 1992.
170. Roth, L. E. and G. Stacey. Bacterium release into host cells of nitrogen-fixing soybean nodules: the symbiosome membrane comes from three sources. *Eur. J. Cell Biol.* 49:13-23, 1989.

171. Russa, R., M. Bruneteau, A. S. Shashkov, T. Urbanik-Sypniewska, and H. Mayer. Characterization of the lipopolysaccharides from *Rhizobium meliloti* strain 102F51 and its nonnodulating mutant WL113. *Arch. Microbiol.* 165:26-33, 1996.

172. Russa, R., O. Luderitz, and E. T. Rietschel. Structural analysis of lipid A from lipopolysaccharide of nodulating and non-nodulating *Rhizobium trifolii*. *Arch. Microbiol.* 141:284-289, 1985.

173. Russa, R., T. Urbanik-Sypniewska, A. Choma, and H. Mayer. Identification of 3-deoxy-*lyxo*-2-heptulosaric acid in the core region of lipopolysaccharides from Rhizobiaceae. *FEMS Microbiol. Lett.* 84:337-344, 1991.

174. Russa, R., T. Urbanik-Sypniewska, K. Lindström, and H. Mayer. Chemical characterization of two lipopolysaccharide species isolated from *Rhizobium loti* NZP2213. *Arch. Microbiol.* 163:345-351, 1995.

175. Schletter, J., H. Holger, A. J. Ulmer, and E. T. Rietschel. Molecular mechanisms of endotoxin activity. *Arch. Microbiol.* 164:383-389, 1995.

176. Schnaitman, C. A. and J. D. Klena. Genetics of lipopolysaccharide biosynthesis in enteric bacteria. *Microbiol. Rev.* 57:655-682, 1993.

177. Schwedock, J. S. and S. R. Long. *Rhizobium meliloti* genes involved in sulfate activation: The two copies of *nodPQ* and a new locus, *saa*. *Genetics* 132:899-909, 1992.

178. Segovia, L., J. P. W. Young, and E. Martínez-Romero. Reclassification of American *Rhizobium leguminosarum* biovar phaseoli type I strains as *Rhizobium etli sp. nov. Int. J. Syst. Bacteriol.* 43:374-377, 1993.

179. Selbitschka, W., W. Arnold, U. B. Priefer, T. Rottschäfer, M. Schmidt, R. Simon, and A. Pühler. Characterization of *recA* genes and *recA* mutants of *Rhizobium meliloti* and *Rhizobium leguminosarum* biovar *viciae*. *Mol. Gen. Genet.* 229:86-95, 1991.

180. Sindhu, S. S., N. J. Brewin, and E. L. Kannenberg. Immunochemical analysis of lipopolysaccharides from free-living and endosymbiotic forms of *Rhizobium leguminosarum*. *J. Bacteriol.* 172:1804-1813, 1990.

181. Sirisena, D. M., K. A. Brozek, P. R. MacLachlan, K. E. Sanderson, and C. R. H. Raetz. The *rfaC* gene of *Salmonella typhimurium*. Cloning, sequencing, and enzymatic function in heptose transfer to lipopolysaccharide. *J. Biol. Chem.* 267:18874-18884, 1992.

182. Spaink, H. P. The molecular basis of infection and nodulation by rhizobia: the ins and outs of sympathogenesis. *Annu. Rev. Phytopathol.* 33:345-368, 1995.

183. Stacey, G., J.-S. So, L. E. Roth, S. K. Bhagya Lakshmi, and R. W. Carlson. A lipopolysaccharide mutant from *Bradyrhizobium japonicum* that uncouples plant from bacterial differentiation. *Mol. Plant-Microbe Interact.* 4:332-340, 1991.

184. Stevenson, G., A. Kessler, and P. R. Reeves. A plasmid-borne O-antigen chain length determinant and its relationship to other chain length determinants. *FEMS Microbiol. Lett.* 125:23-30, 1995.

185. Stevenson, T. T., A. G. Darvill, and P. Albersheim. 3-deoxy-D-*lyxo*-2-heptulosaric acid, a component of the plant cell-wall polysaccharide rhamnogalacturonan-II. *Carbohydr. Res.* 179:269-288, 1988.

186. Streeter, J. G., S. O. Salminen, R. E. Whitmoyer, and R. W. Carlson. Formation of novel polysaccharides by *Bradyrhizobium japonicum* bacteroids in soybean nodules. *Appl. Environ. Microbiol.* 58:607-613, 1992.

187. Takada, H. and S. Kotani. Structure-function relationships of lipid A, in *Bacterial Endotoxic Lipopolysaccharides*, Vol. I, *Molecular Biochemistry and Cellular Biology*. D.C. Morrison and J.L. Ryan (Eds.), CRC Press, Boca Raton, 1992.

188. Tao, H., N. J. Brewin, and K. D. Noel. *Rhizobium leguminosarum* CFN42 lipopolysaccharide antigenic changes induced by environmental conditions. *J. Bacteriol.* 174:2222-2229, 1992.

189. Terefework, Z., G. Nick, S. Suomalainen, L. Paulin, and K. Lindström. Phylogeny of *Rhizobium galegae* with respect to other rhizobia and agrobacteria. *Int. J. Syst. Bacteriol.* 48:349-356, 1998.

190. Urbanik-Sypniewska, T., U. Seydel, M. Greck, J. Weckesser, and H. Mayer. Chemical studies on the lipopolysaccharide of *Rhizobium meliloti* 10406 and its lipid A region. *Arch. Microbiol.* 152:527-532, 1989.

191. VandenBosch, K. A., N. J. Brewin, and E. L. Kannenberg. Developmental regulation of a rhizobium cell-surface antigen during growth of pea root-nodules. *J. Bacteriol.* 171:4537-4542, 1989.

192. Verma, D. P. S. and Z. Hong. Biogenesis of the peribacteroid membrane in root nodules. *Trends Microbiol.* 4:364-368, 1996.

193. Wang, Y. and R. I. Hollingsworth. The structure of the O-antigenic chain of the lipopolysaccharide of *Rhizobium trifolii* 4s. *Carbohydr. Res.* 260:305-317, 1994.

194. Whitfield, C. and M. A. Valvano. Biosynthesis and expression of cell-surface polysaccharides in gram-negative bacteria. *Adv. Microb. Physiol.* 35:136-246, 1993.

195. Williams, M. N. V., R. I. Hollingsworth, P. M. Brzoska, and E. R. Signer. *Rhizobium meliloti* chromosomal loci required for suppression of exopolysaccharide mutations by lipopolysaccharide. *J. Bacteriol.* 172:6596-6598, 1990.

196. Williams, M. N. V., R. I. Hollingsworth, S. Klein, and E. R. Signer. The symbiotic defect of *Rhizobium meliloti* exopolysaccharide mutants is suppressed by *lpsZ*⁺, a gene involved in lipopolysaccharide biosynthesis. *J. Bacteriol.* 172:2622-2632, 1990.

197. Woese, C. R. Bacterial evolution. *Microbiol. Rev.* 51:221-271, 1987.

198. Wolpert, J. S. and P. Albersheim. *Biochem. Biophys. Res. Commun.* 70:729, (Abstract), 1976.

199. Wood, E. A., G. W. Butcher, N. J. Brewin, and E. L. Kannenberg. Genetic derepression of a developmentally regulated lipopolysaccharide antigen from *Rhizobium leguminosarum* 3841. *J. Bacteriol.* 171:4549-4555, 1989.

200. Yanagi, M. and K. Yamasato. Phylogenetic analysis of the family Rhizobiaceae and related bacteria by sequencing of 16S rRNA gene using PCR and DNA sequencer. *FEMS Microbiol. Lett.* 107:115-120, 1993.

201. Young, J. P. W. and K. E. Haukka. Diversity and phylogeny of rhizobia. *New Phytol.* 133:87-94, 1996.

202. Zevenhuizen, L. P. T. M., A. Van Veldhuizen, and R. H. Fokkens. Re-examination of cellular cyclic β-1,2-glucans of Rhizobiaceae: Distribution of ring sizes and degrees of glycerol-1-phosphate substitution. *Antonie Van Leeuwenhoek* 57:173-178, 1990.

203. Zhan, H., J. X. Gray, S. B. Levery, B. G. Rolfe, and J. A. Leigh. Functional and evolutionary relatedness of genes for exopolysaccharide synthesis in *Rhizobium meliloti* and *Rhizobium* sp. strain NGR234. *J. Bacteriol.* 172:5245-5253, 1990.

204. Zhan, H. and J. A. Leigh. Two genes that regulate exopolysaccharide production in *Rhizobium meliloti*. *J. Bacteriol.* 172:5254-5259, 1990.

205. Zhan, H., S. B. Levery, C. C. Lee, and J. A. Leigh. A second exopolysaccharide of *Rhizobium meliloti* strain SU47 that can function in root nodule invasion. *Proc. Natl. Acad. Sci. U.S.A.* 86:3055-3059, 1989.

206. Zhang, Y., R. I. Hollingsworth, and U. B. Priefer. Characterization of structural defects in the lipopolysaccharides of symbiotically impaired *Rhizobium leguminosarum* biovar *viciae* VF-39 mutants. *Carbohydr. Res.* 231:261-271, 1992.

chapter four

The genetics of capsule and lipooligosaccharide biosynthesis in Haemophilus influenzae

Andrew Preston and Michael A. Apicella

Contents

Introduction 91
Capsule 92
    Genetics of capsule production 94
    Amplification of the capsule locus..94
    Population biology of capsulated *H. influenzae* 95
H. influenzae LOS..95
    H. influenzae LOS structure...96
    Genes of LOS biosynthesis...99
    Genetic organization...101
    Genetic regulation of LOS biosynthesis..101
Summary 103
References..103

Introduction

Haemophilus influenzae is a Gram-negative bacterium that is a commensal of the human nasopharynx. On occasion, the balance between colonization and asymptomatic carriage breaks down and disease occurs.[81] Disease caused by *H. influenzae* is significantly affected by the capsulation state of the organism (see below). Both nontypable *H. influenzae*, which does not express a capsule, and *H. influenzae* type b are a significant cause of upper respiratory tract infections.[81,84] Systemic disease caused by *H. influenzae*, the most serious manifestation of which involves crossing the blood–brain barrier leading to meningitis, is caused predominantly by type b organisms,[81] and occurs primarily in neonates and infants during the time after maternal antibody protection is lost and before the developing immune system is capable of protecting the infant.[13] Recently, implementation of vaccines effective against type b *H. influenzae* has dramatically reduced the number of invasive disease episodes in vaccinated populations.[47,51,58,77,78]

Haemophilus influenzae is capable of expressing two major surface polysaccharides. These are capsule and lipooligosaccharide (LOS). Capsule is a polymeric carbohydrate structure that forms the outermost layer of the bacterium. LOS contains both lipid and carbohydrate and is the major component of the outer leaflet of the outer membrane of *H. influenzae*. Both capsule and LOS have been shown to be important contributors to *H. influenzae*'s ability to colonize and infect the human host. In this chapter we attempt to highlight the impact that genetic research into the biosynthesis of these two structures has made on understanding the pathogenesis of *H. influenzae*.

Capsule

A capsule is a polyanionic, well-hydrated structure that provides a physical defense barrier around the bacterium and may also determine access of molecules and ions to the bacterial cell outer membrane.[48] *H. influenzae* can synthesize one of six capsules that confer serotype a to f specificity.[48,62] Type a and b capsules both contain the five-carbon sugar ribitol, type a consisting of a polymer of glucose ribitol phosphate, while type b consists of poly-ribose-ribitol-phosphate.[5,8] Type c and f capsules contain O-acetylated 2-acetimido-2-deoxyhexose, and type d and e capsules contain 2-acetimido-2-deoxy-D-mannose uronic acid.[9,10,79,80]

Although all capsular types share similarities in structure, over 95% of invasive *H. influenzae* disease cases are caused by type b strains, suggesting that this capsule confers an increased invasive capability, or that type b capsule is a marker for strains with an increased invasive capability. Paradoxically, type b capsule appears to inhibit interaction of *H. influenzae* with epithelial cells and thus perhaps inhibit colonization of the nasopharynx.[38,72,73] However, type b capsule does increase serum resistance, enhancing survival of the bacterium in the bloodstream.[50,89] This may be due to the charged, hydrophilic capsule inhibiting interaction with phagocytic cells and inhibiting deposition of complement. However, isogenic transformants of each capsular type demonstrated that capsular type a and e organisms were as resistant to the action of complement as type b transformants,[75] so this alone does not explain the increased disease-causing ability of type b *H. influenzae*. A possible explanation for this may be provided by analysis of the genetics of capsule production (Figure 4.1).

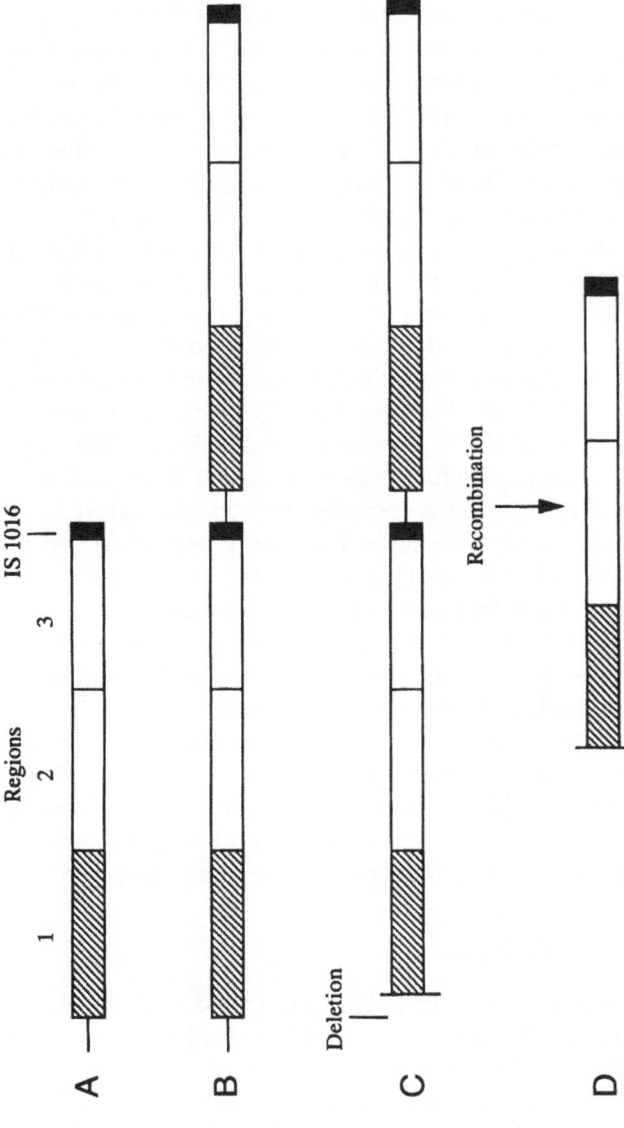

Figure 4.1   The modular arrangement of the capsule locus of *H. influenzae*. (A) The single copy state as found in non-type b strains and some type b strains. In Division I strains, the capsule locus is flanked by copies of the IS1016 sequence (shaded black), leading to frequent amplification of the locus as found in division I type b isolates. (B) The tandem repeat arrangement following duplication of the capsule locus. In division I type b strains, a mutation has deleted 1.2Kb of region 1, resulting in just a single copy of the *bexA* gene remaining (C). Recombination between IS1016 elements can reduce the capsule locus copy number in type b strains, which in the case of reduction to a single copy, results in the absence of a *bexA* gene and thus a defect in capsule export and a capsule minus phenotype (D).

Genetics of capsule production

The genes for capsule production are contained within a single locus of approximately 18Kb[19] (Figure 4.1). The locus is described as comprising three regions.[32,82] Regions 1 and 3 are common to all capsular types: region 1 encoding the "common" functions of capsule export, while region 3 probably encodes control functions. Region 2 differs between the different capsule types and encodes functions of synthesis of the particular capsule.[82] This modular arrangement is also found in *Escherichia coli* K1 and *Neisseria meningitidis* serogroup B, which along with *H. influenzae* type b are the predominant serotypes associated with sepsis and meningitis in infants. The common biochemical properties of these capsules and genetic organization of their production has led to their classification as type II capsules.[14]

For type b *H. influenzae*, the capsule gene arrangement has important consequences. The type b capsule locus is often found as a duplication, creating a direct repeat of the 18Kb locus.[19] However, it is not a true tandem repeat because a deletion has occurred at one end, removing 1.2Kb of the 3′ end of region 1 (Figure 4.1). Thus the duplicated locus contains two repeats of approximately 17Kb, which are separated by a 1.2Kb bridge region containing the remaining copy of the 3′ end of region 1. This 1.2Kb region contains the gene *bexA*, which is the fourth gene in the *bex* operon of region 1, encoding an ABC-type transporter responsible for capsule export.[28]

The duplicated arrangement is unstable. It has long been observed that *H. influenzae* type b capsulation is unstable; capsulated strains give rise to noncapsulated variants at high frequency *in vitro*.[12] Analysis of noncapsulated variants of type b strain Eagan revealed that they had reduced to a single copy of the capsule locus.[19] This event deletes the bridge region containing the only copy of *bexA*, and thus these organisms are capsule-export deficient and noncapsulated. The capsule export deficiency leads to accumulation of intracellular capsule material, a deleterious event which produces small colonies *in vitro*. This situation is often relieved by secondary mutations in the capsule locus that abrogates capsule synthesis and produces large, smooth, noniridescent class II noncapsulated mutant colonies.[4]

Amplification of the capsule locus

An insertion element-like sequence, IS1016, is associated with the type b capsule locus; a single copy is located at each end.[29] Recombination between this repeated element leads to amplification of the capsule locus beyond the duplicated state; up to five copies of the locus have been observed.[7] Interestingly, a survey revealed that approximately 35% of clinical isolates had amplified capsule loci.[7] Amplification also produces a gene dosage effect in that a strain with five copies of the locus synthesized approximately five times the level of capsule as a strain with the duplicated locus.[7] The biological significance of this increased capsule production is unclear, but it is possible the apparent protective effects of capsule are increased accordingly. In support of this theory, an isogenic set of bacteria which contained 2, 3, or 4 copies of the capsule locus were found to differ in their resistance to complement-mediated bacteriolysis, with 4 copies of the capsule locus providing greater resistance than 3 copies, which in turn provided greater protection than did 2 copies.[56] The amplified arrangement is unstable *in vitro*. A reduction in copy number is observed readily after few laboratory passages, and thus there may be a selective pressure that maintains the amplified state *in vivo*.[7] The increased resistance to complement-mediated killing conferred by capsulation may contribute to such a selective pressure.

Population biology of capsulated H. influenzae

Capsulated *H. influenzae* belongs to one of twelve major lineages, designated A–L, each of which belongs to one of two primary divisions. Division one contains all serotype c and d isolates studied: one lineage of serotype a isolates and 95% of type b electrophoretic types. Division II contains all serotype f isolates studied, a second lineage of a isolates, and a second group of serotype b isolates. A serotype belongs to only one or a few lineages, and these lineages are not shared between serotypes.[52]

There is an interesting distinction between the capsule locus of division I and division II type b isolates. In division I isolates, the IS1016 element is associated with the capsule locus. In division II isolates, the element is present, but not associated with the capsule locus. IS1016 is also found at either end of the capsule locus of type a–d strains which belong to division I, although these capsule loci are present as a single copy.[29] Division II type b strains also contain a single copy cap locus. These strains are less virulent in an infant rat model of pathogenesis and are rarely responsible for invasive disease cases.[30] Thus, it appears that the 1.2Kb deletion of the bridge region is an ancestral mutation that causes a selective pressure for maintenance of the duplicated capsule locus state.[31] Reduction to a single copy state in these strains deletes the only copy of the export gene *bexA*, and thus renders the organism capsule deficient and poorly able to survive *in vivo* compared to capsule-proficient organisms. This requirement for a duplicated locus state has the consequence of maintaining organisms that produce more capsule and probably contributes to the increased virulence of these strains, which is suggested by their occurrence as the prime invasive disease causing strains.

H. influenzae LOS

Lipopolysaccharide (LPS) is the major component of the outer membrane of Gram-negative bacteria and is a major distinguishing feature between Gram-negative and Gram-positive bacteria. LPS is a glycolipid in which an acylated diglucosamine backbone (Lipid A) forms a membrane anchor and is linked through 2-keto-3-deoxyoctulosonic acid (KDO) to an oligosaccharide.[39] LPS, particularly the lipid A component, is attributed to mitogenicity, pyrogenicity, platelet aggregation, and cytokine activation, and is thus responsible for much of the inflammation and toxicity associated with Gram-negative bacterial infections.[15,65] *H. influenzae*, along with other bacteria that inhabit human mucosal surfaces, including members of the genera *Neisseria*, *Bordetella*, and *Branhamella*, express shorter, more branched LPS structures than the LPS of enteric bacteria. Also these structures do not contain the repetitive O-antigen side chains of enteric bacteria. To highlight this distinction, these structures are termed lipooligosaccharides (LOSs), although this term is not universally used.

It is now evident that LOS is of fundamental importance to *H. influenzae* for survival and its ability to inhabit and infect the human host. These data are reviewed extensively elsewhere.[63]

H. influenzae *LOS structure*

The structure of LOS from *H. influenzae* strain A2 is shown in Figure 4.2.[61] All *H. influenzae* LOSs share this common structure in terms of an oligosaccharide, consisting mainly of heptose, glucose, and galactose, linked through KDO to lipid A. Sialic acid is also often present, as a terminal component linked to galactose.[40] The oligosaccharide chains display variability between strains. Some structures observed in *H. influenzae* LOS are depicted in Table 4.1.

The LOS is highly variable. A single strain simultaneously synthesizes multiple structures.[42,59-61] These structures are often present in unequal amounts such that a majority of LOS structures reported represent the major LOS structure of that strain. The structures of the full repertoire of LOS molecules of which a strain is capable of producing is unknown but may include truncated versions of the full-length molecule as well as alternative full-length structures.

A feature of *H. influenzae* LOS is phase variation of LOS structures, that is the reversible loss or gain of LOS structures, often described in terms of loss/gain of anti-LOS monoclonal antibody epitopes.[27] Phase variation occurs at high frequency (often 10^{-2} per bacterium per generation) and undoubtedly contributes to the heterogeneity of LOS. The role of phase variation in the biology of *H. influenzae* infection is unclear. It is possible that high frequency switching of exposed structures, and generation of a repertoire of structures is a way of avoiding host defense mechanisms.

Understanding of *H. influenzae* LOS biosynthesis has increased dramatically over the last decade, and recent completion of the whole-genome sequencing of *H. influenzae* strain Rd theoretically allows identification of all the genetic loci responsible for LOS biosynthesis in this organism.[11] Research has now reached the stage, accelerated by advances such as the whole-genome sequencing, at which many of the genes involved in *H. influenzae* LOS biosynthesis have been identified. However, it is evident that simply identifying these genes does not produce a complete understanding of the LOS structures nor, in particular, how the observed variation in these structures is achieved. Research is shifting to understanding the genetic regulation of LOS genes in order to answer these questions.

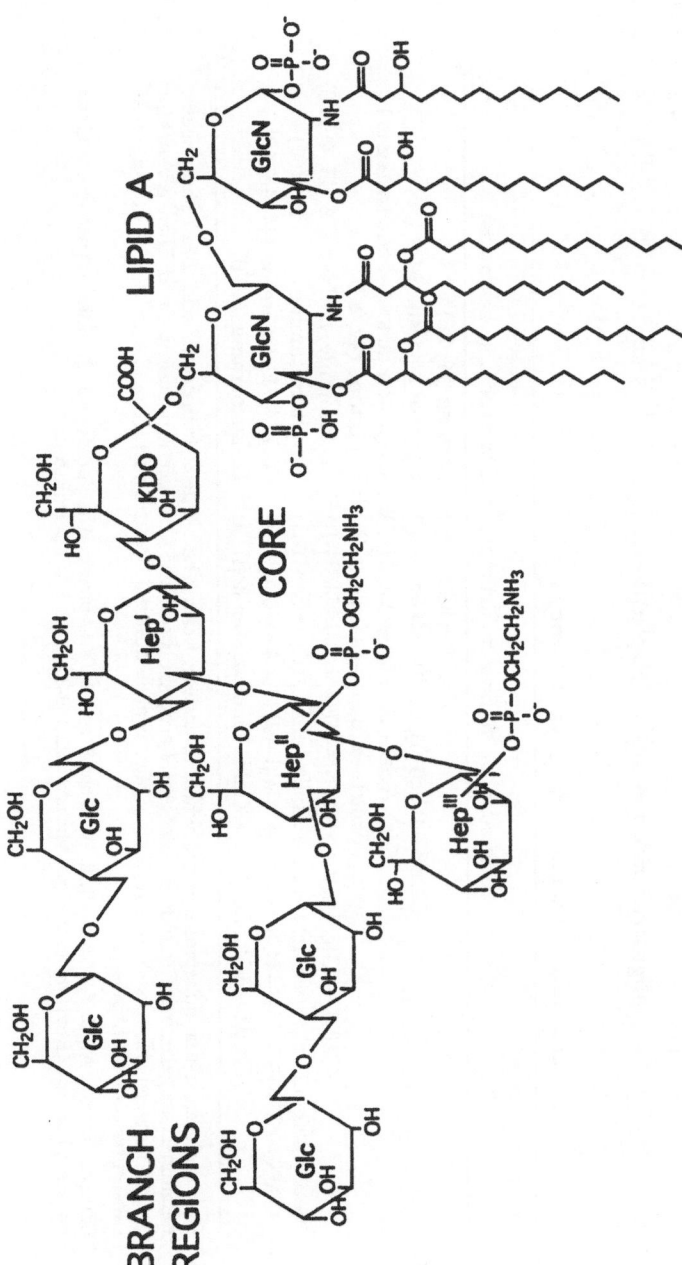

Figure 4.2 Structure of *H. influenzae* A2 LOS. GlcN: glucosamine; KDO: 2-keto-3-deoxyoctulosonic acid; Hep: L-glycero-D-mannoheptose; Glc: glucose.

Table 4.1   Some *Haemophilus* LOS Branch Structures

$$\text{oligosaccharide branch I} \rightarrow 4\text{Hep}^{I}\alpha1 \rightarrow 5\text{KDO(P)*-Lipid A}$$

$$\overset{3}{\uparrow}$$

$$\text{oligosaccharide branch II} \rightarrow 3\text{Hep}^{II}\alpha1(\text{PEA})_{0-1}$$

$$\overset{2}{\uparrow}$$

$$\text{oligosaccharide branch III} \rightarrow \text{Hep}^{III}\alpha1(\text{PEA})_{0-1}$$

Oligosaccharide Branch Structures	Comments, (References)
Galβ1→4Glcβ1-(I)	Lactose (59, 70)
Glcβ1→4Glcβ-(I), Glcβ1→4Glcα1-(II)	Glucose disaccharides (60)
GlcNAcβ1→3Galβ1→4Glcβ1-(I)	Truncated lacto-N-neotetraose (44)
Galα1→4Galβ1→4Glcβ1-(I)(II)	P^k epitope (42, 83)
Galβ1→4GlcNAcβ1→3Galβ1→3Hepα1→6Glcβ1-(I)	Lactosamine, DD-Hep (26, 45, 46, 70)
Neu5Acα2→3Galβ1→4GlcNAcβ1→3Galβ1→3Hepα1→6Glcβ1-(I)	Sialyllactosamine+ DD-Hep (69)
GlcNAcβ1→3Galβ1→4GlcNAcβ1→3Galβ→3Hepα1→6Glcβ1-(I)	GlcNAc-lactosamine (44, 46, 69)
(Galβ1→4GlcNAcβ1)₂→3Galβ→3Hepα1→6Glcβ1-(I)	Di-lactoamine (6, 70)

Hep: L-glycero-D-mannoheptose (except where indicated); KDO: 2-keto-3-deoxyoctulosonic acid; PEA: phosphoethanolamine; Gal: galactose; Glc: glucose; GlcNAc: N-acetylglucose; Neu5Ac: N-acetyl neuraminic acid. There is variability in LOS between strains such that different strains contain different branch structures and some strains do not contain substitutions on all heptoses.

Source: Preston, A., R. E. Mandrell, B. W. Gibson, and M. A. Apicella. The lipooligosaccharides of pathogenic Gram-negative bacteria [Review]. *Crit. Revs. Microbiol.* 22:139-80, 1996. With permission.

An extensive review has appeared recently covering the LOSs of Gram-negative bacteria in general.[63] Here, we shall attempt to highlight some of the recent advances made pertaining to *H. influenzae.*

Genes of LOS biosynthesis

Genes that are involved in LOS biosynthesis are listed in Table 4.2. Some of these genes were identified through mutagenesis, with a mutant producing an altered LOS phenotype on SDS–PAGE analysis and/or in reactivity to anti-LOS monoclonal antibodies. Other genes were cloned through complementation of enteric LPS mutants, and recently a large number of putative LOS genes were identified through analysis of amino acid sequences of ORFs from the whole genome sequence, thus identifying ORFs homologous to previously identified LPS/LOS functions.[20] The list in Table 4.2 clearly demonstrates that a large number of genes are involved in LOS biosynthesis. This is not unexpected because LOS is a multicomponent molecule, the synthesis and assembly of which obviously requires numerous functions. However, *H. influenzae* contains several putative LOS genes for which there is no obvious function. Examples are genes *rfbB, rfbP, cld,* and OrfO, whose products are homologous to proteins involved in O-antigen biosynthesis in enteric bacteria.[20] *H. influenzae* does not produce O-antigen. It is possible that in enteric bacteria these genes have functions in addition to O-antigen biosynthesis and that these additional functions are required by *H. influenzae.* It is also possible that the terminal sugars of *H. influenzae* LOS are assembled in a fashion similar to O-antigen. *Bordetella pertussis* LOS contains a terminal trisaccharide which has been proposed as a single O-antigen unit-like structure and is probably assembled on an acyl-carrier component as is enteric O-antigen.[2] Finally, it is also possible that these genes are nonfunctional in *H. influenzae,* at least under the conditions under which they have been studied. Interestingly, *Neisseria* contain several *rfb* homologs for which no function can be ascribed.[66]

Differences in LOS biosynthesis between closely related strains is also evident. Mutations in several genes of type b strain RM7004 did not produce any detectable change in LOS phenotype, whereas in the type b strain RM153, mutation of these genes did alter LOS as judged by reactivity to anti-LOS monoclonal antibodies.[20] This suggests a high level of heterogeneity of LOS between *H. influenzae* strains even when monoclonal antibody data suggests the presence of the same epitopes. Furthermore, some LOS genes may be strain specific, because the *lex2* locus identified in type b strains is not present in the Rd strain from which the whole genome was sequenced.[20]

Recent studies have begun to further define the genes of lipid A biosynthesis. The *lpx* genes responsible for biosynthesis of the diglucosamine backbone of *E. coli* lipid A were identified some time ago,[68] however, it is only now that genes involved in the acylation of the backbone have been characterized. These studies are providing valuable information on the role of the acyl chains of LPS/LOS in pathogenesis, particularly regarding the toxic effects of lipid A.

Table 4.2 Genes of *H. influenzae* LOS Biosynthesis

Gene/Locus	Function	Reference
lpxA	Transfer of β-hydroxymyristic acid from ACP to UDP-GlcNAc	11
lpxB	Condensation of UDP-2,3-dihydroxymyristoyl-glucosamine with lipid X to form lipid IV$_A$	11
lpxC (*envA*)	Deacetylation of UDP-3-hydroxymyristoyl-GlcNAc	11
lpxD (*fir*)	UDP-2,3-dihydroxymyristoyl-glucosamine formation	11
kdsA	KDO-8-P biosynthesis	11
kdsB	CMP-KDO biosynthesis	11
kdtA	KDO transferase	11
kdtB	KDO transferase	11
msbB	Acyltransferase of lipid A biosynthesis	11/71
htrB	Acyltransferase of lipid A biosynthesis	35
isn/gmhA	Phosphoheptose isomerase	64/3
rfaD	ADP-L-glycero-D-mannoheptose epimerase	54
rfaE	ADP-L-glycero-D-mannoheptose synthetase	36
rfaF	Heptosyltransferase II	54
lsg	Expression of 6E4 epitope	1
lic1	Expression of 12D9 and 6A2 epitopes	85
lic2	Expression of 4C4 epitope	18
lic3	Contains *galE*	41
lex2	Expression of 5G8 epitope	22
lgtC	Putative glycosyltransferase	21
opsX	Heptosyltransferase	20
pgmA-C	Phosphoglucose mutase	20
lpsA	Glycosyltransferase	20
lgtA	Glycosyltransferase	20
nusG (*rfaH*)	Transcriptional terminator	20
OrfH	Heptosyltransferase	20
siaB	CMP-NANA synthetase	11
basR/S	Putative regulator of LOS microheterogeneity	11
OrfM, OrfZ, OrfY, OrfE, OrfO, OrfE, *rfbB*, *rfe*, *kfiC*, *lsg1*, *xylR*, *cld*, *rfbP*	Homologs of previously identified LOS/LPS genes from other organisms	20

Downstream of *rfaE* in *H. influenzae* is an analogue of *E. coli htrB*.[35] HtrB in *E. coli* had been mistakenly identified as a heat shock protein as it is required for growth at 37°C. Several phenotypes were observed in an *E. coli htrB* mutant, including a change in color of the LPS on silver staining, following SDS–PAGE, from black to reddish brown.[24,25] *H. influenzae htrB* mutants have some phenotypes which are similar to those of the *E. coli htrB* mutant, including temperature sensitivity as well as the LOS color modification mentioned above. In addition, induction of suppressors of the mutation occurs in *H. influenzae* in a fashion analogous to *E. coli*. Studies have shown *H. influenzae htrB* to be a KDO-dependent acyltransferase. *htrB* mutants have a lipid A structure which contains predominantly tetraacyl species, compared to wild-type, which has a hexaacyl structure, and enzymatic studies indicate that HtrB is responsible for the 3′ substitution of hydroxymyristic acid in *H. influenzae* lipid A by myristic acid.[24,35]

The predominant tetraacyl lipid A structure of the *H. influenzae htrB* mutant is identical to lipid IV_A. This form of lipid A binds to the LPS receptor, CD14, but does not initiate the signaling which leads to a cytokine response from macrophages and also is nontoxic in animal models. This suggests that the *htrB* lipid A might have reduced toxicity in animal systems and *in vitro* models. Studies of an *H. influenzae* type b strain A2 *htrB* mutant indicate that nasal infectivity in the neonatal rat model is ablated (0 of 30 animals developed bacteremia compared to 8 of 30 of animals infected with the wild type). In addition, limulus amebocyte lysate coagulability and human macrophage TNFα release are reduced by one log when exposed to LOS isolated from the *htrB* mutant when compared to release induced by wild-type LOS.[55]

Genetic organization

The genes of *H. influenzae* LOS biosynthesis are dispersed around the chromosome. Outside of two clusters, the *lsg* locus[1,43] and a cluster of O-antigen related genes (see above), these genes are not found as part of large contiguous gene clusters. Although the consequence of the dispersed nature of the LOS genes is unknown, it is in contrast to the situation in enteric bacteria in which most of the genes for biosynthesis of the oligosaccharide component of LPS are contained in two clusters, *rfa* (core biosynthesis) and *rfb* (O-antigen biosynthesis),[68] and may reflect a complexity of regulation in *H. influenzae* that prohibits clustering of genes under the control of common promoters. Very little is known of transcriptional organization of *H. influenzae* LOS genes.

Genetic regulation of LOS biosynthesis

Phase variation

A striking feature of several LOS biosynthesis genes is the presence of tandem repeats of a tetranucleotide DNA motif in the 5′ region of the predicted coding region of the gene. Association of tandem repeats with LOS biosynthetic genes was made several years ago with identification of the three *lic* loci. *lic1* contains four ORFs, A–D. Initial observations were that *lic1C* and *lic1D* were involved in expression of the epitopes recognized by MAbs 12D9 and 6A2.[85] Recently, the function of the *lic1* locus has been identified as decoration of the LPS with phosphorylcholine. While phosphorylcholine is the major phospholipid of the eukaryotic cell membrane, it has rarely been observed in prokaryotes. *H. influenzae* takes up choline from the growth medium, and *lic1* is proposed as a locus involved in its uptake, phosphorylation, and transfer to the LPS. *lic1A* is a putative choline kinase and contains multiple tandem repeats of 5'-CAAT-3'. Mutation of *lic1A* reduces expression of the 6A2 epitope and also results in the absence of phosphorylcholine from the LPS.[88] *lic2* contains a single gene involved in expression of the 4C4 and 5G8 epitopes and also contains 5'-CAAT-3' repeats in its 5' region.[18,87] This gene probably encodes a glycosyltransferase that adds a sugar to LOS. *lic3* was also identified as a locus containing CAAT repeats and contains four ORFs.[87] *lic3A* contains the repeats and, although its relation to LOS biosynthesis is unknown, *lic3B* encodes GalE, a metabolic enzyme crucial to balancing levels of the LOS precursors UDP-glucose and UDP-galactose.[41] The association of tandem tetrameric DNA repeat motifs and LOS biosynthesis has been strengthened by the recent identification of other LOS biosynthetic loci containing tetrameric repeats, namely *lex2* (containing 5'-GCAA-3')[22] and *lgtC* (containing 5'-GACA-3'),[21] both of which are probably glycosyltransferases.

The repeat motifs are responsible for phase variation of LOS structures. The number of repeats varies due to addition or deletion of single repeat units through slip-strand mispairing during replication.[37] Variation in repeat number moves the coding reading frame downstream of the repeats in or out of frame with the 5' region. A repeat number which leads to an out-of-frame situation prevents synthesis of the affected protein and thus leads to an absence of that LOS biosynthetic function.[49] The repeats are found in glycosyltransferase genes (although the function of the genes containing the repeats in *lic1* and *lic3* is unknown), and thus phase variation probably occurs through switching on and off synthesis of enzymes that directly add components to the oligosaccharide of LOS.

A clear association between repeat number (and thus the corresponding in- or out-of-frame situation) and switching on or off expression of LOS epitopes has been shown for *lic1*, *lic2*, and *lex2*.[18,22,86]

Phase variation through slip-strand mispairing of repetitive DNA elements appears to be a common mechanism of variation of surface components of Gram-negative bacteria; homopolymeric tracts are found in the 5' region of glycosyltransferase genes in pathogenic *Neisseria*,[16,23] although the occurrence of tetrameric repeats has thus far only been identified in *H. influenzae*. Analysis of the whole genome sequence of strain Rd identified twelve open reading frames containing tetrameric repeats, of which four were in obvious LOS biosynthetic loci (the three *lic* loci and *lgtC*).[21] Thus, this strain has multiple loci at which phase variation can alter LOS structure. This highlights the complexity of the LOS biosynthetic process and goes some way to explaining the high-level variability in LOS structures that is observed in a single strain.

Other mechanisms of regulation

Phase variation through alteration of tetrameric repeats is unlikely to be the only mechanism of genetic regulation of LOS biosynthesis. In *lic2*, where 16 tetrameric repeats are associated with strong expression of the 4C4 epitope, 16 repeats are also found in organisms that do not express the epitope.[18]

Also, phase variation in *lic3* occurred even after deletion of the repeats, suggesting that other mechanisms may be operating.[76]

LOS contains multiple sugar residues that are also central molecules of the bacterium's metabolism. It is possible that LOS biosynthesis is under the control of factors that regulate central metabolism, such as signaling through cAMP/CRP. It is known that LOS phenotype changes depending on growth culture conditions, although the exact nature of these changes and the mechanisms by which they occur are unknown.[33,34]

It is also possible that LOS is affected by specific environmental conditions. PmrA/B is a two-component system in *Salmonella typhimurium*, and BasR/S is an analogous system in *Escherichia coli*.[53,67] These systems regulate the pattern of phosphorylation of the LPS in these organisms.[17,57] Although the environmental signal to which these systems react is unknown, mutation of the systems affects resistance to antimicrobial peptides, and for *S. typhimurium* mutation also affects survival within macrophages.[74] A system with considerable homology to these two enteric systems has been identified in *H. influenzae*.[11] Thus it is possible that *H. influenzae* LOS substitutions are subject to regulation through a two-component system in response to an environmental signal.

Summary

Capsule and LOS are two surface polysaccharides that are critical to the infectious capability of *H. influenzae*. Understanding the production of these polysaccharides is important to understanding how *H. influenzae* is able to colonize and invade its human host. This understanding has also been of great value to research of surface polysaccharides in other pathogenic bacteria. A significant contribution to this work has been the elucidation of the genetics of capsule and LOS biosynthesis. We are now reaching a stage of full knowledge of the identity of genes of these processes. Future research is likely to concentrate on the expression of these genes, in particular regulation of the gene expression, because simply knowing the genes involved is insufficient to fully understanding the role of surface polysaccharides in the biology of *H. influenzae*.

References

1. Abu, K. Y., R. E. McLaughlin, M. A. Apicella, and S. M. Spinola. Analysis of *Haemophilus influenzae* type b lipooligosaccharide-synthesis genes that assemble or expose a 2-keto-3-deoxyoctulosonic acid epitope. *Mol. Microbiol.* 5:2475-80, 1991.
2. Allen, A. and D. J. Maskell. The identification, cloning and mutagenesis of a genetic locus required for lipopolysaccharide biosynthesis in *Bordetella pertussis*. *Mol. Microbiol.* 19:37-52, 1996.
3. Brooke, J. S. and M. A. Valvano. Molecular cloning of the *Haemophilus influenzae gmhA* (*lpcA*) gene encoding a phosphoheptose isomerase required for lipooligosaccharide biosynthesis. *J. Bacteriol.* 178:3339-41, 1996.

4. Brophy, L. N., J. S. Kroll, D. J. Ferguson, and E. R. Moxon. Capsulation gene loss and "rescue" mutations during the Cap+ to Cap– transition in *Haemophilus influenzae* type b. *J. Gen. Microbiol.* 137:2571-6, 1991.

5. Byrd, R. A., W. Egan, and M. F. Summers. New N.M.R.-spectroscopic approaches for structural studies of polysaccharides: application to the *Haemophilus influenzae* type a capsular polysaccharide. *Carbohydr. Res.* 166:47-58, 1987.

6. Campagnari, A. A., R. Karalus, M. Apicella, W. Melaugh, A. J. Lesse, and B. W. Gibson. Use of pyocin to select a *Haemophilus ducreyi* variant defective in lipooligosaccharide biosynthesis. *Infect. Immun.* 62:2379-2386, 1994.

7. Corn, P. G., J. Anders, A. K. Takala, H. Kayhty, and S. K. Hoiseth. Genes involved in *Haemophilus influenzae* type b capsule expression are frequently amplified. *J. Infect. Dis.* 167:356-364, 1993.

8. Crisel, R. M., R. S. Baker, and D. E. Dorman. Capsular polymer of *Haemophilus influenzae*, type. I. Structural characterisation of the capsular polymer of strain Eagan. *J. Biol. Chem.* 250:4926-4930, 1975.

9. Egan, W., F.-P. Tsui, P. A. Climenson, and R. Schneerson. Structural and immunological studies of the *Haemophilus influenzae* type c capsular polysaccharide. *Carbohydr. Res.* 80:305-316, 1980.

10. Egan, W., F.-P. Tsui, and R. Schneerson. Structural studies of the *Haemophilus influenzae* type f capsular polysaccharide. *Carbohydr. Res.* 79:271-277, 1980.

11. Fleischmann, R. D., M. D. Adams, O. White, R. A. Clayton, E. F. Kirkness, A. R. Kerlavage, C. J. Bult, J. F. Tomb, B. A. Dougherty, J. M. Merrick, K. McKenney, G. Sutton, W. FitzHugh, C. Fields, J. D. Gocayne, J. Scott, R. Shirley, L. I. Liu, A. Glodek, J. M. Kelley, J. F. Weidman, C. A. Phillips, T. Spriggs, and E. Hedblom. Whole-genome random sequencing and assembly of *Haemophilus influenzae* Rd. *Science.* 269:496-512, 1995.

12. Fothergill, L. and C. A. Chandler. Observations on the dissociation of meningitic strains of *Haemophilus influenzae.* *J. Immunol.* 31:401-15, 1936.

13. Fothergill, L. D. and J. Wright. Influenzal meningitis. The relation of age incidence to the bactericidal power of blood against the causal organism. *J. Immunol.* 24:273-284, 1933.

14. Frosch, M., U. Edwards, K. Bousset, B. Krausse, and C. Weisgerber. Evidence for a common molecular origin of the capsule gene loci in Gram-negative bacteria expressing group II capsular polysaccharides. *Mol. Microbiol.* 5:1251-63, 1991.

15. Galanos, C., O. Luderitz, E. T. Rietschel, O. Westphal, H. Brade, et al. Synthetic and natural *Escherichia coli* lipid A express identical endotoxic activities. *Eur. J. Biochem.* 145:1-5, 1985.

16. Gotschlich, E. C. Genetic locus for the biosynthesis of the variable portion of *Neisseria gonorrhoeae* lipooligosaccharide. *J. Exp. Med.* 180:2181-90, 1994.

17. Helander, I. M., I. Kilpelainen, and M. Vaara. Increased substitution of phosphate groups in lipopolysaccharides and lipid A of the polymyxin-resistant *pmrA* mutants of *Salmonella typhimurium*: a 31P-NMR study. *Mol. Microbiol.* 11:481-7, 1994.

18. High, N. J., M. E. Deadman, and E. R. Moxon. The role of a repetitive DNA motif (5'-CAAT-3') in the variable expression of the *Haemophilus influenzae* lipopolysaccharide epitope αGal(1-4)βGal. *Mol. Microbiol.* 9:1275-82, 1993.

19. Hoiseth, S. K., E. R. Moxon, and R. P. Silver. Genes involved in *Haemophilus influenzae* type b capsule expression are part of an 18-kilobase tandem duplication. *Proc. Nat. Acad. Sci. U.S.A.* 83:1106-10, 1986.

20. Hood, D. W., M. E. Deadman, T. Allen, H. Masoud, A. Martin, J. R. Brisson, R. Fleischmann, J. C. Venter, J. C. Richards, and E. R. Moxon. Use of the complete genome sequence information of *Haemophilus influenzae* strain Rd to investigate lipopolysaccharide biosynthesis. *Mol. Microbiol.* 22:951-65, 1996.

21. Hood, D. W., M. E. Deadman, M. P. Jennings, M. Bisercic, R. C. Fleischmann, J. C. Venter, and E. R. Moxon. DNA repeats identify novel virulence genes in *Haemophilus influenzae*. *Proc. Nat. Acad. Sci. U.S.A.* 93:11121-11125, 1996.

22. Jarosik, G. P. and E. J. Hansen. Identification of a new locus involved in expression of *Haemophilus influenzae* type b lipooligosaccharide. *Infect. Immun.* 62:4861-7, 1994.

23. Jennings, M. P., D. W. Hood, I. R. A. Peak, M. Virji, and E. R. Moxon. Molecular analysis of a locus for the biosynthesis and phase-variable expression of the lacto-N-neotetraose terminal lipopolysaccharide structure in *Neisseria meningitidis. Mol. Microbiol.* 18:729-740, 1995.

24. Karow, M., O. Fayet, and C. Georgopoulos. The lethal phenotype caused by null mutations in the *Escherichia coli htrB* gene is suppressed by mutations in the *accBC* operon, encoding two subunits of acetyl coenzyme A carboxylase. *J. Bacteriol.* 174:7407-18, 1992.

25. Karow, M. and C. Georgopoulos. The essential *Escherichia coli msbA* gene, a multicopy suppressor of null mutations in the *htrB* gene, is related to the universally conserved family of ATP-dependent translocators. *Mol. Microbiol.* 7:69-79, 1993.

26. Kelly, J. F., P. Thibault, H. Massoud, M. B. Perry, and J. C. Richards. Presented at the Abs 41st Conf. Mass Spectrom, 1993.

27. Kimura, A. and E. J. Hansen. Antigenic and phenotypic variations of *Haemophilus influenzae* type b lipopolysaccharide and their relationship to virulence. *Infect. Immun.* 51:69-79, 1986.

28. Kroll, J. S., B. Loynds, L. N. Brophy, and E. R. Moxon. The *bex* locus in encapsulated *Haemophilus influenzae*: a chromosomal region involved in capsule polysaccharide export. *Mol. Microbiol.* 4:1853-62, 1990.

29. Kroll, J. S., B. M. Loynds, and E. R. Moxon. The *Haemophilus influenzae* capsulation gene cluster: a compound transposon. *Mol. Microbiol.* 5:1549-60, 1991.

30. Kroll, J. S. and E. R. Moxon. Capsulation and gene copy number at the cap locus of *Haemophilus influenzae* type b [published erratum appears in *J. Bacteriol.* 1988. 170:2418]. *J. Bacteriol.* 170:859-64, 1988.

31. Kroll, J. S., E. R. Moxon, and B. M. Loynds. An ancestral mutation enhancing the fitness and increasing the virulence of *Haemophilus influenzae* type b. *J. Infect. Dis.* 168:172-6, 1993.

32. Kroll, J. S., S. Zamze, B. Loynds, and E. R. Moxon. Common organization of chromosomal loci for production of different capsular polysaccharides in *Haemophilus influenzae. J. Bacteriol.* 171:3343-7, 1989.

33. Langford, P. R. and E. R. Moxon. The dilution rate affects the outer membrane protein and lipopolysaccharide composition of *Haemophilus influenzae* type b grown under iron limitation. *J. Bacteriol.* 175:2462-4, 1993.

34. Langford, P. R. and E. R. Moxon. Growth of *Haemophilus influenzae* type b in continuous culture: effect of dilution rate on outer-membrane protein and lipopolysaccharide expression. *FEMS Microbiol. Letts.* 72:43-7, 1992.

35. Lee, N.-G., M. G. Sunshine, J. J. Engstrom, B. W. Gibson, and M. A. Apicella. Mutation of the *htrB* locus of *Haemophilus influenzae* nontypable strain 2019 is associated with modifications of lipid A and phosphorylation of the lipooligosaccharide. *J. Biol. Chem.* 270:27151-59, 1995.

36. Lee, N. -G., M. G. Sunshine, and M. A. Apicella. Molecular cloning and characterization of the nontypable *Haemophilus influenzae* 2019 *rfaE* gene required for lipopolysaccharide biosynthesis. *Infect. Immun.* 63:818-24, 1995.

37. Levinson, G. and G. A. Gutman. Slipped-strand mispairing: a major mechanism for DNA sequence evolution. *Mol. Biol. Evol.* 4:203-21, 1987.

38. LiPuma, J. J. and J. R. Gilsdorf. Role of capsule in adherence of *Haemophilus influenzae* type b to human buccal epithelial cells. *Infect. Immun.* 55:2308-10, 1987.

39. Luderitz, P., K. Tanamato, C. Galanos, G. R. McKenzie, H. Brade, U. Zahringer, E. T. Rietschel, S. Kusumoto, and T. Shiba. Lipopolysaccharides: structural principles and biologic activities. *Rev. Infect. Dis.* 6:428-431, 1984.

40. Mandrell, R. E., R. McLaughlin, Y. Aba Kwaik, A. Lesse, R. Yamasaki, B. Gibson, S. M. Spinola, and M. A. Apicella. Lipooligosaccharides (LOS) of some *Haemophilus* species mimic human glycosphingolipids, and some LOS are sialylated. *Infect. Immun.* 60:1322-8, 1992.

41. Maskell, D. J., M. J. Szabo, P. D. Butler, A. E. Williams, and E. R. Moxon. Molecular analysis of a complex locus from *Haemophilus influenzae* involved in phase-variable lipopolysaccharide biosynthesis. *Mol. Microbiol.* 5:1013-22, 1991.

42. Masoud, H., E. R. Moxon, A. Martin, D. Krajcarski, and J. C. Richards. Structure of the variable and conserved lipopolysaccharide oligosaccharide epitopes expressed by *Haemophilus influenzae* serotype b strain Eagan. *Biochem.* 36:2091-2103, 1997.

43. McLaughlin, R., S. M. Spinola, and M. A. Apicella. Generation of lipooligosaccharide mutants of *Haemophilus influenzae* type b. *J. Bacteriol.* 174:6455-9, 1992.

44. Melaugh, W., B. W. Gibson, and A. A. Campagnari. The lipooligosaccharides of *Haemophilus ducreyi* are highly sialylated. *J. Bacteriol.* 178:564-70, 1996.

45. Melaugh, W., N. Phillips, A. A. Campagnari, R. Karalus, and B. A. Gibson. Partial characterization of the major lipooligoasccharide from a strain of *Haemophilus ducreyi*, a genital ulcer disease. *J. Biol. Chem.* 267:13434-13439, 1992.

46. Melaugh, W., N. J. Phillips, A. A. Campagnari, M. V. Tullius, and B. W. Gibson. Structure of the major oligosaccharide from the lipooligosaccharide of *Haemophilus ducreyi* strain 35000 and evidence for additional glycoforms. *Biochem.* 33:13070-8, 1994.

47. Mohle Boetani, J. C., G. Ajello, E. Breneman, K. A. Deaver, C. Harvey, B. D. Plikaytis, M. M. Farley, D. S. Stephens, and J. D. Wenger. Carriage of *Haemophilus influenzae* type b in children after widespread vaccination with conjugate *Haemophilus influenzae* type b vaccines. *Pediatr. Infect. Dis. J.* 12:589-93, 1993.

48. Moxon, E. R. and J. S. Kroll. The role of bacterial polysaccharide capsules as virulence factors. *Curr. Top. Microbiol. Immunol.* 150:65-85, 1990.

49. Moxon, E. R. and D. J. Maskell. *Haemophilus influenzae* lipopolysaccharide: the biochemistry and biology of a virulence factor, in C. E. Hormaeche, C. W. Penn, and C. J. Smyth (Eds.), *Molecular Biology of Bacterial Infection: Current Status and Future Perspectives.* Cambridge University Press, Cambridge, 1992.

50. Moxon, E. R. and K. A. Vaughn. The type b capsular polysaccharide as a virulence determinant of *Haemophilus influenzae*: studies using clinical isolates and laboratory transformants. *J. Infect. Dis.* 143:517-24, 1981.

51. Murphy, T. V., P. Pastor, F. Medley, M. T. Osterholm, and D. M. Granoff. Decreased *Haemophilus* colonization in children vaccinated with *Haemophilus influenzae* type b conjugate vaccine. *J. Pediatr.* 122:517-23, 1993.

52. Musser, J. M., J. S. Kroll, D. M. Granoff, E. R. Moxon, B. R. Brodeur, J. Campos, H. Dabernat, W. Frederiksen, J. Hamel, G. Hammond, et al. Global genetic structure and molecular epidemiology of encapsulated *Haemophilus influenzae*. *Rev. Infect. Dis.* 12:75-111, 1990.

53. Nagasawa, S., K. Ishige, and T. Mizuno. Novel members of the two-component signal transduction genes in *Escherichia coli*. *J. Biochem.* 114:350-57, 1993.

54. Nichols, W. A., B. W. Gibson, W. Melaugh, N. G. Lee, M. Sunshine, and M. A. Apicella. Identification of the ADP-L-glycero-D-manno-heptose-6-epimerase (*rfaD*) and heptosyltransferase II (*rfaF*) biosynthesis genes from nontypeable *Haemophilus influenzae* 2019. *Infect. Immun.* 65:1377-86, 1997.

55. Nichols, W. A., C. R. H. Raetz, T. Clementz, A. L. Smith, J. A. Hanson, M. R. Ketterer, M. Sunshine, and M. A. Apicella. *htrB* of *Haemophilus influenzae*: determination of biochemical activity and effects on virulence and lipooligosaccharide toxicity. *J. Endotox. Res.* 4:163-72, 1997.

56. Noel, G. J., A. Brittingham, A. A. Granato, and D. M. Mosser. Effect of amplification of the Cap-B locus on complement-mediated bacteriolysis and opsonization of type-B *Haemophilus influenzae*. *Infect. Immun.* 64:4769-75, 1996.

57. Nummila, K., I. Kilpelaeinen, U. Zaehringer, M. Vaara, and I. M. Helander. Lipopolysaccharides of polymyxin B-resistant mutants of *Escherichia coli* are extensively substituted by 2-aminoethyl pyrophosphate and contain aminoarabinose in lipid A. *Mol. Microbiol.* 16:271-278, 1995.

58. Peltola, H., T. Kilpi, and M. Anttila. Rapid disappearance of *Haemophilus influenzae* type b meningitis after routine childhood immunisation with conjugate vaccines. *Lancet.* 340:592-4, 1992.

59. Phillips, N. J., M. A. Apicella, J. M. Griffiss, and B. W. Gibson. Structural characterization of the cell surface lipooligosaccharides from a nontypable strain of *Haemophilus influenzae*. *Biochem.* 31:4515-26, 1992.

60. Phillips, N. J., M. A. Apicella, J. M. Griffiss, and B. W. Gibson. Structural studies of the lipooligosaccharides from *Haemophilus influenzae* type b strain A2. *Biochem.* 32:2003-12, 1993.

61. Phillips, N. J., R. McLaughlin, T. J. Miller, M. A. Apicella, and B. W. Gibson. Characterization of two transposon mutants from *Haemophilus influenzae* type b with altered lipooligosaccharide biosynthesis. *Biochem.* 35:5937-47, 1996.

62. Pittman, M. Variation and type specificity in the bacterial species *Haemophilus influenzae*. *J. Exp. Med.* 53:471-92, 1931.

63. Preston, A., R. E. Mandrell, B. W. Gibson, and M. A. Apicella. The lipooligosaccharides of pathogenic Gram-negative bacteria [Review]. *Crit. Revs. Microbiol.* 22:139-80, 1996.

64. Preston, A., D. Maskell, A. Johnson, and E. R. Moxon. Altered lipopolysaccharide characteristic of the I69 phenotype in *Haemophilus influenzae* results from mutations in a novel gene, *isn*. *J. Bacteriol.* 178:396-402, 1996.

65. Raetz, C. R., K. A. Brozek, T. Clementz, J. D. Coleman, S. M. Galloway, D. T. Golenbock, and R. Y. Hampton. Gram-negative endotoxin: a biologically active lipid. *Cold Spring Harb. Symp. Quant. Biol.* 2:973-82, 1988.

66. Robertson, B. D., M. Frosch, and J. P. van Putten. The identification of cryptic rhamnose biosynthesis genes in *Neisseria gonorrhoeae* and their relationship to lipopolysaccharide biosynthesis. *J. Bacteriol.* 176:6915-20, 1994.

67. Roland, K. L., L. E. Martin, C. R. Esther, and J. K. Spitznagel. Spontaneous *pmrA* mutants of *Salmonella typhimurium* LT2 define a new two-component regulatory system with a possible role in virulence. *J. Bacteriol.* 175:4154-64, 1993.

68. Schnaitman, C. A. and J. D. Klena. Genetics of lipopolysaccharide biosynthesis in enteric bacteria. *Microbiol. Rev.* 57:655-82, 1993.
69. Schweda, E. K., J. A. Jonasson, and P. E. Jansson. Structural studies of lipoolig osaccharides from *Haemophilus ducreyi* ITM 5535, ITM 3147, and a fresh clinical isolate, ACY1: evidence for intrastrain heterogeneity with the production of mutually exclusive sialylated or elongated glycoforms. *J. Bacteriol.* 177:5316-21, 1995.
70. Schweda, E. K., A. C. Sundstrom, L. M. Eriksson, J. A. Jonasson, and A. A. Lindberg. Structural studies of the cell envelope lipopolysaccharides from *Haemophilus ducreyi* strains ITM 2665 and ITM 4747. *J. Biol. Chem.* 269:12040-8, 1994.
71. Somerville, J. E., Jr., L. Cassiano, B. Bainbridge, M. D. Cunningham, and R. P. Darveau. A novel *Escherichia coli* lipid A mutant that produces an antiinflammatory lipopolysaccharide. *J. Clin. Invest.* 97:359-65, 1996.
72. St. Geme, J. W. I. and S. Falkow. Capsule loss by *Haemophilus influenzae* type b results in enhanced adherence to and entry into human cells. *J. Infect. Dis.* 165:S117, 1992.
73. St. Geme, J. W. I. and S. Falkow. Loss of capsule expression by *Haemophilus influenzae* type b results in enhanced adherence to and invasion of human cells. *Infect. Immun.* 59:1325-33, 1991.
74. Stinavage, P., L. E. Martin, and J. K. Spitznagel. O Antigen and lipid A phosphoryl groups in resistance of *Salmonellae typhimurium* LT-2 to nonoxidative killing in human polymorphonuclear neutrophils. *Infect. Immun.* 57:3894-900, 1989.
75. Swift, A. J., E. R. Moxon, A. Zwahlen, and J. A. Winkelstein. Complement-mediated serum activities against genetically defined capsular transformants of *Haemophilus influenzae. Microb. Pathog.* 10:261-9, 1991.
76. Szabo, M., D. Maskell, P. Butler, J. Love, and R. Moxon. Use of chromosomal gene fusions to investigate the role of repetitive DNA in regulation of genes involved in lipopolysaccharide biosynthesis in *Haemophilus influenzae. J. Bacteriol.* 174:7245-52, 1992.
77. Takala, A. K., J. Eskola, M. Leinonen, H. Kayhty, A. Nissinen, E. Pekkanen, and P. H. Makela. Reduction of oropharyngeal carriage of *Haemophilus influenzae* type b (Hib) in children immunized with an Hib conjugate vaccine. *J. Infect. Dis.* 164:982-6, 1991.
78. Takala, A. K., M. Santosham, H. J. Almeido, M. Wolff, W. Newcomer, R. Reid, H. Kayhty, E. Esko, and P. H. Makela. Vaccination with *Haemophilus influenzae* type b meningococcal protein conjugate vaccine reduces oropharyngeal carriage of *Haemophilus influenzae* type b among American Indian children. *Pediatr. Infect. Dis. J.* 12:593-9, 1993.
79. Tsui, F.-P., R. Schneerson, R. A. Boykins, A. B. Karpas, and W. Egan. Structural and immunological studies of the *Haemophilus influenzae* type d capsular polysaccharide. *Carbohydr. Res.* 97:293-306, 1981.
80. Tsui, F.-P., R. Schneerson, and W. Egan. Structural studies of the *Haemophilus influenzae* type e capsular polysaccharide. *Carbohydr. Res.* 88:85-92, 1981.
81. Turk, D. C. The pathogenicity of *Haemophilus influenzae. J. Med. Microbiol.* 18:1-16, 1984.
82. van Eldere, J., L. Brophy, B. Loynds, P. Celis, I. Hancock, S. Carman, J. S. Kroll, and E. R. Moxon. Region II of the *Haemophilus influenzae* type b capsulation locus is involved in serotype-specific polysaccharide synthesis. *Mol. Microbiol.* 15:107-18, 1995.

83. Virji, M., J. N. Weiser, A. A. Lindberg, and E. R. Moxon. Antigenic similarities in lipopolysaccharides of *Haemophilus* and *Neisseria* and expression of a digalactoside structure also present on human cells. *Microb. Pathog.* 9:441-50, 1990.

84. Wallace, R. J., C. J. Baker, F. J. Quinones, et al. Non-typeable *Haemophilus influenzae* (biotype 4) as a neonatal, maternal and genital pathogen. *Rev. Infect. Dis.* 5:123-36, 1983.

85. Weiser, J. N., A. A. Lindberg, E. J. Manning, E. J. Hansen, and E. R. Moxon. Identification of a chromosomal locus for expression of lipopolysaccharide epitopes in *Haemophilus influenzae. Infect. Immun.* 57:3045-52, 1989.

86. Weiser, J. N., J. M. Love, and E. R. Moxon. The molecular mechanism of phase variation of *H. influenzae* lipopolysaccharide. *Cell.* 59:657-65, 1989.

87. Weiser, J. N., D. J. Maskell, P. D. Butler, A. A. Lindberg, and E. R. Moxon. Characterization of repetitive sequences controlling phase variation of *Haemophilus influenzae* lipopolysaccharide. *J. Bacteriol.* 172:3304-9, 1990.

88. Weiser, J. N., M. Shchepetov, and S. T. H. Chong. Decoration of lipopolysaccharide with phosphorylcholine: A phase-variable characteristic of *Haemophilus influenzae. Infect. Immun.* 65:943-950, 1997.

89. Zwahlen, A., J. S. Kroll, L. G. Rubin, and E. R. Moxon. The molecular basis of pathogenicity in *Haemophilus influenzae*: comparative virulence of genetically-related capsular transformants and correlation with changes at the capsulation locus cap. *Microb. Pathog.* 7:225-35, 1989.

chapter five

The genetics of LPS synthesis by the gonococcus

Susie Y. Minor and Emil C. Gotschlich

Contents

Introduction 111
Generic structure of neisserial LOS 112
Genetics of core biosynthesis 113
Genetics of α-chain biosynthesis 114
Genetics of high-frequency variation 116
Sialylation of LOS 118
Genetics of β-chain biosynthesis 119
Genes with secondary effects on LOS biosynthesis 119
Biosynthetic genes not yet identified 120
Selective advantage of LOS in pathogenesis 120
    Pili 120
    Opacity proteins 121
    LOS and the mucosal infection 122
Conclusion 123
Acknowledgments 123
References 124

Introduction

Gonococci (GC) and meningococci (MC), like typical Gram-negative bacteria, contain lipopolysaccharide (LPS) in their outer membranes. Unlike enterobacterial LPS, GC and MC LPS is modest in molecular size and therefore has been referred to as lipooligosaccharide (LOS). Yet work over the past 15 years has shown that within the constraints of the small size lurks a surprising degree of antigenic complexity and size heterogeneity. The technological

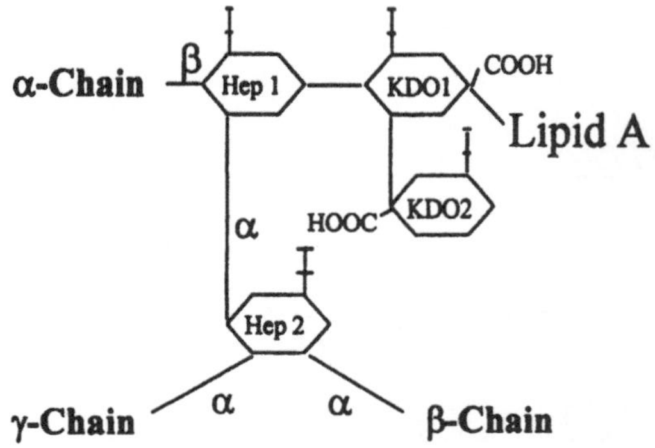

Figure 5.1   The generic structure of neisserial LOS. The LOS of MC and GC has a triantennary structure. This contrasts with the structure of enterobacterial LPS which is substituted only at the 3 position of heptose 2.

breakthroughs needed for this understanding were the ability to characterize LOS by SDS–PAGE visualized with a highly sensitive silver stain,[1] the ability to relate the bands seen on SDS–PAGE with reactivity of a panoply of monoclonal antibodies (MAb) by Western blots,[2,3] and detailed structures of the LOS of a number of GC and MC strains. It is only recently that full structures of several GC LOS have become available.[4] It was found that they resembled the structure of MC LOS that had been determined earlier.[5-9] The structural chemistry of MC LOS has been reviewed extensively by others[10] and is presented here only to provide a context for the genetic studies.

Generic structure of neisserial LOS

Abbreviations used in the text and figures are: Glc = glucose; Gal = galactose; GlcNAc = *N*-acetyl glucosamine; GalNAc = *N*-acetyl galactosamine; KDO = 2-keto-3-deoxy-manno-octulosonic acid; Hep = heptose; NANA = *N*-acetyl neuraminic acid. The inner core region is quite similar to enterobacterial LPS and is illustrated in Figure 5.1. It consists of lipid A moiety which is linked to 2 KDO units and these in turn to two heptose residues. At this point the neisserial structure diverges substantially from enterobacterial LPS in that three positions on the Hep residues may be substituted, and these have been referred to as the α, β, and γ chains. In GC the γ-chain is always present and consists of GlcNAc. The β-chain in MC may be phosphoryl ethanolamine (PE) (L1, L3, L9), Glc (L5), or absent L2, L4, L6. It should be noted that L2, L4 and L6 carry PE residues on C6 or C7 or both, on the Hep2 residue. In GC the β-chain in strains 15253 and FA1090 is a lactosyl group.[11,12] It is not present in strains F62, MS11 or 1291.[4,13,14] There is evidence that suggests that certain MAb, such as 2C7 and 3G9, may recognize the lactosyl group as part of their specificity.[11] If this is correct then the majority of GC strains carry this substitution.[15]

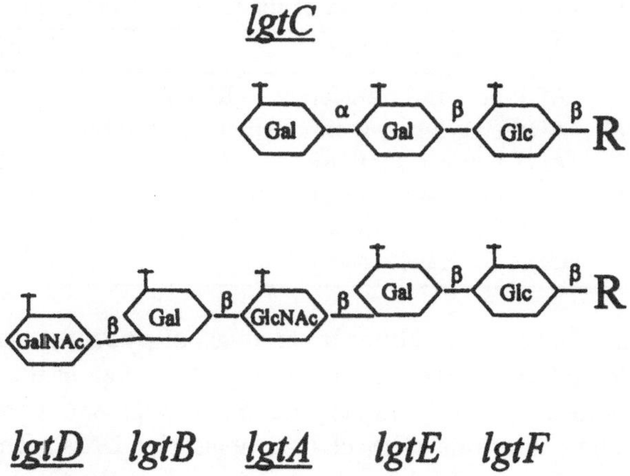

lgtC

lgtD   *lgtB*   *lgtA*   *lgtE*   *lgtF*

Figure 5.2   Structure and genetics of the α-chain of GC LOS. The LOS structure in the lower portion of the figure has been reported for strain F62[4] and for strain MS11.[13] It consists of lacto-*N*-neotetraose with a GalNAc at the reducing end. The glycosyl transferases responsible for each addition are indicated next to the sugar residue they add. The *lgt* genes underlined are those that contain poly-G tracts and are susceptible to high-frequency variation. A sialic acid unit will replace the GalNAc if *lgtA* is in frame, *lgtD* is out of frame, and the organism is grown in the presence of CMP–NANA. The alternative α-chain shown in the upper part of the figure is produced when *lgtA* is out of frame and *lgtC* is in frame.

The α-chain is the most complex and variable and is shown in the lower half of Figure 5.2. In some strains, such as F62, the α-chain consists of a GalNAcβ1-3Galβ1-4GlcNAcβ1-3Galβ1-4Glc pentasaccharide,[4] which is identical to the pentasaccharide found on the human red cell X_2 antigen.[16] However, in many strains the terminal GalNAc is not added, and the resultant tetrasaccharide is then identical to lacto-*N*-neotetraose, a human glycolipid which is the core structure of the blood group substances.[17] In other strains the α-chain is limited to a disaccharide lactosyl unit and is then identical to lactosylceramide. Occasionally, an alternative structure (Galα1-4Galβ1-4Glc) is produced where the lactosyl unit bears an additional α-linked Gal.[14] This structure, referred to below as the alternative α-chain, is identical to that found on the p^k blood group antigen globotriglycosyl ceramide[17] and is shown in the upper part of Figure 5.2. In MC the substitution with a terminal GalNAc has not been reported; however, the lacto-*N*-neotetraose structure is found in the L2, L3, L4, L7 immunotypes, L1 carries the globotriose structure, and L8 contains a lactosyl residue.

Genetics of core biosynthesis

A number of genes involved in the biosynthesis of the core region have recently been identified, and these findings are summarized in Table 5.1.

Table 5.1 Biosynthesis Genes of the GC LOS-Core Region

Gene	Biochemical Activity	Reference
rfaC	ADP-heptosyl transferase to KDO1	18
rfaD (lsi-6)	ADP-L-glycero-D-mannoheptose epimerase	22
rfaE (lsi-7)	ADP-heptose synthase	22
rfaF (lsi1)	ADP-heptosyl transferase to Hep1	19–21
rfaK	UDP-GlcNAc transferase to Hep2	24
lgtF	UDP-Glc transferase to Hep1	25

These genes have been found either by analysis of pyocin-selected mutants with defective core biosynthesis, or by taking advantage of the similarity of the core LPS structures of *Neisseria* and *Salmonella* and complementing known *Salmonella* mutations with cloned neisserial DNA. Zhou et al. by complementation of an *rfaC Salmonella* mutant cloned the GC homologue of the *rfaC* (heptosyl 1 transferase) gene.[18] The *rfaF* gene was identified by analysis of cloned GC DNA able to complement another pyocin-mutant GC strain FA5100, and this locus was first named *lsi1*.[19] Subsequently it was shown that *lsi1* is a homologue of *rfaF*, both genes able to complement the *Salmonella* mutation; the deduced proteins have very convincing homology to the enterobacterial *rfaF*.[20,21] Levin and Stein,[22] by complementing a transformable derivative of pyocin-mutant WS1,[23] identified GC homologues of *rfaE* and *rfaD*, demonstrated by complementation in *Salmonella* mutants that these did complement known mutations, and showed that *rfaD–rfaE* form a single transcriptional unit. In the case of MC, analysis of Tn916 mutants has proved very informative. A mutant producing a truncated LOS molecule was found to have the Tn916 insertion in a gene with a deduced amino acid sequence clearly similar to *Salmonella rfaK*, and the chemistry of the mutant's LPS lacks GlcNAc on Hep2.[24] Immediately upstream of the MC *rfaK* gene, Kahler et al. found another gene which they named *lgtF* and demonstrated that it catalyzes the addition of Glc to Hep1.[25]

Genetics of α-chain biosynthesis

The genes responsible for the biosynthesis of the α-chain of GC have been identified and are summarized in Figure 5.2, which shows the structure of the α-chain reported for F62[4] and for MS11 variant C.[13] The most recently identified gene has been named *lgtF* and is the glycosyl transferase responsible for the addition of the Glc to the Hep1.[25] This gene has been identified in MC[25] and GC[25a] and is immediately upstream of the *rfaK* gene. The two genes are co-transcribed and this operon has been named *ice*, (inner core extension).[25] The *lgtA–E* genes are located in a single cluster that is not adjacent to other LOS synthesis genes. This gene cluster was first identified by serendipity and was carefully analyzed when it was noted that two of its genes, *lgtB* and *lgtE*, are homologous to a *Haemophilus influenzae* gene (called *lex-1*[26] or *lic2A*[27]) that is known to be involved in LPS synthesis of

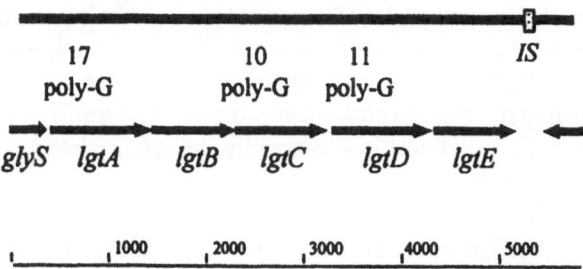

Figure 5.3 Structure of the *lgt* locus. The gene organization has been observed in the majority of GC strains. The sequence is almost identical to that of FA1090, except that in FA1090 the poly-G regions are, respectively, 11, 14, 14; (Gonococcal Genome Sequencing Project, and B.A. Roe, S. Clifton and D.W. Dyer; this project is supported by USPHS/NIH grant #AI38399). This is in accord with the known structure of FA1090 LOS with lacto-*N*-neotetraose structure without a GalNAc substitution. In strain 15253 an internal recombination has occurred between the homologous portions of *lgtB* and *lgtE*, thereby excising *lgtC* and *lgtD*.[12]

that species. Subsequently, Danaher et al.[28] cloned a portion of this locus based on its ability to alter expression of the α-chain. Four of the genes are the glycosyl transferases responsible for the synthesis of the α-chain from the Glc residue outward. The fifth gene is the α-galactosyl transferase responsible for the alternative α-chain.[29] The structure of the locus is shown in Figure 5.3. Following the *glyS* gene, there are five closely spaced open reading frames (orf). A sequence typical of a ρ-independent termination signal is located 46 bp downstream of the termination codon of the last orf. Subsequently, there is an area of approximately 100 bp that has striking homology to the IS1106 neisserial insertion sequence.[30] Searches for internal homology within this locus indicated that the DNA coding for the first two genes (*lgtA*, *lgtB*) is homologous to the fourth and fifth genes (*lgtD*, *lgtE*) and that interposed is an additional open reading frame, *lgtC*. The homologous genes encode proteins that are nearly identical at the N-termini and increasingly divergent toward the COOH-termini. The *lgtC* sequence interposed between the repeated portions of the locus is not repeated within the locus nor elsewhere in the *Neisseria gonorrhoeae* genome. It is homologous to *E. coli rfaI* or *rfaJ* genes, which are two very closely related genes encoding, respectively, a galactosyl and a glucosyl transferase in core LPS biosynthesis.[31] GC *lgtC* is also homologous to the subsequently discovered *lgtC* gene of *Haemophilus influenzae*, which is one of the phase variable genes of the latter species by virtue of a tract of repeating $(GACA)_n$ in its coding frame.[32]

The function of the *lgt* genes was defined by introducing into the GC genome insertions or deletions of the *lgt* locus and determining the effect on the LOS purified from each of the mutants. The size of the LOS by SDS–PAGE together with the reactivity with MAbs and the chemical data allowed assignment of the function of each of the coding frames (Figure 5.2). More recently the chemical activities of the enzymes encoded by *lgtA*, *lgtB*, and *lgtE* with synthetic acceptors have been determined.[33] It is noteworthy that *lgtB* and *lgtE*, which are structurally very closely related, also perform a very similar biosynthetic task, i.e., the addition of Galβ1→4 to GlcNAc or Glc, respectively. Similarly, the strongly homologous *lgtA* and *lgtD* add GlcNAc or GalNAcβ1→3 to a Gal residue, respectively. *lgtC*, which is unrelated to the other genes in the locus, is responsible for the addition of a Gal α1→4.[29]

We have studied the genomic organization of five additional strains of GC that were isolated at very different times and places: R10, MS11, FA1090, M94, and 15253. Southern blots with probes derived from *lgtA*, *lgtB*, and *lgtC* were performed with genomic DNA digested with *Bsa*BI and, with one exception, two signals were obtained with the first two probes (*lgtA* and *lgtB* probes also detect *lgtD* and *lgtE*, respectively) and a single signal with the *lgtC* probe. The exception proved to be strain 15253 where only a single signal was seen with *lgtA* and *lgtB* and none with *lgtC*. Sequence analysis of the *lgt* locus cloned from this strain indicated that it had undergone an internal recombination within the *lgt* locus in an N-terminal conserved portion of *lgtB* and *lgtE*, leaving a functional *lgtE* gene and excising the C-terminal portion of *lgtB* as well as the intervening *lgtC* and *lgtD* genes.[12] This is in accord with the published structure of the LOS of this strain showing that its α-chain is limited to a lactosyl group.[11] The study of the genetics of LOS synthesis by MC has shown that it has great similarities to that described for GC. Jennings et al. characterized the *lgt* locus of MC strain MC58 and found that it contains only three genes, *lgtA*, *lgtB*, and *lgtE*.[33,34] In the case of strain M1080, it was found that *lgtC* was also part of the *lgt* locus.[12]

Genetics of high-frequency variation

Studies with MAb have revealed the remarkable ability of GC to alter at high frequency the structure of their LOS. Thus far, most of the changes observed have been in the α-chain, but there is recent evidence of variation of the expression of the β-chain.[35] Regarding the α-chain, Schneider et al. demonstrated by a colony immunoblot technique that the expression of the lacto-*N*-neotetraose epitope (recognized by MAb 3F11) is lost/regained *in vitro* at a frequency of 10^{-3} per generation.[36] *In vivo* selection of the 3F11 epitope was seen when volunteers were challenged with strain MS11$_{mk}$ variant A, which produces an LOS with only a lactosyl α-chain. Some of the clones recovered from the urine of the infected individuals produced a complete LOS with a pentasaccharide α-chain; a representative clone was named variant C.[13,37]. van Putten[38] has reported a high frequency variation of the reaction with MAb 1-1-M, reflecting the addition of the terminal GalNAc.

When the *lgt* locus was sequenced, it was found that three of the genes (underlined in Figure 5.2) contain runs of guanosines coding for stretches of glycines. In strain F62, these poly-G regions are found in *lgtA* (17 bp), *lgtC* (10 bp), and *lgtD* (11 bp). In each case, the number of G residues is one that maintains an intact reading frame coding for glycines, and in each of the three genes a change of one or two G bases would cause premature termination of translation (see Figure 5.3). Thus, these poly-G tracts provide a mechanism for high-frequency variation of expression of these genes.[29] Slippage in such poly-G tracts is well documented to control the expression of the GC *pilC* genes, for example, with resultant effects on pilus adhesiveness to human epithelial cells.[39] Three aspects of LOS α-chain biosynthesis are subject to high frequency variation. The first is the addition of the terminal GalNAc (*lgtD*), causing an alteration of reactivity with monoclonal antibody 1-1-M; this phase variation has been reported by van Putten.[38] Second, a phase change in *lgtA* prevents the addition of GlcNAc to the growing chain and truncates the LOS at the β-lactosyl level. This is a very common form of LOS in GC, seen on SDS–PAGE as a 3.6 kD band[40] and in MC as immunotype L8. Finally, the variable addition of Galα1→4 to the β-lactosyl (p^k-like globo-triose) is under the control of the expression of *lgtC*.

That high frequency antigenic variation is in fact due to shifts in these poly-G regions has been documented by a number of investigators. Yang and Gotschlich subjected an F62 *lgtAΔ* mutant (F62Δ1) to colony blotting with MAb 17-1-L1, which recognizes the globotriose structure, and three independent L1^{minus} variants were selected. SDS–PAGE indicated that the LOS of the variants was smaller. The poly-G region in *lgtC* of F62 Δ1 and of the three variants was sequenced. It was found that F62 Δ1 (like F62 wt) has 10 Gs, which is in frame. The three variants have 11, 17, and 17 poly-G tracts, which are out of frame. Thus the variable expression of the Galα1-4Galβ1-4Glc-R epitope is dependent on shifts in *lgtC*.[41] To document the role of *lgtA* in variation, we obtained from Dr. H. Schneider strain MS11$_{mk}$ variant A (LOS with only a lactosyl α-chain) and variant C (complete α-chain) recovered from a volunteer infected with variant A.[37] The *lgtA* and the *lgtC* regions were sequenced, and it was found that *lgtC* contains eight Gs in both strains (off) and that the *lgtA* region changed from 12 Gs (off) in variant A to 11 Gs (on) in variant C. In variant A, in which both *lgtC* and *lgtA* are off, the LOS α-chain is limited to the lactosyl group. Thus, the loss of a single G residue in *lgtA* of variant C accounts for the regained ability to synthesize the full LOS.[41] The effect of alterations in poly-G in the *lgtA* gene on antigenic shifts in strain 1291 was documented by Danaher et al.[28] Furthermore, Burch et al. recently reported that GC strain FA19 produces a mixture of LOS with a lacto-*N*-neotetraose and a lactosyl α-chain and reacts with two MAbs that recognize these epitopes. They demonstrated that this occurs when *lsi2* (identical to *lgtA*) has a poly-G tract frame-shifted 2+ and ascribed the low level *lsi2* transferase activity to either transcriptional or translational frame-shifting.[42] Interestingly, in Schneider's challenge experiment (described above) another MS11 variant was frequently reisolated from the urine of infected volunteers and, like FA19, expresses an LOS with both lacto-*N*-neotetraose and lactosyl epitopes.[37] This clone, named MS11 variant B, was sequenced in our laboratory and had a 2+ frame-shifted *lgtA* poly-G tract of 13 Gs (unpublished results), supporting the observation of Burch et al. A shift in the poly-G region of *lgtA* from 15 (off) to 14 (on) was also documented in strain 15253 which, as described above, is deleted in *lgtB–D*. In this strain, the activation of *lgtA* causes a shift from the expression of a lactosyl group to the expression of a very unusual α-chain, i.e., GlcNAcβ1-3Galβ1-4Glc.[12] It is significant that this type of α-chain is also seen in MC of the L6 immunotype, suggesting that these strains may have undergone a similar internal recombination in their *lgt* loci eliminating *lgtB*. In addition, Jennings et al. have demonstrated that changes in MC *lgtA* accounts for shifts between expression of the lacto-*N*-neotetraose or lactosyl α-chain.[34]

Sialylation of LOS

The correlation of LOS structure with function is still in it early stages. It is known that LOS greatly influences the bacteriolytic effects of normal human serum (NHS) on GC. It has been shown that a large proportion of human bactericidal antibodies are directed to LOS epitopes. GC strains grown *in vitro* vary widely in their ability to resist the bacteriolytic effect of NHS. At one pole are strains that resist the bactericidal activity of most normal human sera. These are referred to as Ser[R] and are commonly found among isolates from patients with disseminated GC infection (DGI).[43] The convalescent sera of these patients may develop antibodies that are lethal to these strains. At the other pole are GC strains that are sensitive to most NHS when grown on ordinary media. The structural basis for the difference between Ser[R] and Ser[S] has recently been ascribed to the ability of a conserved exposed sequence present on PorA protein to bind factor H.[44] It has been proposed that the LOS with the terminal GalNAc added is a particularly good bacteriolytic target.[45] It has also been reported that GC that carry only a lactosyl α-chain tend to resist killing by NHS.[40] In addition, GC strains lacking the α-chain (e.g., 1291 d/e; phosphoglucomutase defective) and strains with mutations in the inner core region become highly serum sensitive.[43]

Among GC Ser[S] strains there are a large number that become serum-resistant when incubated for a period of three hours or more in the presence of CMP–NANA. GC possess a sialyl transferase that is capable of using exogenous cytidine-5′-monophospho-*N*-acetyl-neuraminic acid (CMP–NANA) to sialylate its LOS.[46] The concentrations of CMP–NANA found *in vivo* are sufficient to support this reaction,[47] which was discovered when the effect of *in vivo* growth in subcutaneous chambers on the ability of GC to resist the bactericidal action of serum was studied. However, the reaction depends on the LOS being a competent acceptor; it is well established that the GC lacto-*N*-neotetraose chain is the main substrate for sialylation.[48] The α2, 3-sialyltransferase (*lst* gene) has recently been cloned from both MC and GC.[49] It was accomplished by testing a large number of plasmid clones of MC DNA by an exquisitely sensitive assay for sialyltransferase activity. This membrane-bound enzyme, which can use α- as well as β-linked Gal residues as acceptors,[50] has a clear preference for sialylating *N*-acetyl lactosamine in comparison to lactose or galactose.[49] There is a correlation between the expression of free lacto-*N*-neotetraose on MC LOS and the inhibition of serum bactericidal activity.[51] The mechanism by which sialylation renders the organisms phenotypically Ser[R] has been attributed to a general defect of effective complement deposition[52,53] or an inability of antibodies to bind to the modified LOS.[54] Researchers have found that factor H binds to sialic acid on sialylated GC, increasing the conversion of C3b to iC3b and thus blocking complement pathway activation;[55] data suggest a similar mechanism in MC.[56]

Genetics of β-chain biosynthesis

The α1-3Glc transferase that catalyzes the first step in the addition of the β-chain, a reaction chemically analogous to that performed by *Salmonella rfaG*, has been cloned.[35] In addition, Erwin et al.[12] made an observation that bears upon the next biosynthetic step, the addition of the Gal residue. Among the strains studied, 15253 and FA1090 LOS react with MAb 2C7. Deleton of *lgtA* through *lgtD* (Δ5) of FA1090 results in production of an LOS with mobility very similar to that of 15253, and this LOS retains 2C7 reactivity. However, deletion of *lgtE* in GC strains that are reactive with MAb 2C7 abrogates this activity. This observation suggests that *lgtEΔ* has an effect on β-chain synthesis. The sugar composition of the LOS of GC 15253 and 15253(*lgtEΔ*) was determined. In the parent strain, Glc:Gal:GlcNAc was found to be approximately in the expected ratio of 2:2:1. In the *lgtEΔ* mutant, galactose is entirely absent; the same is true for strain FA1090. Thus, the deletion of *lgtE* not only has the expected effect on the addition of Gal to the α-chain, but also prevents addition of Gal to the β-chain. Recent data suggest that the *lgtE* product is in fact bifunctional, attaching Gal to the initial glucose of the β-chain as well as the α-chain (Stein, D.C., personal communication). It is known that certain steps in LOS synthesis occur in a specific order. Apparently, addition of the α-chain requires the presence of heptose 2 and its substituent γ-chain, since the α-chain is absent from the LOS of an MC *rfaK* mutant (lacking the γ-chain)[24] and a GC *rfaF* mutant (lacking heptose 2).[21] Furthermore, the addition of the first glucose of the β-chain by *lgtG* possibly requires the presence of the first glucose of the α-chain, because MC strain NMB (β-chain positive in the wild type) disrupted in *lgtF* lacks glucose on HepII as well as HepI.[25]

Genes with secondary effects on LOS biosynthesis

The genes *pgm* and *galE*, which act indirectly upon LOS biosynthesis, are of considerable interest because mutants of these genes existed before the glycosyl transferase genes were identified; these mutants were used in very informative functional studies that are summarized below. Sandlin et al.[57] studied a set of pyocin mutants which previously had been derived from GC strain 1291 and structurally characterized.[14] They cloned a DNA fragment that repairs the defect of 1291d and 1291e[57] and subsequently demonstrated that the defect is a mutated phosphoglucomutase (*pgm*), which prevents the synthesis of UDP-glucose and hence the addition of the Glc residue.[58] The *pgm* gene of MC was identified by determining the gene affected in a Tn*916* transposon mutant with a truncated LOS.[59] Robertson et al.[60] cloned and inactivated the *galE* (UDP-galactose-4-epimerase) gene of GC thus generating a mutant expressing Glcβ1→4Hep, which is a severely truncated α-chain.[60] Similarly, Jennings et al.[61] identified this gene in MC and showed that its deletion affects LOS biosynthesis.

Biosynthetic genes not yet identified

Neisserial LOS shows considerable microheterogeneity. It is well established that SDS–PAGE analysis of a given strain's LOS can yield multiple uncharacterized bands. Some of these, such as the L1 (globotriose) LOS, possibly represent alternative structures that are not direct precursors of the larger and predominant LOS forms. Lee et al. have used mass spectrometry to demonstrate the presence of repetitive hexoses in certain LOS forms of MC strain NMB. They have also found LOS structures with similar repeats in a Tn*916*-generated NMB *galE* mutant. The hexoses are probably glucose residues, since the mutant is unable to synthesize UDP-galactose.[62] The glycosyltransferases responsible for this alternative polymerization are still unknown. Other unidentified genes include those responsible for additional LOS modifications such as acetylation and ethanolamine substitutions. Furthermore, the complete structures of the L10 and L11 immunotypes of Group A MC have not yet been determined. Doubtless, these remain areas for further investigation.

Selective advantage of LOS in pathogenesis

GC possess surface structures that are relevant to inter-gonococcal adhesion, adherence to eukaryotic cells, invasion of host mucosa, and immune evasion. These surface structures include pili, opacity (Opa) proteins, and LOS, and share the property of being subject to antigenic variation. We next focus briefly on pili and opacity proteins to provide context for a consideration of the physiological relevance of LOS variation.

Pili

GC pili are associated with virulence,[63] mediate inter-gonococcal adhesion,[64] and are necessary for transformation[65,66] (and reviewed in Reference 67). Pili also mediate attachment of GC and MC to host epithelial and endothelial cells via the pilus-tip adhesion PilC.[39,68-70] However, there is some evidence that pili may be inhibitory to the invasion step, *in vitro*.[71,72] Pilin, the subunit protein of the GC pilus, is usually encoded by one complete expression locus termed *pilE1*. The genome also contains several promoter-less (silent) pilin sequences, designated *pilS1*, *pilS2*, etc. These sequences are homologous to *pilE*. A given strain of GC can express many different sizes of pilin subunits.[73] Pilus phase transition occurs at a frequency of ~10⁻³;[73,74] pilin antigenic variation occurs at frequencies of 10^{-2} to 10^{-4} [75] and is generally *recA*-dependent.[76] Studies on this variation have shown that a combination of mechanisms may be responsible. These include transformation-mediated recombination between *pilS* sequences released by GC autolyzed in culture and *pilE* genes in live GC,[77,78] intragenomic reciprocal recombination between *pilE* and *pilS*,[77] and gene conversion.[79] The end result is the rearrangement of the variable cassettes of the pilin expression locus and expression of antigenically distinct pili. In contrast, *pilC* varies at high frequency due to a slipped-strand mechanism involving a stretch of guanine residues early in the coding frame.[80,81] Pilus variation contributes to GC evasion of host immune defenses and could possibly mediate tissue tropism.

Note that the *pgm* and *galE* genes (mentioned above as they relate to LOS biosynthesis) are most likely involved in other housekeeping functions, including the modification of pili by glycosylation. An early study revealed the presence of galactose and glucose residues on GC pili by qualitative gas chromatography.[82] More recently the presence of carbohydrate groups on the MC pilus has been established. Indeed, α1,3-galactosyl antibodies can bind to MC pili, blocking complement-dependent lysis.[83] Virji et al. demonstrated that the glycosylation status of MC pili accounts for the differential mobility of pilin on SDS–PAGE, and variation in the putative N-glycosylation motifs of PilE modulates MC adherence to host cells.[84] Furthermore, Stimson et al. have shown that the MC *galE* mutation results in the simultaneous decrease of pilin and LOS M_r as analyzed by SDS–PAGE.[85] Thus, future studies—particularly clinical trials—of the specific role of LOS structure in natural infection, should be done using mutants of the specific glycosyl transferases.

Opacity proteins

Both GC and MC Opa proteins have been clearly implicated in adherence and epithelial cell invasion. The GC Opa protein repertoire consists of up to 11 opacity loci, each constitutively transcribed.[86,87] GC expressing certain Opa proteins assume an opaque and ground-glass appearance on translucent, solid media.[88] Opacity protein variation, which occurs at a frequency of ~10^{-3},[89] is *recA*-independent and, due to slipped-strand mispairing of pentameric (CTCTT) coding, repeats within the opacity genes.[90] Accordingly, control of Opa protein expression is at the level of translation.[91] Phase variation (on/off switching) results in the expression of zero, one, or more Opa proteins by a single GC. Antigenic variation thus occurs due to the random expression of a particular locus or loci.[87]

Opacity proteins mediate GC adherence to and invasion of host cells and may target the bacteria to specific cell types such as epithelium[71,92-94] or PMNs.[93,95,96] The 30 kD OpaA mediates invasion of GC strain MS11 into Chang conjunctiva cells[71,93] by binding to heparan sulfate proteoglycans on the eukaryotic cell surface;[96a,97] there is evidence that this interaction also contributes to serum resistance.[98] Interestingly, vitronectin has been found to promote the OpaA-heparin interaction in Chinese hamster ovary cells[99] and HeLa cells.[100] Furthermore, recent studies show that several of the GC and MC Opa proteins bind to various members of the carcinoembryonic antigen (CEA) family that are expressed on both phagocytic and epithelial cells,[101-105] indicating that Opa antigenic variation leads to changes in tissue tropism.

LOS and the mucosal infection

Much interest centers on the role of LOS in GC infection of its obligate human host. LOS varies independently of other GC surface structures, such as Opa proteins and pili, at a frequency of ~10^{-3} in the absence of selective pressures.[36,106] There is evidence that the specific structure of LOS affects GC invasiveness. It has been shown that LOS is stably expressed during adherence, invasion, and internal processing of the GC by epithelial cells.[107] Indeed, data suggest that LOS interacts with eukaryotic cell membranes[107] and can bind to human epithelial glycosphingolipids (reviewed in Reference 17). The effect of sialylation in the local infection is under study by a number of groups. van Putten has shown that sialylation of LOS has a marked inhibitory effect on epithelial cell invasion, without greatly altering adhesion.[38] Variants that incorporate low amounts of sialic acid show high rates of invasion into human mucosal epithelial cells and are susceptible to killing by complement and anti-LOS monoclonal antibodies. By contrast, when incubated with CMP–NANA, GC bearing highly sialylated LOS are equally adhesive, but less invasive and more serum resistant. In the absence of sialylation, there is no significant difference in invasion between the LOS variants.[38] van Putten's studies suggest that in the mucosal infection, α-chain structures that cannot be sialylated may be important for efficient cell invasion. This finding has been supported by the demonstration that GC with sialylated LOS are less able to cause infection in human volunteers.[108]

The *galE* mutant of Robertson et al. exhibits wild-type levels of adhesion to and invasion of epithelial cells, as well as unaltered inter-gonococcal adhesion.[60] Likewise, the results of van Putten's experiments, as well as our own measurements of the invasion rates of OpaA+ MS11 *lgt* mutants,[25a] suggest that the terminal GalNAcβ1→3Galβ1→4GlcNAcβ1→3Gal residue may not be necessary for invasion. However, the proximal glucose residue and the heptosyl core is more likely essential. Schwan et al. have shown that a disruption of the GC *lsi-1/rfaF* gene leads to production of LOS that is truncated in the inner-core region, abolishing GC invasion of Chang conjunctiva cells despite expression of OpaA in these mutants.[21] This is the first instance where a clear defect in interaction with epithelial cells is linked to an LOS mutation, and it is remarkable that it is dominant over the OpaA phenotype. The MC *lgtF* gene, which encodes the glucosyl transferase of the α-chain, has recently been identified.[25] Cloning and inactivation of the GC homologue and the generation of GC *lgtF* mutants was subsequently done in our laboratory.[25a] Interestingly, we found a GC MS11 *lgtF* mutant to be unaffected in adherence but significantly impaired in the invasion of Chang conjunctiva cells despite expression of OpaA,[25a] suggesting the importance of the α-chain's proximal glucose residue.

There is evidence of a possible interaction between Opa proteins and LOS which may be important in the association with host cells. For instance, *E. coli* expressing recombinant OpaA adhere to but do not invade Chang conjunctiva cells, demonstrating that this particular Opa–protein interaction is insufficient to mediate invasion;[109,110] this contrasts with findings that expression of OpaI is sufficient for internalization of *E. coli* by HeLa cells expressing the CEA antigen CGM1a.[101] In addition, investigators have reported the direct binding of GC lacto-*N*-neotetraose LOS to the asialoglycoprotein receptor in HepG2 cells,[111] to a 70 kD HepG2 protein which is recognized by an anti-Opa monoclonal immunoglobulin,[112] and to GC Opa proteins themselves.[113,114] Doubtless, the specific contribution of LOS to invasion, either alone or via a possible interaction or synergism with Opa, needs to be dissected in future studies.

Conclusion

In sum, GC possess an elegant genetic system for varying the structure of their LOS. The fact that each of the various α-chain structures produced by GC is a mimic of a host sugar structure has raised the question of what the role of this mimicry may be. Immune evasion is one possibility but is rendered less credible by the observation that the majority of bactericidal antibodies in human serum are directed to LOS.[115] The host uses a very large number of ligand binding proteins, recognizing the rich array of carbohydrate structures on glycolipids and glycoproteins for its own homeostatic purposes. Among these, the C-lectins, the galectins, and the sialoadhesins may be particularly significant because their binding specificities would include structures found on LOS. Thus, it is possible that mimics of these host structures on the GC would also be recognized and may contribute in important ways to the mucosal infection. *Haemophilus influenzae*, an organism that produces chemically similar LOS, also has frame-shift mechanisms for varying LOS structures[27] and sialylates its LOS.[116] The existence of this genetic capability argues that the process is biologically important, and that organisms with phase-variable LOS have a selective advantage. It strongly suggests that specific LOS structures afford an advantage in one biological niche but a disadvantage in another host environment, and that the organism negotiates this dilemma by phase variation.

Acknowledgments

This work was supported by PHS grants AI10615 and AI26558. S.Y.M. was supported by GE Foundation Academic Fellowship and NIH training grants GM15317 and GM07739. We are grateful to Drs. Asesh Banerjee and Daniel Stein for sharing their unpublished data and to Mr. James Parker and Ms. Clara Eastby for technical assistance.

References

1. Hitchcock, P.J. and Brown, T.M., Morphological heterogeneity among *Salmonella* lipopolysaccharide chemotypes in silver-stained polyacrylamide gels, *J. Bacteriol.*, 154, 269, 1983.
2. Mandrell, R.E., Schneider, H., Zollinger, W., Apicella, M.A., and Griffiss, J.M., Characterization of mouse monoclonal antibodies specific for gonococcal lipooligosaccharides, in Schoolnik, G.K., Ed., *The Pathogenic Neisseriae*. Washington, American Society for Microbiology; 1985:379-384.
3. Mandrell, R., Schneider, H., Apicella, M., Zollinger, W., Rice, P.A., and Griffiss, J.M., Antigenic and physical diversity of *Neisseria gonorrhoeae* lipooligosaccharides, *Infect. Immun.*, 54, 63, 1986.
4. Yamasaki, R., Bacon, B.E., Nasholds, W., Schneider H., and Griffiss, J.M., Structural determination of oliogosaccharides derived from lipooligosaccharide of *Neisseria gonorrhoeae* F62 by chemical, enzymatic, and two-dimensional NMR methods, *Biochemistry*, 30, 10566, 1991.
5. Jennings, H.J., Johnson, K.G., and Kenne, L., The structure of an R-type oligosaccharide core obtained from some lipopolysaccharides of *Neisseria meningitidis*, *Carbohydr. Res.*, 121, 233, 1983.
6. Dell, A., Azadi, P., Tiller, P., et al., Analysis of oligosaccharide epitopes of meningococcal lipopolysaccharides by fast-atom-bombardment mass spectrometry, *Carbohydr. Res.*, 200, 59, 1990.
7. Michon, F., Beurret, M., Gamian, A., Brisson, J.R., and Jennings, H.J., Structure of the L5 lipopolysaccharide core oligosaccharides of *Neisseria meningitidis*, *J. Biol. Chem.*, 265, 7243, 1990.
8. Gamian, A., Beurret, M., Michon, F., Brisson, J.R., and Jennings, H.J., Structure of the L2 lipopolysaccharide core oligosaccharides of *Neisseria meningitidis*, *J. Biol. Chem.*, 267, 922, 1992.
9. Pavliak, V., Brisson, J.R., Michon, F., Uhrin, D., and Jennings, H.J., Structure of the sialylated L3 lipopolysaccharide of *Neisseria meningitidis*, *J. Biol. Chem.*, 268, 14146, 1993.
10. Verheul, A.F.M., Snippe, H., and Poolman, J.T., Meningococcal lipopolysaccharides: virulence factor and potential vaccine component, *Microbiol. Rev.*, 57, 34, 1993.
11. Yamasaki, R., Kerwood, D.E., Schneider, H., Quinn, K.P., Griffiss, J.M., and Mandrell, R.E., The structure of lipooligosaccharide produced by *Neisseria gonorrhoeae*, strain 15253, isolated from a patient with disseminated infection. Evidence for a new glycosylation pathway of the gonococcal lipooligosaccharide, *J. Biol. Chem.*, 269, 30345, 1994.
12. Erwin, A.L., Haynes, P.A., Rice, P.A., and Gotschlich, E.C., Conservation of the lipooligosaccharide synthesis locus *lgt* among strains of *Neisseria gonorrhoeae*: requirement for *lgtE* in synthesis of the 2C7 epitope and of the β chain of strain 15253, *J. Exp. Med.*, 184, 1233, 1996.
13. Kerwood, D.E., Schneider, H., and Yamasaki, R., Structural analysis of lipooligosaccharide produced by *Neisseria gonorrhoeae*, strain MS11mk (variant A): A precursor for a gonococcal lipooligosaccharide associated with virulence, *Biochemistry*, 31, 12760, 1992.
14. John, C.M., Griffiss, J.M., Apicella, M.A., Mandrell, R.E., and Gibson, B.W., The structural basis for pyocin resistance in *Neisseria gonorrhoeae* lipooligosaccharides, *J. Biol. Chem.*, 266, 19303, 1991.

15. Mandrell, R., Apicella, M., Boslego, J., Chung, R., Rice, P., and Griffiss, J.M., Human immune response to monoclonal antibody-defined epitopes of *Neisseria gonorrhoeae* lipooligosaccharides, in Pollman, J.T., Zanen, H.C., Meyer, T.F., et al. (Eds.): *Gonococci and Meningococci: Epidemiology, Genetics, Immunochemistry, and Pathogenesis.* Dordrecht, Kluwer Academic Publishers, 1988:569-574.
16. Kannagi, R., Fukuda, M.N., and Hakomori, S., A new glycolipid antigen isolated from human erythrocyte membranes reacting with antibodies directed to globo-N-tetraosylceramide (globoside), *J. Biol. Chem.*, 257, 4438, 1982.
17. Mandrell, R.E. and Apicella, M.A., Lipo-oligosaccharides (LOS) of mucosal pathogens: Molecular mimicry and host-modification of LOS, *Immunobiology,* 187, 382, 1993.
18. Zhou, D., Lee, N., and Apicella, M.A., Lipooligosaccharide biosynthesis in *Neisseria gonorrhoeae*: cloning, identification and characterization of the α1, 5 heptosyltransferase I gene (*rfaC*), *Mol. Microbiol.*, 14, 609, 1994.
19. Petricoin, E.F., III, Danaher, R.J., and Stein, D.C., Analysis of the *lsi* region involved in lipooligosaccharide biosynthesis in *Neisseria gonorrhoeae, J. Bacteriol.*, 173, 7896, 1991.
20. Sandlin, R.C., Danaher, R.J., and Stein, D.C., Genetic basis of pyocin resistance in *Neisseria gonorrhoeae, J. Bacteriol.*, 176, 6869, 1994.
21. Schwan, T.E., Robertson, B.D., Brade, H., and van Putten, J.P.M., Gonococcal *rfaF* mutants express Rd$_2$ chemotype LPS and do not enter epithelial host cells, *Molec. Microbiol.*, 15, 267, 1995.
22. Levin, J.C. and Stein, D.C., Cloning, complementation, and characterization of an rfaE homolog from *Neisseria gonorrhoeae, J. Bacteriol.*, 178, 4571, 1996.
23. Shafer, W.M., Onunka, V., and Hitchcock, P.J., A spontaneous mutant of *Neisseria gonorrhoeae* with decreased resistance to neutrophil granule proteins, *J. Infect. Dis.*, 153, 910, 1986.
24. Kahler, C.M., Carlson, R.W., Rahman, M.M., Martin, L.E., and Stephens, D.S., Inner core biosynthesis of lipooligosaccharide (LOS) in *Neisseria meningitidis* serogroup B: identification and role in LOS assembly of the alpha 1,2 N-acetylglucosamine transferase (RfaK), *J. Bacteriol.*, 178, 1265, 1996.
25. Kahler, C.M., Carlson, R.W., Rahman, M.M., Martin, L.E., and Stephens, D.S., Two glycosyl transferase genes, *lgtF* and *rfaK*, constitute the lipooligosaccharide *ice* (inner core extension) biosynthesis operon of *Neisseria meningitidis, J. Bacteriol.*, 178, 6677, 1996.
25a. Minor, S.Y., Banerjee, A., and Gotschlich, E.C., Effect of α-oligosaccharide phenotype of *Neisseria gonorrhoeae* on its invasion of Chang conjunctiva cells, manuscript in preparation.
26. Cope, L.D., Yogev, R., Mertsola, J., et al., Molecular cloning of a gene involved in lipooligosaccharide biosynthesis and virulence expression by *Haemophilus influenzae* type b, *Mol. Microbiol.*, 5, 1113, 1991.
27. High, N.J., Deadman, M.E., and Moxon, E.R., The role of a repetitive DNA motif (5'-CAAT-3') in the variable expression of the *Haemophilus influenzae* lipopolysaccharide epitope αGal(1-4)βGal, *Mol. Microbiol.*, 9, 1275, 1993.
28. Danaher, R.J., Levin, J.C., Arking, D., Burch, C.L., Sandlin, R., and Stein, D.C., Genetic basis of *Neisseria gonorrhoeae* lipooligosaccharide antigenic variation, *J. Bacteriol.*, 177, 7275, 1995.
29. Gotschlich, E.C., Genetic locus for the biosynthesis of the variable portion of *Neisseria gonorrhoeae* lipooligosaccharide, *J. Exp. Med.*, 180, 2181, 1994.

30. Knight, A.I., Ni, H., Cartwright, K.A., and McFadden, J.J., Identification and characterization of a novel insertion sequence, IS1106, downstream of the porA gene in B15 *Neisseria meningitidis, Mol. Microbiol.*, 6, 1565, 1992.

31. Pradel, E., Parker, C.T., and Schnaitman, C.A., Structures of the *rfaB, rfaI, rfaJ,* and *rfaS* genes of *Escherichia coli* K-12 and their roles in assembly of the lipopolysaccharide core, *J. Bacteriol.*, 174, 4736, 1992.

32. Hood, D.W., Deadman, M.E., Jennings, M.P., et al., DNA repeats identify novel virulence genes in *Haemophilus influenzae, Proc. Nat. Acad. Sci. U.S.A.*, 93, 11121, 1996.

33. Wakarchuk, W., Martin, A., Jennings, M.P., Moxon, E.R., and Richards, J.C., Functional relationships of the genetic locus encoding the glycosyltransferase enzymes involved in expression of the lacto-N-neotetraose terminal lipopolysaccharide structure in *Neisseria meningitidis, J. Biol. Chem.*, 271, 19166, 1996.

34. Jennings, M.P., Hood, D.W., Peak, I.R.A., Virji, M., and Moxon, E.R., Molecular analysis of a locus for the biosynthesis and phase-variable expression of the lacto-N-neotetraose terminal lipopolysaccharide structure in *Neisseria meningitidis, Mol. Microbiol.*, 18, 729, 1995.

35. Banerjee, A., Wang, R., Uljohn, S., Rice, P.A., Gotschlich, E.C., Stein, C.A., Identification of the gene (lgtG) encoding the lipooligosaccharide β-chain synthesizing glucosyl transferase from *Neisseria gonorrhoeae, Proc. Natl. Acad. Sci. USA*, 95, 10872-10877, 1998.

36. Schneider, H., Hammack, C.A., Apicella, M.A., and Griffiss, J.M., Instability of expression of lipooligosaccharides and their epitopes in *Neisseria gonorrhoeae, Infect. Immun.*, 56, 942, 1988.

37. Schneider, H., Griffiss, J.M., Boslego, J.W., Hitchcock, P.J., Zahos, K.M., and Apicella, M.A., Expression of paragloboside-like lipooligosaccharides may be a necessary component of gonococcal pathogenesis in men, *J. Exp. Med.*, 174, 1601, 1991.

38. van Putten, J.P.M., Phase variation of lipopolysaccharide directs interconversion of invasive and immuno-resistant phenotypes of *Neisseria gonorrhoeae, EMBO J.*, 12, 4043, 1993.

39. Rudel, T., van Putten, J.P.M., Gibbs, C.P., Haas, R., and Meyer, T.F., Interaction of two variable proteins (PilE and PilC) required for pilus-mediated adherence of *Neisseria gonorrhoeae* to human epithelial cells, *Mol. Microbiol.*, 6, 3439, 1992.

40. Schneider, H., Griffiss, J.M., Mandrell, R.E., and Jarvis, G.A., Elaboration of a 3.6-kilodalton lipooligosaccharides, antibody against which is absent from human sera, is associated with serum resistance of *Neisseria gonorrhoeae, Infect. Immun.*, 50, 672, 1985.

41. Yang, Q-L. and Gotschlich, E.C., Variation of gonococcal LOS structure is due to alterations in poly-G tracts in *lgt* genes encoding glycosyl transferases, *J. Exp. Med.*, 183, 323, 1996.

42. Burch, C.L., Danaher, R.J., and Stein, D.C., Antigenic variation in *Neisseria gonorrhoeae*: production of multiple lipooligosaccharides, *J. Bacteriol.*, 179, 982, 1997.

43. Rice, P.A., Molecular basis for serum resistance in *Neisseria gonorrhoeae, Clin. Microbiol. Rev.*, 2, S112, 1989.

44. Ram, S., Binding of complement factor H to loop 5 of porin protein IA: a molecular mechanism of serum resistance of the non-sialylated *Neisseria gonorrhoeae, J. Exp. Med.*, 188, 671-680, 1998.

45. Griffiss, J.M., Jarvis, G.A., O'Brien, J.P., Eads, M.M., and Schneider, H., Lysis of *Neisseria gonorrhoeae* initiated by binding of normal human IgM to a hexosamine-containing lipooligosaccharide epitope(s) is augmented by strain-specific, properdin-binding-dependent alternative complement pathway activation, *J. Immunol.*, 147, 298, 1991.

46. Parsons, N.J., Andrade, J.R., Patel, P.V., Cole, J.A., and Smith, H., Sialylation of lipopolysaccharide and loss of absorption of bactericidal antibody during conversion of gonococci to serum resistance by cytidine 5'-monophospho-N-acetyl neuraminic acid, *Microb. Pathol.*, 7, 63, 1989.

47. Apicella, M.A., Mandrell, R.E., Shero, M., et al., Modification by sialic acid of *Neisseria gonorrhoeae* lipooligosaccharide epitope expression in human urethral exudates: an immunoelectron microscopic analysis, *J. Infect. Dis.*, 162, 506, 1990.

48. Mandrell, R.E., Lesse, A.J., Sugai, J.V., et al., In vitro and in vivo modification of *Neisseria gonorrhoeae* lipooligosaccharide epitope structure by sialylation, *J. Exp. Med.*, 171, 1649, 1990.

49. Gilbert, M., Watson, D.C., Cunningham, A.M., Jennings, M.P., Young, N.M., and Wakarchuk, W., Cloning of the lipooligosaccharide alpha-2, 3-sialyltransferase from the bacterial pathogens *Neisseria meningitidis* and *Neisseria gonorrhoeae*, *J. Biol. Chem.*, 271, 28271, 1996.

50. Gilbert, M., Cunningham, A.M., Watson, D.C., Martin, A., Richards, J.C., and Wakarchuk, W., Characterization of a recombinant *Neisseria meningitidis* alpha-2,3-sialyltransferase and its acceptor specificity, *Eur. J. Biochem.*, 249, 187, 1997.

51. Estabrook, M.M., Griffiss, J.M., and Jarvis, G.A., Sialylation of *Neisseria meningitidis* lipooligosaccharide inhibits serum bactericidal activity by masking lacto-N-neotetraose, *Infect. Immun.*, 65, 4436, 1997.

52. Wetzler, L.M., Barry, K., Blake, M.S., and Gotschlich, E.C., Gonococcal lipooligosaccharide sialylation prevents complement-dependent killing by immune sera, *Infect. Immun.*, 60, 39, 1992.

53. Elkins, C., Carbonetti, N.H., Varela, V.A., Stirewalt, D., Klapper, D.G., and Sparling, P.F., Antibodies to N-terminal peptides of gonococcal porin are bactericidal when gonococcal lipopolysaccharide is not sialylated, *Mol. Microbiol.*, 6, 2617, 1992.

54. de la Paz, H., Cooke, S.J., and Heckels, J.E., Effect of sialylation of lipopolysaccharide of *Neisseria gonorrhoeae* on recognition and complement-mediated killing by monoclonal antibodies directed against different outer-membrane antigens, *Microbiology*, 141, 913, 1995.

55. Ram, S., Sharma, A.K., Simpson, S.D, Gulati, S., McQuillen, D.P., and Pangburn, M.K., A novel sialic acid binding site on factor H mediates serum resistance of sialylated *Neisseria gonorrhoeae*, *J. Exp. Med.*, 187, 743, 1998.

56. Vogel, U., Weinberger, A., Frank, R., Muller, A., Kohl, J., and Atkinson, J.P., Complement factor C3 deposition and serum resistance in isogenic capsule and lipooligosaccharide sialic acid mutants of serogroup B *Neisseria meningitidis*, *Infect. Immun.*, 65, 4022, 1997.

57. Sandlin, R.C., Apicella, M.A., and Stein, D.C., Cloning of a gonococcal DNA sequence that complements the lipooligosaccharide defects of *Neisseria gonorrhoeae* 1291d and 1291e, *Infect. Immun.*, 61, 3360, 1993.

58. Sandlin, R.C. and Stein, D.C., Role of phosphoglucomutase in lipopolysaccharide biosynthesis in *Neisseria gonorrhoeae*, *J. Bacteriol.*, 176, 2930, 1994.

59. Zhou, D., Stephens, D.S., Gibson, B.W., et al., Lipooligosaccharide biosynthesis in pathogenic *Neisseria*. Cloning, identification, and characterization of the phosphoglucomutase gene, *J. Biol. Chem.*, 269, 11162, 1994.

60. Robertson, B.D., Forsch, M., and van Putten, J.P.M., The role of galE in the biosynthesis and function of gonococcal lipopolysaccharide, *Mol. Microbiol.*, 8, 891, 1993.

61. Jennings, M.P., van der Ley, P., Wilks, K.E., Maskell, D.J., Poolman, J.T., and Moxon, E.R., Cloning and molecular analysis of the *galE* gene of *Neisseria meningitidis* and its role in lipopolysaccharide synthesis, *Mol. Microbiol.*, 10, 361, 1993.

62. Lee, F.K.N., Stephens, D.S., Gibson, B.W., Engstrom, J.J., Zhou, D., and Apicella, M.A., Microheterogeneity of Neisseria lipooligosaccharide: analysis of a UDP-glucose 4-epimerase mutant of *Neisseria meningitidis* NMB, *Infect. Immun.*, 63, 2508, 1995.

63. Kellogg, D.S., Jr., Peacock, W.L., Jr., Deacon, W.E., Brown, L., and Pirkle, C.I., *Neisseria gonorrhoeae*. I. Virulence genetically linked to colonial variation, *J. Bacteriol.*, 85, 1274, 1963.

64. Swanson, J.L., Kraus, S.J., and Gotschlich, E.C., Studies on gonococcus infection. I. Pili and zones of adhesion: their relation to gonococcal growth patterns, *J. Exp. Med.*, 134, 886, 1971.

65. Rudel, T., Facius, D., Barten, R., Scheuerpflug, I., Nonnenmacher, E., and Meyer, T.F., Role of pili and the phase-variable PilC protein in natural competence for transformation of *Neisseria gonorrhoeae*, *Proc. Nat. Acad. Sci. U.S.A.*, 92, 7986, 1995.

66. Sparling. P.F., Genetic transformation of *Neisseria gonorrhoeae* to streptomycin resistance, *J. Bacteriol.*, 92, 1364, 1966.

67. Fussenegger, M., Rudel, T., Barten, R., Ryll, R., and Meyer, T.F., Transformation competence and type-4 pilus biogenesis in *Neisseria gonorrhoeae*: a review. [Review] [93 refs], *Gene*, 192, 125, 1997.

68. Rudel, T., Scheurerpflug, I., and Meyer, T.F., Neisseria PilC protein identified as type-4 pilus tip-located adhesion, *Nature*, 373, 357, 1995.

69. Virji, M., Kayhty, H., Ferguson, D.J., Alexandrescu, C., Heckels, J.E., and Moxon, E.R., The role of pili in the interactions of pathogenic Neisseria with cultured human endothelial cells, *Mol. Microbiol.*, 5, 1831, 1991.

70. Pron, B., Taha, M.K., Rambaud, C., et al., Interaction of *Neisseria meningitidis* with the components of the blood-brain barrier correlates with an increased expression of PilC, *J. Infect. Dis.*, 176, 1285, 1997.

71. Makino, S., van Putten, J.P.M., and Meyer, T.F., Phase variation of the opacity outer membrane protein controls invasion by *Neisseria gonorrhoeae* into human epithelial cells, *EMBO J.*, 10, 1307, 1991.

72. Ilver, D., Kallstrom, H., Normark, S., and Jonsson, A.B., Transcellular passage of *Neisseria gonorrhoeae* involves pilus phase variation, *Infect. Immun.*, 66, 469, 1998.

73. Swanson, J.L. and Barrera, O., Gonococcal pilus subunit size heterogeneity correlates with transitions in colony piliation phenotype, not with changes in colony opacity, *J. Exp. Med.*, 158, 1459, 1983.

74. Swanson, J.L., Bergstrom, S., Barrera, O., Robbins, K., and Corwin, D., Pilus-gonococcal variants: evidence for multiple forms of piliation control, *J. Exp. Med.*, 162, 729, 1985.

75. Serkin, C.D. and Seifert, H.S., Frequence of pilin antigenic variation in *Neisseria gonorrhoeae*, *J. Bacteriol.*, 180, 1955, 1998.
76. Koomey, M., Gotschlich, E.C., Robbins, K., Bergstrom, S., and Swanson, J.L., Effects of *recA* mutations on pilus antigenic variation and phase transitions in *Neisseria gonorrhoeae*, *Genetics*, 117, 391, 1987.
77. Gibbs, C.P., Reimann, B.-Y., Schultz, E., Kaufmann, A., Haas, R., and Meyer, T.F., Reassortment of pilin genes in *Neisseria gonorrhoeae* occurs by two distinct mechanisms, *Nature*, 338, 651, 1989.
78. Seifert, H.S., Ajioka, R.S., Marchal, C., Sparling, P.F., and So, M., DNA transformation leads to pilin antigenic variation in *Neisseria gonorrhoeae*, *Nature*, 336, 392, 1988.
79. Zhang, Q.Y., DeRychere, D., Lauer, P., and Koomey, M., Gene conversion in *Neisseria gonorrhoeae*: Evidence for its role in pilus antigenic variation, *Proc. Natl. Acad. Sci. U.S.A.*, 89, 5366, 1992.
80. Jonsson, A.-B., Nyberg, G., and Normark, S., Phase variation of gonococcal pili by frameshift mutation in *pilC*, a novel gene for pilus assembly, *EMBO J.*, 10, 477, 1991.
81. Jonsson, A.-B., Pfeifer, J., and Normark, S., *Neisseria gonorrhoeae* PilC expression provides a selective mechanism for structural diversity of pili, *Proc. Natl. Acad. Sci. U.S.A.*, 89, 3204, 1992.
82. Robertson, J.P., Vincent, P., and Ward, M.E., The preparation and properties of gonococcal pili, *J. Gen. Microbiol.*, 102, 169, 1977.
83. Hamadeh, R.M., Estabrook, M.D., Zhou, P., Jarvis, G.A., and Griffiss, J.M., Anti-Gal binds to pili of *Neisseria meningitidis*: the immunoglobin A isotype blocks complement-mediated killing, *Infect. Immun.*, 63, 4900, 1995.
84. Virji, M., Saunders, J.R., Sims, G., Makepeace, K., Maskell, D., and Ferguson, D.J., Pilus-facilitated adherence of *Neisseria meningitidis* to human epithelial and endothelial cells: modulation of adherence phenotype occurs concurrently with changes in primary amino acid sequence and the glycosylation status of pilin, *Mol. Microbiol.*, 10, 1013, 1993.
85. Stimson, E., Virji, M., Makepeace, K., et al., Meningococcal pilin: a glycoprotein substituted with digalactosyl 2,4-diacetamido-2,4,6-trideoxyhexose, *Mol. Microbiol.*, 17, 1201, 1995.
86. Meyer, T.F., Gibbs, C.P., and Haas, R., Variation and control of protein expression in *Neisseria*, *Ann. Rev. Microbiol.*, 44, 451, 1990.
87. Stern, A., Brown, M., Nickel, P., and Meyer, T.F., Opacity genes in *Neisseria gonorrhoeae*: control of phase and antigenic variation, *Cell*, 47, 61, 1986.
88. Swanson, J.L., Studies on gonococcus infection. XII. Colony color and opacity variants of gonococci, *Infect. Immun.*, 19, 320, 1978.
89. Mayer, L.W., Rates of in vitro changes of gonococcal opacity phenotypes, *Infect. Immun.*, 37, 481, 1982.
90. Murphy, G.L., Connell, T.D., Barritt, D.S., Koomey, M., and Cannon, J.G., Phase variation of gonococcal protein II: regulation of gene expression by slipped-strand mispairing of a repetitive DNA sequence, *Cell*, 56, 539, 1989.
91. Belland, R.J., Morrison, S.G., van der Ley, P., and Swanson, J., Expression and phase variation of gonococcal P.II genes in *Escherichia coli* involves ribosomal frameshifting and slipped-strand mispairing, *Mol. Microbiol.*, 3, 777, 1989.
92. Bessen, D. and Gotschlich, E.C., Interaction of gonococci with HeLa cells: attachment, detachment, replication, penetration, and the role of protein II, *Infect. Immun.*, 54, 154, 1986.

93. Kupsch, E.-M., Knepper, B., Kuroki, T., Heuer, I., and Meyer, T.F., Variable opacity (Opa) outer membrane proteins account for the cell tropisms displayed by *Neisseria gonorrhoeae* for human leukocytes and epithelial cells, *EMBO J.*, 12, 641, 1993.

94. Simon, D. and Rest, R.F., *Escherichia coli* expressing a *Neisseria gonorrhoeae* opacity-associated outer membrane protein invade human cervical and endometrial epithelial cell lines, *Proc. Natl. Acad. Sci. U.S.A.*, 89, 5512, 1992.

95. Belland, R.J., Chen, T., Swanson, J., and Fischer, S.H., Human neutrophil response to recombinant neisserial Opa proteins, *Mol. Microbiol.*, 6, 1729, 1992.

96. Elkins, C. and Rest, R.F., Monoclonal antibodies to outer membrane protein PII block interactions of *Neisseria gonorrhoeae* with human neutrophils, *Infect. Immun.*, 58, 1078, 1990.

96a. Chen, T., Belland, R.J., Wilson, J., and Swanson, J., Adherence of pilus⁻ Opa⁺ gonococci to epithelial cells *in vitro* involves heparan sulfate, *J. Exp. Med.*, 182, 511-517, 1995.

97. van Putten, J.P.M. and Paul, S.M., Binding of syndecan-like surface proteoglycan receptors is required for *Neisseria gonorrhoeae* entry into human mucosal cells, *EMBO J.*, 14, 2144, 1995.

98. Chen, T., Swanson, J., Wilson, J., and Belland, R.J., Heparin protects opa+ *Neisseria gonorrhoeae* from the bacterial action of normal human serum, *Infect. Immun.*, 63, 1790, 1995.

99. Duensing, T.D. and van Putten, J.P., Vitronectin mediates internalization of *Neisseria gonorrhoeae* by Chinese hamster ovary cells, *Infect. Immun.*, 65, 964, 1997.

100. Gomez-Duarte, O.G., Dehio, M., Guzman, C.A., Chatwal, G.S., Dehio, C., and Meyer, T.F., Binding of vitronectin to opa-expressing *Neisseria gonorrhoeae* mediates invasion of HeLa cells, *Infect. Immun.*, 65, 3857, 1997.

101. Chen, T. and Gotschlich, E.C., CGM1a antigen of neutrophils, a receptor of gonococcal opacity proteins, *Proc. Natl. Acad. Sci. U.S.A.*, 93, 14851, 1996.

102. Virji, M., Makepeace, K., Ferguson, D.J., and Watt, S.M., Carcinoembryonic antigens (CD66) on epithelial cells and neutrophils are receptors for Opa proteins of pathogenic neisseriae, *Mol. Microbiol.*, 22, 941, 1996.

103. Virji, M., Watt, S.M., Barker, S., Makepeace, K., and Doyonnas, R., The N-domain of the human CD66a adhesion molecule is a target for Opa proteins of *Neisseria meningitidis* and *Neisseria gonorrhoeae*, *Mol. Microbiol.*, 22, 929, 1996.

104. Chen, T., Grunert, F., Medina-Marino, A., and Gotschlich, E.C., Several carcinoembryonic antigens (CD66) serve as receptors for gonococcal opacity proteins, *J. Exp. Med.*, 185, 1557, 1997.

105. Bos, M.P., Grunert, F., and Belland, R.J., Differential recognition of members of the carcinoembryonic antigen family by opa variants of *Neisseria gonorrhoeae*, *Infect. Immun.*, 65, 2353, 1997.

106. Apicella, M.A., Shero, M., Jarvis, G.A., Griffiss, J.M., Mandrell, R.E., and Schneider, H., Phenotypic variation in epitope expression of the *Neisseria gonorrhoeae* lipooligosaccharide, *Infect. Immun.*, 55, 1755, 1987.

107. Weel, J.F.L., Hopman, C.T.P., and van Putten, J.P.M., Stable expression of lipooligosaccharide antigens during attachment, internalization, and intracellular processing of *Neisseria gonorrhoeae* in infected epithelial cells, *Infect. Immun.*, 57, 3395, 1989.

108. Schneider, H., Schmidt, K.A., Skillman, D.R., et al., Sialylation lessens the infectivity of *Neisseria gonorrhoeae* MS11mkC, *J. Infect. Dis.*, 173, 1422, 1996.

109. Grassme, H.U., Ireland, R.M., and van Putten, J.P., Gonococcal opacity protein promotes bacterial entry-associated rearrangements of the epithelial cell actin cytoskeleton, *Infect. Immun.*, 64, 1621, 1996.
110. van Putten, J.P.M., Duensing, T.D., and Cole, R.L., Entry of OpaA⁺ gonococci into Hep-2 cells requires concerted action of glycosaminoglycans, fibronectin, and integrin receptors, *Mol. Microbiol.*, 29, 369-379, 1998.
111. Porat, N., Apicella, M.A., and Blake, M.S., Neisseria gonorrhoeae utilizes and enhances the biosynthesis of the asialoglycoprotein receptor expressed on the surface of the hepatic HepG2 cell line, *Infect. Immun.*, 63, 1498, 1995.
112. Porat, N., Apicella, M.A., and Blake, M.S., A lipooligosaccharide-binding site on HepG2 cells similar to gonococcal opacity-associated surface protein opa, *Infect. Immun.*, 63, 2164, 1995.
113. Blake, M.S., Functions of the outer membrane proteins of *Neisseria gonorrhoeae*, in Jackson, G.G., Thomas, H., (Eds.) *The Pathogenesis of Bacterial Infections.* Berlin, Springer Verlag KG, 1985:51-66.
114. Blake, M.S., Blake, C.M., Apicella, M.A., and Mandrell, R.E., Gonococcal opacity: lectin-like interactions between Opa proteins and lipooligosaccharide, *Infect. Immun.*, 63, 1434, 1995.
115. Rice, P.A., McCormack, W.M., and Kasper, D.L., Natural serum bactericidal activity against *Neisseria gonorrhoeae* isolates from disseminated, locally invasive, and uncomplicated disease, *J. Immunol.*, 124, 2105, 1980.
116. Mandrell, R.E., McLaughlin, R., Kwaik, Y.A., et al., Lipooligosaccharides (LOS) of some *Haemophilus* species mimic human glycosphingolipids, and some LOS are sialylated, *Infect. Immun.*, 60, 1322, 1992.

chapter six

Genetics of Vibrio cholerae O1 and O139 surface polysaccharides

Uwe H. Stroeher and Paul A. Manning

Contents

Introduction 133
Vibrio cholerae O1 O-antigen 134
    Genetics of *Vibrio cholerae* O1 O-antigen 136
    GDP–perosamine biosynthesis in *V. cholerae* O1 137
    O-antigen transport in *Vibrio cholerae* O1....................................... 139
    Tetronate biosynthesis in *Vibrio cholerae* O1 141
    Ogawa–Inaba serotype switching.. 142
    rfa genes linked to the *rfb* operon in *Vibrio cholerae* O1............... 143
    Virulence of *V. cholerae* O1 and O139 O-antigen (*rfb*) mutants.............. 144
    Insertion sequences associated with the *rfb* region in *Vibrio cholerae*.... 144
    Surface polysaccharides of *Vibrio cholerae* O139 145
    Genetics of *Vibrio cholerae* O139 capsule and O-antigen
        biosynthesis 147
    Evolution of *V. cholerae* O139 polysaccharide genes 151
Conclusion 152
Acknowledgments 152
References... 153

Introduction

Cholera is characterized by massive diarrhea due to the secretion of electrolytes and fluids (often up to 12 liters a day) into the stool. This sudden loss of fluid can cause shock and lead to death due to organ failure. Since the beginning of the 19th century there have been seven recorded cholera pandemics across the globe. The most recent of these began in 1961 in the Philippines. *Vibrio cholerae* comprises over 150 serogroups based upon the antigenicity of the surface polysaccharides, and until the recent identification of the O139 serogroup, only the O1 strains were associated with epidemic cholera. In O1 strains the major surface polysaccharide is the lipopolysaccharide, and the O-antigen component has been shown to extend over the cell surface and the flagellum.[1] *V. cholerae* O1 can be divided into two biotypes: classical strains, which are thought to have been responsible for the first six recorded cholera pandemics, and the El Tor biotype, thought to have been responsible for the seventh. It has been assumed that only the *V. cholerae* O1 serogroup could give rise to pandemic cholera,[2] but in 1992 the picture changed considerably. In the region around the Bay of Bengal a new serogroup termed O139, with pandemic potential, was isolated.[3,4,5,6a] This novel O-serogroup began to replace traditional cholera strains in Bangladesh and the Indian subcontinent, and it may have arisen by selection because although there was significant immunity to *V. cholerae* of the O1 serogroup in the population, it was naive to the O139 strains. The O-antigen of the LPS also appears to be the major protective antigen, and consequently a change in the O-antigen, and thus the serogroup of the organism, can potentially lead to strains with new pandemic capabilities. Interestingly, a nonpathogenic *V. cholerae* serogroup O139 strain has also arisen in Argentina, but its origin appears to be very different from that of the O139 Bengal strains.[5,6]

In this chapter we will review the genetics of biosynthesis of both the *V. cholerae* O1 and O139 lipopolysaccharides using the nomenclature described in the original publications. However, in Tables 6.1 and 6.2 we also present the new uniform nomenclature according to the system of Reeves et al.[7] as recently reported.[8]

Vibrio cholerae *O1 O-antigen*

V. cholerae O1 strains of both biotypes have been further subdivided into three serotypes, depending on the presence and amount of particular antigens on the O-antigen of the lipopolysaccharide. The three serotypes are designated Inaba, Ogawa, and Hikojima, which all share a common antigen referred to as the A antigen. There are two specific antigens, B and C, which are present in varying amounts on the different serotypes. Inaba strains express only C and A, while Ogawa strains express all three antigens A, B, and C, although C is present in much reduced amounts compared to Inaba.[9-11] The third serotype, termed Hikojima, is extremely rare and unstable and expresses elevated levels of all three antigens.[12,13] It would appear to be just an Ogawa variant in which the level of expression of the enzyme (RfbT) responsible for the B antigen synthesis is altered.

The O-antigen of *V. cholerae* O1 consists of a homopolymer of 4-amino-4,6-dideoxy-mannose (perosamine) which is substituted with 3-deoxy-L-*glycero*-tetronic acid (tetronate)[11,14,15] (Figure 6.1). This basic structure is repeated 18 times on average and is joined to lipid A via a linker core oligosaccharide

Table 6.1 Possible Functions of *V. cholerae* O1 *rfb* Gene Products

Old ORF designation	New ORF designation	G+C content	No. of a.a.	Putative function	Homology to	Accession number
RfaD	GmhD	46.1	314	core biosynthesis	*Escherichia coli* RfaD	M33577
RfbA	ManC	42.1	465	PMI/GMP activity	*Salmonella enterica* RfbM	P26404
RfbB	ManB	38.6	463	PMM activity	*Escherichia coli* K-12 CpsG	P24175
RfbD	Gmd	39.6	373	Oxidoreductase	*Vibrio cholerae* O139 RfbD	Y07786
RfbE	WbeE	40	367	Pyridoxal binding protein	*Bacillus subtilis* Ipa65	X73124
RfbG	WbeG	35.1	463	unknown	no homolog	
RfbH	Wzm	35.1	257	O-antigen export (channel?)	*Yersinia enterocolitica* RfbD	S28580
RfbI	Wzt	36.5	250	O-antigen export (energizer?)	*Yersinia enterocolitica* RfbE	S28581
RfbK	WbeK	33.3	80	acyl carrier protein	*Escherichia coli* ACP	P02901
RfbL	WbeL	37.7	471	fatty acid ligase	*Escherichia coli* FadE	P29212
RfbM	WbeM	42.5	374	Alcohol dehydrogenase	*Clostridium acetbutylicum* ADHI	P13604
RfbN	WbeN	42.4	825	fatty acid reductase	*Vibrio harveyi* LuxC/E	P08639
RfbO	WbeO	39	188	acetyl Co-A transferase	*Rhizobium meliloti* NodL	P28266
RfbP	WbeP	43.1	73	unknown	no homolog	
RfbQ,R,S	IS1358d1	42.7	65,234,68	Insertion sequence	*Vibrio cholerae* O139 IS1358	X91246
RfbT	WbeT	31.1	287	Ogawa determination	no homolog	
RfbU	WbeU	35.1	370	Mannosyl transferase	*Eschericia coli* MtfC	D43637
RfbV	WbeV	43.5	621	LPS biosynthesis	*Yersinia enterocolitica* TrsG	S51266
ORF35.7	ORF35.7	45.2	323	dTDP-glucose dehydratase	*Neisseria gonorrhoeae* RfbB	P37761
RfbW	WbeW	44.1	184	Galactosyl transferase	*Vibrio cholerae* O139 ORF7	U47057

Only the best homologs are shown. The old ORF designations are those used in the original references. The new nomenclature is based on that described by Reeves et al.[7]

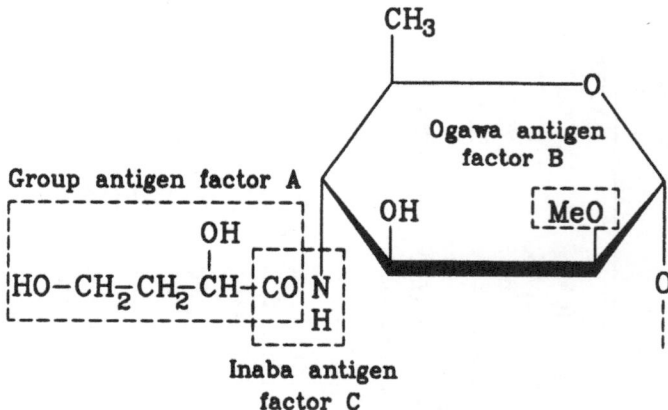

Figure 6.1 The structure of the A, B, and C antigens on the O-antigen subunit is shown. The B-antigen is formed essentially by the addition of the methyl group on the number-two carbon in the perosamine sugar. The C-antigen is the linkage between the perosamine and the tetronate, whereas the A-antigen corresponds to the tetronate. (Based on Hisatsune, K., S. Kondo, T. Iguchi, T. Ito, and K. Hiramatsu. Lipopolysaccharide of *Escherichia coli* K-12 strains that express cloned genes for the Ogawa and Inaba antigens of *Vibrio cholerae* O1; Identification of O-antigenic factors, in *The Thirty-Second Joint Conference U.S.–Japan Cooperative Medical Science Program Cholera and Related Diarrheal Diseases Panel*, 1996. U.S.–Japan Cooperative Medical Science Program.)

region. The exact chemical nature of the A antigen has not yet been elucidated, but it is likely to be the tetronate substitution on the perosamine backbone.[16] The gene (*rfbT*) responsible for the chemical modification leading to the B antigen has been determined.[17,18] The modified repeat unit is 2-O-methyl-D-perosamine. It is thought that the C antigen is the linkage between the perosamine backbone and the tetronate substitution and that in Ogawa strains this antigen is partially sterically masked by the presence of the B antigen[16] (Figure 6.1). The O-antigen also contains another sugar known as quinovosamine. This sugar is found in a ratio of approximately 1 to 20 compared to perosamine, but it is not known whether quinovosamine either caps the distal end of the O-antigen or the core at its site of attachment to the O-antigen. The sugar quinovosamine is a 4-amino-2,6-dideoxy-glucose and may be derived from a perosamine precursor.

Genetics of Vibrio cholerae O1 O-antigen

Several years ago Manning et al.[19] cloned and expressed the genes for both the Inaba and Ogawa serotype of the *V. cholerae* O-antigen in *Escherichia coli* K-12. From heteroduplex analysis and independent cosmid clones, this region was initially thought to be approximately 18 to 20 kb in size, was referred to as the *rfb* region,[20,21] and corresponds to the *oag* locus first described by Bhaskaran and Gorrill in 1957. The *rfb* operon of *V. cholerae* O1 can be split into five major regions: (1) perosamine biosynthesis, (2) O-antigen transport, (3) tetronate biosynthesis, (4) O-antigen modification, and (5) accessory *rfb* genes (Figure 6.2). This region also contains a defective form of the putative insertion sequence designated IS*1358*d1, and downstream of *rfbT* lies a recently described region containing the accessory *rfb* genes involved in O-antigen biosynthesis in *V. cholerae* but not required for expression in a heterologous host such as *E. coli* K-12.[23]

Upstream of the perosamine biosynthetic genes lies *rfaD*, which is involved in core biosynthesis: RfaD converts ADP-D-glycero-D-mannoheptose to ADP-L-*glycero*-D-mannoheptose.[24] In *E. coli* and *Salmonella enterica* (serovar Typhimurium) the *rfaD* gene is found as part of the *rfa* gene cluster,[25,26] but this does not appear to be the case for *V. cholerae* (see Figure 6.2). The proximity of the *rfb* operon and *rfaD* with divergent promoters is suggestive of some form of common regulation. Just upstream of the start of the *rfb* operon lies a short sequence of approximately 40 base pairs which corresponds to the JUMPstart sequence[27] that has also been described as a σ54-dependent promoter-like sequence.[28] This sequence is found in many polysaccharide operons and is thought to be involved in the initiation and stabilization of the transcription complex mediated by RfaH homologs.[29]

The genetic organization of the *rfb* region of *V. cholerae* O1 suggests that the genes are translationally coupled, because in most cases there are only a few bases separating the genes. The exception to this is the gap of approximately 70 bp between *rfbN* and *rfbO*. Extensive sequence analysis has found neither a promoter region nor any errors in the sequence that could close this gap. Since *rfbO* is the last of the true *rfb* genes in this region and it is thought to condense the tetronate and GDP-perosamine into the O-antigen subunit (see below), it could have ancestrally been associated with *rfbA* to *rfbE* in the perosamine pathway. Thus, *rfbA* through *rfbO* forms the major *rfb* operon of *V. cholerae* O1. As mentioned, the DNA downstream from this operon, including IS*1358*d1, *rfbT*, and the recently described *rfb* genes, are separate transcriptional units.[6,17,23]

GDP–perosamine biosynthesis in V. cholerae O1

The perosamine biosynthetic genes in *V. cholerae* O1 are thought to comprise *rfbA*, *rfbB*, *rfbD*, and *rfbE*,[30] and their corresponding protein products all show homology to proteins involved in polysaccharide biosynthesis from a number of other bacterial species (Table 6.1). The biosynthesis of perosamine parallels that of alginate in *Pseudomonas aeruginosa*.[31-33] The first enzyme in the pathway is RfbA, which is thought to be bifunctional, having both phosphomannose isomerase (PMI) and guanosine pyrophosphorylase activity (GMP).[34] Based on studies with AlgA, RfbA contains all the regions and residues critical for PMI–GMP function.[35] The RfbA protein is predicted to catalyze the first and third step in the biosynthesis of perosamine[30] (Figure 6.3). The second enzyme in the pathway requires RfbB, which is thought to be a phospho-manno-mutase (PMM) that converts mannose-

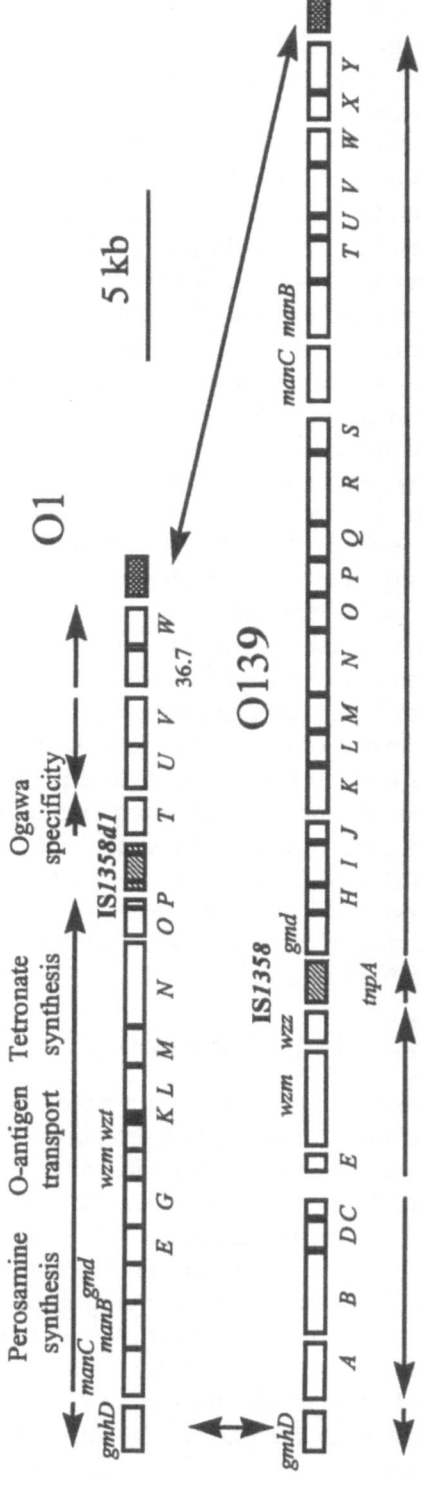

Figure 6.2 Comparison of the *rfb* regions of *V. cholerae* serotypes O1 and O139. The ORF designations are based on the scheme proposed by Reeves et al.,[7] for the nomenclature of Gram-negative lipopolysaccharide and capsular genes; see Tables 6.1 and 6.2 for old nomenclature. The ORFs are represented by boxes. The regions of homology between *V. cholerae* O1 and O139 are indicated by vertical arrows. The directions of transcription of the genes within these regions are represented by horizontal arrows.

6-phosphate to mannose-1-phosphate (Figure 6.3) but also shows homology to a number of phospho-gluco-mutases (PGM).[30] It is not known whether both activities are seen in *V. cholerae* O1 RfbB. The fourth enzymatic step is likely to be carried out by RfbD, which is an oxidoreductase/dehydratase. In *V. cholerae* O1 GDP-mannose is predicted to be converted to GDP-4-keto-6-deoxymannose by the action of RfbD[30] (Figure 6.3). The final step is the conversion of GDP-4-keto-6-deoxymannose to GDP-4-amino-4,6-dideoxymannose (GDP-perosamine)[30] (Figure 6.3), which is proposed to be performed by RfbE that shows homology to a number of pyridoxal-binding proteins from a wide range of organisms. Based on these data, it is likely that RfbE is the amino-transferase or perosamine synthetase.[30] As yet we have not identified an O-antigen polymerase or perosamine transferase. However, since no function has yet been attributed to RfbG, which appears to have no homology to any known proteins, this could in fact be the perosamine transferase, but it could equally well be involved in the biosynthesis of quinovosamine. Alternatively, as mentioned quinovosamine is a 4-amino-2,6-dideoxy-glucose and may be derived from a perosamine precursor by the function of *rfbU*, which is the last complete gene found in the minimal *SacI* region. A combination of genetics and chemical studies are needed to resolve this. Analysis of non-O1 *V. cholerae* strains has shown that the perosamine pathway is also present in O140.[37] This serogroup has an O-antigen which is also a homopolymer of perosamine but substituted with *N*-acetyl instead of tetronate.[36,37]

O-antigen transport in Vibrio cholerae *O1*

A variety of molecules that are synthesized in the bacterial cell are exported to other locations. However, whereas the export of proteins from the cell often involves N-terminal signal sequences and a variety of accessory proteins required in the secretion process, the export of polysaccharides is less well understood. Much is known about the biosynthesis of molecules, such as lipopolysaccharides and capsular polysaccharides, but details of the topology and export of these molecules are far from complete. The lipid carrier, bactoprenol or undecaprenol-phosphate, is critical in the export of all polysaccharides such as LPS, peptidoglycan, and capsule and has been proposed to function by flipping the newly synthesized polysaccharide across the cytoplasmic membrane via a flippase complex.[38] With the cloning and sequencing of *rfb* regions, it has been possible to make generalizations about polysaccharide export. The mechanism appears to involve a pair of proteins. One is an integral membrane protein with a number of membrane spanning domains, and the other has an ATP-binding motif and is thought to be the energizing partner.[39] In *V. cholerae* O1 these correspond to RfbH and RfbI, respectively[40,41] (Table 6.1).

The integral membrane proteins fall into two distinct classes: those for LPS export and those for capsular polysaccharide export. The LPS export

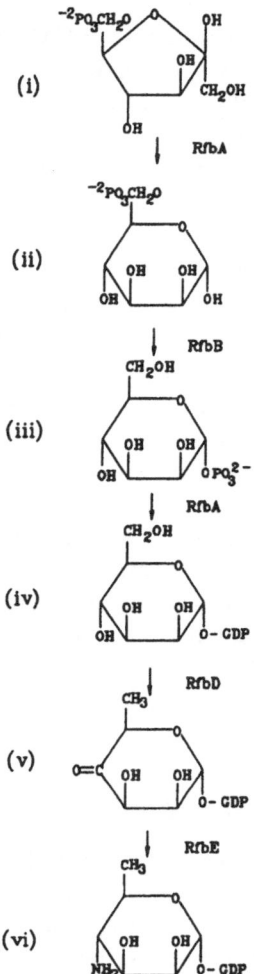

Figure 6.3 Proposed pathway for the biosynthesis of perosamine. The pathway is based solely on the homologies described. The pathway commences with fructose-6-phosphate (i) which is converted to mannose-6-phosphate by RfbA (ii). This is subsequently converted to mannose-1-phosphate by the action of RfbB (iii), which is then made into GDP-mannose by RfbA (iv). GDP-mannose is then further converted by the action of RfbD to GDP-4-keto-6-deoxymannose (v). This substrate is then thought to be converted by RfbE to perosoamine (vi). The substrates for the various enzymes are shown.

proteins are from organisms such as *Yersinia enterocolitica*[42] and *Salmonella enterica* (serovar Typhi),[43] whereas the capsular exporters are found in *Neisseria meningitidis*[44] and *E. coli*.[45,46] The RfbH protein from *V. cholerae* is likely to be the integral membrane protein, since its hydropathy profile suggests six α-helical transmembrane domains.[40,41] Like KpsM of *E. coli*[46] and BexB of *H. influenzae*,[47] RfbH is thought to act as a carrier or pore. RfbI shows homology to export proteins from the same systems as are seen for RfbH. These proteins all contain an adenine nucleotide binding fold.[48] Since these systems usually require two membrane proteins and two ATP-binding domains, it is likely that RfbH and RfbI may be present as homodimers in the export complex.[44] There are other export systems involving proteins, which show no homology to either RfbH or RfbI, such as in *Shigella flexneri,* where the protein designated RfbX is thought to form a pore in the cytoplasmic membrane to allow LPS export.[49,49a]

Tetronate biosynthesis in Vibrio cholerae *O1*

As already mentioned, the perosamine backbone of the *Vibrio cholerae* O1 O-antigen is substituted with 3-deoxy-L-*glycero*-tetronic acid (tetronate).[11,14,15] Within the *rfb* cluster, there are several genes (*rfbK, rfbL, rfbM, rfbN,* and *rfbO*) whose products are predicted to be involved in tetronate biosynthesis. All of these proteins have homologs in other bacterial polysaccharide systems, and their putative functions are based on these homologies (Table 6.1).

The initial step in tetronate biosynthesis is thought to start with the Krebs cycle intermediate malate, which is a 4-C dicarboxylic acid. The first step in the biosynthesis of tetronate is thought to be undertaken by RfbN. RfbN would appear to have evolved by gene fusion since it has homology to LuxE at the N-terminus and to LuxC at the C-terminus.[50] LuxC is the fatty acid reductase, and LuxE is the acyl-protein synthetase. Because of its similarity to both LuxC and LuxE, RfbN presumably has both enzymatic activities. Thus, RfbN is thought to act on a dicarboxylic acid to produce an aldehyde (Figure 6.4). The next enzymatic step is likely to be performed by RfbM, which is predicted to be an iron-containing alcohol dehydrogenase. RfbM has a number of conserved motifs that appear essential for this class of enzyme.[51-53] RfbM may then act on the aldehyde to convert it to the di-hydroxy carboxylic acid (Figure 6.4). The action of RfbL is most likely that of an acetyl-CoA synthetase due to its homology to adenylate-forming enzymes. RfbL contains a highly conserved motif which is thought to correspond with the AMP-binding motif of adenylate-forming enzymes. This group of enzymes works by adenylation of the substrates, which are carboxylic acids (either aliphatic or aromatic). Thus, RfbL would act on the dihydroxy carboxylic acid to activate it to a CoA form (Figure 6.4).[50] This substrate is 3-deoxy-L-*glycero*-tetronyl:CoA which must now be activated to an acyl carrier protein (ACP) form by an enzyme such as acetyl-CoA transacylase. This is likely to be a housekeeping enzyme from the general metabolic pool. RfbK is the ACP and it has all the conserved features of a number of ACPs from different bacterial species (Table 6.1). All ACPs have a common amino acid motif of Asp, Ser, Lys in which the serine is the site of covalent modification with phosphopantetheine.[54,55] The activated precursor could then be condensed with a molecule of GDP-perosamine via its free amino group, by a transferase proposed to be RfbO (Figure 6.4), which shows good homology to acetyl-CoA transferases.[50]

The three components described above, namely perosamine biosynthesis, transport, and tetronate biosynthesis, are capable of synthesizing a com-

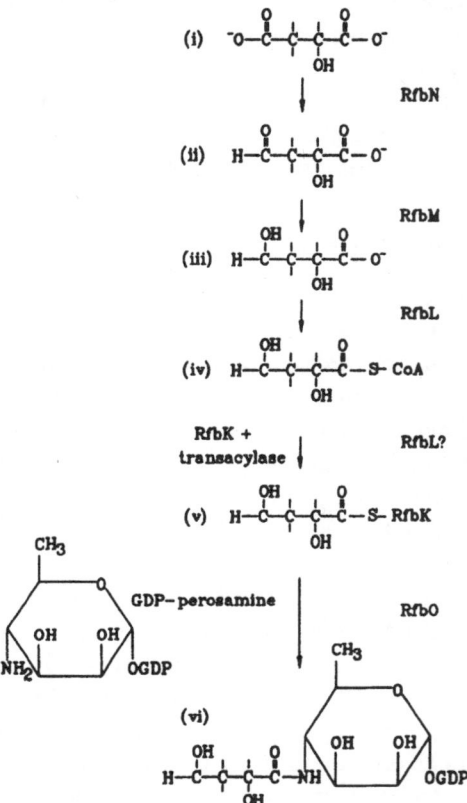

Figure 6.4 Proposed pathway for the biosynthesis of tetronate. The pathway is based solely on the homologies described. The starting substrate is thought to be malate (i) from the Krebs cycle, which is converted first to an aldehyde by the action of RfbN (ii) and then to a di-hydroxy carboxylic acid by RfbM (iii). This is subsequently activated to a Co-A form by the action of RfbL (iv), this substrate is 3-deoxy-ʟ-*glycero*-tetronyl: CoA. This is then likely to be activated to the acyl carrier protein (ACP) form (v) by an as yet unidentified enzyme. This ACP-activated precursor can be condensed with a molecule of perosamine to give rise to a complete O-antigen subunit (vi). The proposed substrates for the various enzymes are shown.

plete O-antigen unit. This basic O-antigen unit can subsequently be modified by the addition of the methyl group in the case of Ogawa serotype[18] however, it is not known whether this modification is added co- or post-synthetically (see below).

Ogawa–Inaba serotype switching

As mentioned above *V. cholerae* O1 is now recognized as two distinct sero-types designated Inaba and Ogawa, depending on the antigens expressed on their O-antigen. *V. cholerae* strains are not fixed in their serotype but can undergo serotype conversion. The switching from Ogawa to Inaba occurs at a frequency of approximately 10^{-5},[10,22] whereas the reverse event is much rarer. The mechanism for serotype switching has been elucidated[17] and may be one that allows the organism to evade the host immune system and therefore persist longer.[13]

The introduction of the *rfb* region from either Inaba or Ogawa strains leads to the expression of the respective serotype in *E. coli* K-12,[19] implying that the gene or genes responsible for the serotype specificity must lie within this region (Figure 6.2). Introduction of distal regions of the *rfb* operon from the Ogawa serotype into Inaba indicated that the serotype specific gene(s) must lie there.[56] Complete sequencing of the *rfb* operons from Inaba and Ogawa strains revealed a single base pair deletion in the *rfbT* gene (Figure 6.2); further analysis of *rfbT* genes from a number of Inaba and Ogawa strains indicated that Inaba strains all appear to have a mutated *rfbT*. The introduction of *rfbT* alone from an Ogawa strain into an Inaba allows the serotype conversion to Ogawa, and the construction of defined mutations in *rfbT* of an Ogawa strain results in the Inaba serotype.[17] This mechanism explains why the conversion from Ogawa occurs at such a high frequency, because this requires only a mutation in the *rfbT* gene, but the switch from Inaba to Ogawa is extremely rare because it requires a precise reversion of the *rfbT* mutation. This appears to have occurred in the switching of the epidemic Inaba strain to the Ogawa form in the 1991 South American cholera epidemic.[57]

The precise modification giving rise to the B antigen has now been determined to be the incorporation of a methyl group into the perosamine backbone of the O-antigen.[16,18] However, the mechanism by which this occurs has not been elucidated. RfbT does not appear to show similarity to any known methylase, and it is unclear if RfbT itself interacts with the O-antigen to methylate it. However, in the absence of a functional RfbT the Ogawa-specific B epitope is not added to the O-antigen, and therefore the C antigen is found in high levels.

The *rfbT* gene, although involved in modifying the O-antigen, is not vital for O-antigen expression in *V. cholerae* and exists as a separate transcriptional unit. It is interesting to note that *rfbT* is preceded by a defective insertion element first identified as three small open reading frames designated RfbQ, RfbR and RfbS but now known as IS1358d1. An apparently active form of this element has also been found linked to the *rfb* regions in *V. cholerae* O139 (see below)[6] and *Vibrio anguillarum* O1 and O2.[58] It would appear that IS1358d1 defines the end of the actual *rfb* operon and that *rfbT* and the newly described *rfb* genes[23] are quite separate from the rest of the *rfb* operon.

rfa genes linked to the rfb operon in Vibrio cholerae *O1*

The linkage of *rfaD* to the *rfb* operon in *V. cholerae* has already been mentioned. This gene is found upstream of and divergently transcribed to *rfbA* and is probably involved in heptose biosynthesis for the *V. cholerae* O1 (and probably O139) core oligosaccharide. To date, it has not been possible to make mutations in this gene, suggesting that a deep rough core may be lethal.[24] Interestingly, *rfaD* is conserved in *V. cholerae* O139, whereas the rest of the *V. cholerae* O1 *rfb* region has been lost.[6,40] Furthermore, this homology between O1 and O139 may well have provided a point of recombination between the various polysaccharide gene clusters (see below). As yet no other *rfa* genes have been reported for *V. cholerae*. The core structure of *V. cholerae* has not been unequivocally determined, possibly due to a lack of (deep) rough mutants which could simplify sugar analyses. It is, however, known that the *V. cholerae* O1 core contains 2-keto-deoxyoctonic acid (KDO) and other sugars such as fructose, heptose, ethanolamine, and *N*-acetyl glucosamine.[59]

Virulence of V. cholerae O1 and O139 O-antigen (rfb) mutants

The LPS represents the most important protective antigen in a number of Gram-negative organisms, including *V. cholerae* O1 and O139.[19,40,60] A variety of *rfb* mutants have been studied in both *V. cholerae* O1 and O139. In *V. cholerae* O1 perturbation of the outer membrane occurs in *rfb* mutants, leading to a change in the composition of the major outer membrane proteins, reduced motility possibly due to an effect on the sheathed flagellum, and trapping in the periplasmic space of the major subunit of an important adhesin known as the toxin coregulated pilus (TCP).[61,62] It is thought that it is these effects that lead to *rfb* strains being avirulent. Similarly, mutants lacking the O-antigen in *V. cholerae* O139 also show reduced virulence.[63,64] Interestingly, unlike *V. cholerae* O1, which is nonencapsulated, in *V. cholerae* O139 two distinct types of surface polysaccharide mutants have been isolated: mutants lacking only the capsule and those lacking both the capsular and the O-antigen polysaccharides. Both of these mutant types show increased serum sensitivity and reduced virulence in the infant mouse cholera model.[63,64]

Insertion sequences associated with the rfb region in Vibrio cholerae

An element designated IS*1358* was identified as being associated with the *rfb* region *of V. cholerae* O1 and O139 as well as the *rfb* region of *V. anguillarum* O1.[6,58] Furthermore, this element is widespread in non-O1 *V. cholerae* strains. Screening of *V. cholerae* serogroups O1 to O155 has revealed that approximately 30% of all strains contain this or a closely-related element, and although it has been found in the related *V. anguillarum*, it has not been found in the more distantly related species, such as *V. parahaemolyticus* and *V. fluvialis*. Upon screening *V. cholerae* O139 it was found that *rfaD* (see above) and a region homologous to IS*1358*d1 was present.[6,40] IS*1358* has several features common to insertion sequences and is similar to IS*10* and IS*50*. It has inverted repeats at its ends (17 bp in length), a small inverted repeat spanning the ribosome binding site, and an intact transposase gene (*tnpA*).[65,66] Furthermore, IS*1358* shows homology to the H-repeat of the RHS element in *E. coli*[6,67,68] and an insertion sequence IS*AS1* in *Aeromonas salmonicida*.[69] Interestingly, an H-repeat element has been found associated with *Salmonella enterica rfb* region, and it has been proposed to play a role in recombination between different *rfb* operons.[70] The H-repeat present as part of the RHS element in *E. coli* has also been proposed to play a role in large chromosomal rearrangements.[67,68] Initial investigations have shown that IS*1358* and IS*1358*d1 are flanked by a 17-bp inverted repeat. Furthermore, sequencing out of the ends of IS*1358* in *V. cholerae* O139 has shown that the right-hand inverted repeat is part of two 31-bp repeats. The role of these direct 31-bp repeats and an additional 8-bp repeat following the two 31-bp repeats has not yet been elucidated, but perhaps these repeats are remnants of transposition and imprecise recombination events which occurred during the formation of *V. cholerae* O139 *rfb* region.[71,72]

Although, the IS*1358* element found in *V. cholerae* O139 appears to be intact and have all the features of a functional IS element, it has not yet been possible to demonstrate transposition. However, the widespread nature of this element and the multiple copies in some strains would suggest that these elements are genetically mobile.[58] Further studies of these elements may provide insights into horizontal gene transfer and genetic evolution of O-antigen specificities in Gram-negative bacteria.

Surface polysaccharides of Vibrio cholerae O139

From the first appearance of the *V. cholerae* O139 Bengal serogroup, numerous studies have attempted to identify the differences between the O1 and O139 serogroups.[4,40,73,74] *V. cholerae* O139 Bengal strains appear to closely resemble *V. cholerae* O1 El Tor strains in most characteristics.[63,75-77] However, without a doubt the most dramatic difference, possibly selected for by immune pressure in the population, was the acquisition of the ability to synthesize a new O-antigen and capsule and the associated novel sugars.[40,78] The additional ability to synthesize a capsule has been linked with increased serum resistance.[78-80] Detailed sequence analyses have now revealed that in O139 strains there appears to have been a relatively precise replacement of the original *V. cholerae* O1 *rfb* region with novel DNA.[40,71,72,80,81]

V. cholerae O139 strains produce two colonial morphologies, either opaque or translucent. This phenotype correlates with the capsule status of *V. cholerae* O139: opaque colonies are encapsulated, whereas translucent colonies are nonencapsulated.[73] Using silver stain analysis after SDS–PAGE, the LPS of O139 *V. cholerae* is markedly different from that of O1 and can best be described as semi-rough[40] (Figure 6.5). Whereas the *V. cholerae* O1 LPS is smooth with long O-antigen chains attached to the lipid A-core, *V. cholerae* O139 appears to have only a single sugar repeat on a more completely substituted core.[82] Furthermore, *V. cholerae* O139 has been reported to have other electrophoretic forms of the surface polysaccharide.[78] There appears to be a rapid migrating form corresponding to the O-antigen/lipid A-core, and a medium migrating form and slow migrating forms which are thought to be capsular material since these forms are not seen in translucent colonies using western blot analysis.[78] This marked change in the LPS suggested the possibility that *V. cholerae* O139 had lost the original LPS genes from O1 and acquired a new set.[40] Probing with the regions for perosamine, tetronate biosynthesis, and O-antigen transport from *V. cholerae* O1 (see above) showed that these genes were missing from *V. cholerae* O139.[6,40] Work in a number of laboratories has now led to the characterization of the novel O139 DNA.[8,30,71,72,80,81,83,85] However, the O139 LPS/capsular region still has *rfaD* linked to it which is identical to the *V. cholerae* O1 *rfaD* gene and has an intact IS*1358* (see above).

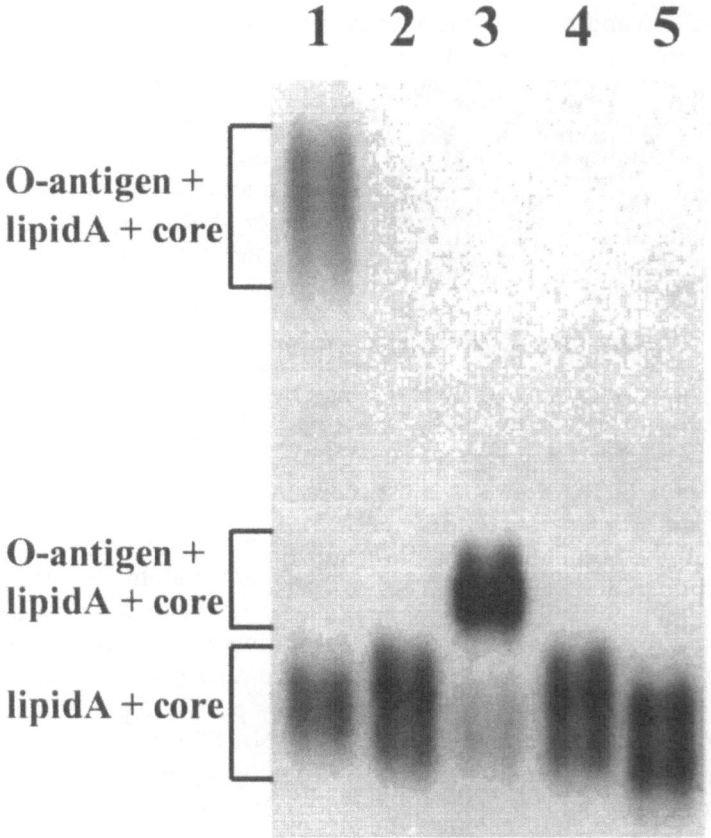

Figure 6.5 Detection of LPS in *V. cholerae* O1 and O139. LPS samples were prepared by treatment of whole-cell lysates with proteinases K, subjected to SDS–PAGE on a 20% polyacrylamide gel, and silver stained. Lanes 1 to 5 correspond to: 1: *V. cholerae* O1 strain O17, 2: *V. cholerae* O1 strain O17 *rfb* rough mutant, 3: *V. cholerae* O139 strain AI-1837, 4: *V. cholerae* O139 strain AI-1837 *rfb* (rough) mutant, 5: *V. cholerae* strain AI-1837 *rfb/rfa* (rough) mutant. The second type of *rfb* mutant in AI-1837 appears to be a deep rough strain and may be an *rfa* mutant. The position of the lipid A core and the O-antigen linked to the lipid A core are indicated.

The sugars found in the LPS of *V. cholerae* serogroup O139 are primarily colitose, glucose, heptose, fructose, glucosamine, and quinovosamine.[86-88] Some of these sugars are found not only in the O-antigen but also in the core. It would appear that the O-antigen and the capsular material are essentially the same except that the single O-antigen is attached to the lipid A-core.[78] The presence of quinovosamine in O139 suggests that it may cap the core, making its synthesis essentially an *rfa* function, and so the same may hold true for O1 (see above). However, quinovosamine synthesis would appear to be encoded in the O1 *rfb* region since it is present in the LPS of *E. coli* K-12 harboring the cloned O1 *rfb* region.[16] It would thus seem likely that the minimal 20kb *Sac*I region contains all the genes required for the expression of the complete O-antigen in *E. coli* K-12. Colitose is 3,6 dideoxy-L-galactose and has not previously been reported in *V. cholerae*, although it is found in the LPS of a number of other enteric bacterial species, such as *Salmonella enterica* and *E. coli* O111.[70,84]

Genetics of Vibrio cholerae *O139 capsule and O-antigen biosynthesis*

The region involved in the biosynthesis of the *V. cholerae* O139 surface polysaccharide is complex and composed of several independent transcriptional units (Figure 6.2). Mutational analysis of this region has clearly shown that some genes are involved in both capsule and O-antigen biosynthesis as well as genes specific for either capsule or O-antigen.[6,64,71,72,78,80,83] The first described region contained two ORFs designated OtnA and OtnB.[81] The OtnA protein shows homology to proteins for O-antigen chain length regulation (Cld or Rol) from a number of bacteria, whereas OtnB is homologous to KspD1 involved in capsule transport in *E. coli* (Table 6.2).[81] Mutations in both *otnA* and *otnB* appear to affect capsule but not O-antigen biosynthesis.[83] The remaining region between IS*1358* and *rfaD* has been reported to contain *otnD*, *otnE*, *otnF*, *otnG*, and *otnH*, none of which show any homology to polysaccharide biosynthesis genes, and their roles, if any, are unknown (Figure 6.2; Table 6.2).[83] The *rfaD* gene of *V. cholerae* O139 Bengal, which lies at the end of the *otn* region, is 99.7% identical to that of *V. cholerae* O1, which is not surprising since O139 and O1 appear to have an identical core.[82] The region downstream of *rfaD* also appears to be identical between the two serogroups, indicating that *rfaD* defines one end of novel O139 DNA.[6,83] However, in the O139 Argentinian isolate[5] there are not only differences within *rfaD* but also downstream.[6,71,72] Furthermore, the exact point of fusion between O139 DNA and *rfaD* is also different. This variation probably reflects the alternate parentage of the O139 Argentinian isolate, while the different fusion point could indicate an independent genesis for the two O139 strains.[6,71] Bik and co-workers[83] also reported an ORF downstream from IS*1358* which shows a high degree of homology to RfbD, the oxidoreductase from *V. cholerae* O1, clearly indicating polysaccharide genes on both sides of IS*1358* in *V. cholerae* O139 (Table 6.2). Comstock et al.[85] have also identified a number of other ORFs which show homology to polysaccharide biosyn-

Table 6.2 Possible Functions of *V. cholerae* O139 *rfb*/Capsule Genes

Old ORF designation	New ORF designation	G+C content	No. of a.a.	Putative function	Homology to	Accession number
RfaD	GmhD	46	314	core biosynthesis	*Vibrio cholerae* O1 RfaD	X59554
OtnH	WbfA	34.5	505	unknown	*Vibrio cholerae* O141 OfoA	
OtnG	WbfB	48.8	730	unknown	*Escherichia coli* K-12 o698	U00006
OtnF	WbfC	49.1	228	unknown	*Escherichia coli* K-12 o245	U00006
OtnE	WbfD	45.7	229	unknown	*Escherichia coli* K-12 o222	U00006
OtnD	WbfE	44.7	196	unknown	no homolog	
OtnA	Wzm	47.3	911	O-antigen export	*Escherichia coli* K1 KpsD	M64977
OtnB	Wzz	45.9	335	O-antigen chain length regulation	*Salmonella enterica* Cld	Z17278
IS1358	IS1358	42.9	375	insertion sequence	*Vibrio cholerae* O1 IS1358d1	X59554
ORF41.9/RfbD	Gmd	46.6	372	oxidoreductase	*Vibrio cholerae* O1 RfbD	X59554
ORF34.6	WbfH	43.6	308	colitose biosynthesis	*Escherichia coli* O111 *rfb* ORF6.7	U13629
ORF43.9	WbfI	41.4	390	colitose biosynthesis	*Escherichia coli* O111 *rbf* ORF7.7	U13629
ORF18.8	WbfJ	35.2	161	colitose biosynthesis	*Escherichia coli* O111 *rbf* ORF3.4	U13629
ORF54.4	WbfK	34.4	478	unknown	no homolog	
ORF40.0	WbfL	30.2	338	unknown	no homolog	

Old ORF	New name	MW	Size	Function	Homolog	Accession
ORF40.1	WbfM	29.3	337	unknown	no homolog	
ORF56.2	WbfN	30.2	485	unknown	no homolog	
ORF39.2	WbfO	29.7	337	Galactosyl transferase	*Neisseria gonorrheae* LgtD	U14554
ORF35.9	WbfP	33.9	310	Galactosyl transferase	*Erwinia amylovora* AmsB	X77921
ORF41.8/ORF1	WbfQ (Wzy)	28	354	Rfc like (O-antigen polymerase)	no homolog	
ORF71.9/ORF2	WbfR	43.1	636	Asparagine synthetase	Wide number of species	
ORF41.3/ORF3	WbfS	41.3	377	Galactosyl transferase	*Proteus morbilis* CpsF	L36823
ORF50.8/ORF4	ManC	46.8	460	PMI/GMP activity	*Salmonella enterica* RfbM	M84642
ORF5	ManB	46.2	483	PMM activity	*Salmonella enterica* RfbK	X56793
ORF6	WbfT	41.6	328	UDP-galactose 4-epimerase	*Neisseria gonorrheae* GalE	Z21508
ORF7	WbfU	42.7	186	Galactosyl transferase	*Streptococcus agalactiae* CpsD	L09116
ORF8	WbfV	45.5	388	Nucleotide sugar dehydrogenase	*Escherichia coli* O111 ORF1 near *rfb*	Z17241
ORF9	WbfW	44.4	334	Nucleotide sugar epimerase	*Escherichia coli* O111 ORF2 near *rfb*	Z17241
ORF10	WbfX	44.6	227	unknown	no homolog	
ORF11	WbfY	46.6	378	unknown	*Vibrio cholerae* RfbU	Y07788

Only the best homologs are shown. The old ORF designation, are those used in the original references. The new nomenclature is based on that described by Reeves et al.[7] Data in this table have been extracted in part from Bik et al.[83] and Comstock et al.[85] Data for the ORFs from IS1358 to ORF50.8 are derived from our laboratory.[71,72]

thetic proteins, as well as defining the other end point of novel O139 *V. cholerae* DNA. Interestingly, the differences between O1 and O139 mean that precisely all of the *V. cholerae* O1 *rfb* region has been deleted, including the additional recently described *rfb* genes, with the end point directly after RfbW[23] (Figure 6.2). This suggests the precise nature of the recombination events occurring at the extremities of this region, or alternatively it could indicate an incompatibility between the incoming O139 capsule/O-antigen genes and those which were already present in *V. cholerae* O1.[40]

Unlike the *V. cholerae* O1 *rfb* operon, the O139 region does not appear to be made up of clearly defined regions corresponding to specific biosynthetic pathways (Figure 6.2; Table 6.2). The *V. cholerae* O139 polysaccharide biosynthetic cluster clearly has a number of redundancies in that there are at least two possible galactosyl transferases of differing classes and a number of UDP-galactose 4-epimerases involved in the synthesis of UDP-galactose (Table 6.2). The revised *V. cholerae* O1 *rfb* region described by Fallarino et al.[23] contains a galactosyl transferase which shows 68% identity to ORF7 of *V. cholerae* O139. A mutation in this gene in *V. cholerae* O1 affects O-antigen synthesis[23] and this may well be the case in O139. Thus, together with the ORFs thought to be homologous to GalE (ORF6 and ORF9) in *V. cholerae* O139, it is possible that the UDP-galactose produced by these enzymes is picked up by one of the putative galactosyl transferases (ORF3 and ORF7) and transferred to the lipid carrier, bactoprenol.

The region encompassing ORF36.6, ORF43.9, and ORF18.8 is most likely involved in colitose biosynthesis since these ORFs show good homology to three genes in the *E. coli* O111 *rfb* region[71,90] to which no function has been defined as yet. Another region to which a function can be assigned is ORF4 (ORF50.8) and ORF5. These proteins are involved in the synthesis of GDP-mannose from fructose-6-phosphate similar to that described for *V. cholerae* O1 RfbA and RfbB (see above). ORF4 is likely to have PMI and GMP activity, whereas ORF5 is the PMM.[85] GDP-mannose is required for the biosynthesis of not only perosamine in *V. cholerae* O1 but also colitose in *V. cholerae* O139.[30,71]

Interestingly, *V. cholerae* O139 has an *rfc* homolog (ORF41.8/ORF1). The Rfc protein is involved in O-antigen polymerization and is not usually found in bacterial strains which have a homopolymer as their O-antigen, such as *V. cholerae* O1.[30,71] Furthermore, in *V. cholerae* O139 there would not appear to be a need for an Rfc if only a single O-antigen subunit sugar is substituted on the core. Thus, the *rfc* found in *V. cholerae* O139 is either nonfunctional or is involved in the polymerization of the capsule which is thought to be composed of material identical to the O-antigen but not linked to lipid A-core.[78] This may be possible since an Rol/Cld homolog is also found in O139 (*otnB*) which probably interacts with the *rfc* product. This is similar to the situation seen with the group I capsular K antigens in *E. coli* O8 and O9.[91]

The *rfb* region of *V. cholerae* O139 contains only one partner of the two-component polysaccharide export system. Unlike *V. cholerae* O1, which has *rfbH* and *rfbI* (described above), O139 contains only the transmembrane protein[81] (*otnA*) but no energizing partner. Mutations in *otnA* eliminate capsule production indicating that *otnA* is an essential gene in *V. cholerae* O139 capsule biosynthesis.[81] It is thus likely that one of the yet unidentified ORFs in this region acts as the energized partner or that it lies somewhere else on the chromosome.

The *V. cholerae* O139 capsule/*rfb* region is not only more complex than that of *V. cholerae* O1 *rfb* but also appears to be less well organized into defined biosynthetic regions. This suggests that the evolution of this region may have occurred via the acquisition of genes from a number of different ancestors to generate the new serogroup.

Evolution of V. cholerae O139 polysaccharide genes

A number of studies have shown that the *V. cholerae* O139 Bengal strains are closely related to *V. cholerae* O1 El Tor. Furthermore, all of the O-antigen biosynthetic genes of the O1 serotype have been lost and replaced with novel O139 surface polysaccharide genes. Bik and colleagues[82,84,91] have shown that at least some of the DNA associated with O139 surface polysaccharide biosynthesis is also found in non-O1 *V. cholerae* serogroups. In addition, work carried out by Comstock et al.[80] has shown that some of the regions downstream of IS*1358* are found in non-O1 and non-cholera *Vibrios*.

The *otn* region is found in part in *V. cholerae* O69 and O141. Furthermore, the *otnAB* region was found associated with IS*1358* in serogroup O69 but not O141.[81,83] Sequence analysis of the *otn* regions of *V. cholerae* O69 and O141 has shown that the O139 DNA has not been directly acquired from these serogroups.[83] Perhaps the O139 Bengal isolate arose from the strain isolated in Argentina, since the evidence is strong that at least the *otn* region is from a *Vibrio* source.[5,81] It has been shown that the exact end point in the vicinity of *rfaD* is different in the two O139 isolates, but this does not rule out the possibility that the Argentinian strain was the ancestor of at least the O139 Bengal surface polysaccharide genes, as opposed to being an example of convergent evolution.

The region downstream of IS*1358* contains an ORF (ORF41.9) which is approximately 80% identical to the *rfbD* gene of *V. cholerae* O1[71,83] (Table 6.2). Southern hybridization with a probe to ORFs 9 to 10 described by Comstock et al.[80] has shown that homologous DNA is present in not only a wide number of non-O1 strains but is also found in a variety of non-cholera-*Vibrios*. A probe to ORF71x8 only hybridizes to *Vibrio damsela* but not to any non-O1 *V. cholerae* strains. Interestingly, insertions in both of these ORFs causes the loss of O-antigen.[80]

Thus, it would appear that the *V. cholerae* O139 surface polysaccharide genes have their origin within the *Vibrionaceae*, but the exact mechanism by which these genes are transferred is unknown. The loss of the recipient *rfb* region can be more readily explained by reciprocal recombination rather than incompatibility between resident and incoming genes. However, the

mode of transfer of DNA is completely unknown, but it has been specu-
lated that the DNA transfers via conjugation since many Bengal isolates
carry a large conjugative plasmid.[93] Alternatively, one of the many phages
found in *V. cholerae* could have transduced the specific regions, which then
became incorporated. The insertion of the donor DNA is likely to involve
homologous recombination at conserved sites such as *rfaD*, IS*1358* and the
region found downstream of *rfbW* in *V. cholerae* O1. An alternative possibility is that the insertion is mediated by IS*1358*, since an element similar
to IS*1358* has been shown to transpose in *Aeromonas salmonicida*.[69] There is
now evidence in *V. anguillarum* that IS*1358* exists as multiple copies, perhaps as many as 20 per chromosome, which would strongly indicate a
mobile nature for this element.[58] No matter what the actual events prove
to be, it is clear that there is tremendous potential for the reassortment of
genes involved in the synthesis of surface polysaccharides to generate new
serogroup specificities.

Conclusion

Many years of research have gone into the elucidation of the *V. cholerae* O1
lipopolysaccharide and, although the structure has been determined and the
mechanism of Inaba–Ogawa serotype switching solved, there are still many
unanswered questions. Much is known about the genes in the *rfb* operon of
the O1 and O139 serogroups and their organization, but little or nothing is
known about how the various Rfb proteins interact and how the LPS is
exported to the cell surface. With the emergence of the *V. cholerae* serogroup
O139, still more questions arise as to how new serogroups evolve, and what
the role of IS*1358* is in this process, and whether some of the other serogroups
have arisen by a similar mechanism to *V. cholerae* O139. It is worth noting
that new cholera-related non-O1, non-O139 serotypes continue to be isolated.
Thus the question arises: are the various known serogroups simply the result
of recombination events from a large pool which have successfully recom-
bined to give a new functional set of O-antigen/polysaccharide genes? Only
a wider and more detailed analysis will provide clues.

Acknowledgments

The authors wish to gratefully acknowledge that the research from their
laboratory presented herein has been supported by grants from the Austra-
lian Research Council, the Clive and Vera Ramaciotti Foundations, and the
Global Vaccines Programme of the Diarrhoeal Diseases Research Committee
of WHO.

References

1. Fuerst, J.A. and J.W. Perry. Demonstration of lipopolysaccharide on sheathed flagella of *Vibrio cholerae* O1 by Protein A-gold immunoelectron microscopy. *J. Bacteriol.* 170, pp. 1488-1494, 1988.
2. Barua, D. History of cholera, in *Cholera,* Barua, D. and W.B. Greenough, III (Eds.) Plenum. New York, pp. 1-36, 1992.
3. International Center for Diarrheal Disease Research, Bangladesh, Cholera Working Group. Large epidemic of cholera-like disease in Bangladesh caused by *Vibrio cholerae* O139 synonym Bengal. *Lancet* 342, pp. 387-390, 1993.
4. Ramamurthy, T., S. Garg, R. Sharma, S.K. Bhattachary, G.B. Nair, T. Shimada, T. Takeda, T. Karasawa, H. Kurazano, A. Pal, and Y. Takeda. Emergence of novel strain of *Vibrio cholerae* with epidemic potential in southern and eastern India. *Lancet* 341, pp. 703-704, 1993.
5. Rivas, M., C. Toma, E. Miliwebsky, M.I. Caffer, M. Galas, P. Varela, M. Tous, A.M. Bru, and N. Binsztein. Cholera isolates in relation to the "eighth pandemic." *Lancet* 342, pp. 926-927, 1993.
6. Stroeher, U.H., K.E. Jedani, B.K. Dredge, R. Morana, M.H. Brown, L.E. Karageorgos, J. M. Albert, and P.A. Manning. Genetic rearrangement of the *rfb* regions of *Vibrio cholerae* O1 and O139. *Proc. Natl. Acad. Sci. U.S.A.* 92, pp. 10374-10378, 1995.
6a. Shimada, T., G.B. Nair, B.C. Deb, M.J. Albert, R.B. Sack, and Y. Takeda. Outbreak of *Vibrio cholerae* non-O1 in India and Bangladesh. *Lancet* 341, pp. 1347, 1993.
7. Reeves, P.R., M. Hobbes, M.A. Valvano, M. Skurnik, C. Whitfield, D. Coplin, N. Kido, J. Klena, D. Maskell, C.R.H. Raetz, and P.D. Rick. Bacterial polysaccharide synthesis and gene nomenclature. *Trends Microbiol.* 4, pp. 495-503, 1996.
8. Stroeher, U.H. and P.A. Manning. *Vibrio cholerae* serotype O139: Swapping genes for surface polysaccharide biosynthesis. *Trends Microbiol.* 5, pp. 178-180, 1997.
9. Burrows, W., A.N. Mathers, V.G. McGann, and S.M. Wagner. Studies on immunity to asiatic cholera. Part I: Introduction. *J. Infect. Dis.* 79, pp. 159-167, 1946.
10. Sakazaki, R. and K. Tamura. Somatic antigen variation in *Vibrio cholerae. Jpn. J. Med. Sci. Biol.* 24, pp. 93-100, 1971.
11. Redmond, J.W. The structure on the O-antigenic side chain of the lipopolysaccharide of *Vibrio cholerae* 569B (Inaba). *Biochem. Biophys. Acta.* 584, pp. 346-352, 1979.
12. Nobechi, K. Immunological studies upon the types of *Vibrio cholerae. Sci. Rep. Inst. Infect. Dis. Tokyo Uni.* 2, pp. 43, 1923.
13. Gangarosa, E. J., A. Sonati, H. Saghari, and J.C. Feely. Multiple serotypes of *Vibrio cholerae* from a case of cholera. *Lancet* 1, pp. 646-648, 1967.
14. Kenne, L., B. Lindberg, P. Unger, B. Gustafson, and T. Holme. Structural studies of the *Vibrio cholerae* O-antigen. *Carbohydr. Res.* 100, pp. 341-349, 1982.
15. Kenne, L., B. Lindberg, P. Unger, T. Holme, and J. Holmgren. Structural studies of the *Vibrio cholerae* O-antigen. *Carbohydr. Res.* 68, pp. C14-C16, 1979.
16. Hisatsune, K., S. Kondo, T. Iguchi, T. Ito, and K. Hiramatsu. Lipopolysaccharide of *Escherichia coli* K-12 strains that express cloned genes for the Ogawa and Inaba antigens of *Vibrio cholerae* O1; Identification of O-antigenic factors, in *The Thirty-Second Joint Conference U.S.–Japan Cooperative Medical Science Program Cholera and Related Diarrheal Diseases Panel,* 1996. U.S.–Japan Cooperative Medical Science Program.

17. Stroeher, U.H., L.E. Karageorgous, R. Morona, and P.A. Manning. Serotype conversion in *Vibrio cholerae* O1. *Proc. Natl. Acad. Sci. U.S.A.* 89, pp. 2566-2570, 1992.

18. Hisatsune, K., S. Kondo, Y. Isshiki, T. Iguchi, and Y. Haishima. Occurrence of 2-O-methyl-N-(3-deoxy-L-*glycero*-tetronyl)-D-perosamine (4-amino-4, 6-dideoxy-D-manno-pyranose) in lipopolysaccharide from the Ogawa but not from Inaba O forms of O1 *Vibrio cholerae*. *Biochem. Biophys. Res. Commun.* 190, pp. 302-307, 1993.

19. Manning, P.A., M.W. Heuzenroeder, J. Yeadon, D.I. Leavesley, P.R. Reeves, and D. Rowley. Molecular cloning and expression in *Escherichia coli* K-12 of the O-antigen of the Ogawa and Inaba serotypes of the lipopolysaccharide of *Vibrio cholerae* O1 and their potential for vaccine development. *Infect. Immun.* 53, pp. 272-277, 1986.

20. Ward, H.M., G. Morelli, M. Kamke, R. Morona, J. Yeadon, J.A. Hackett, and P.A. Manning. A physical map of the chromosomal region determining O-antigen biosynthesis in *Vibrio cholerae* O1. *Gene* 55, pp. 197-204, 1987.

21. Ward, H.M. and P.A. Manning. Mapping of chromosomal loci associated with lipopolysaccharide biosynthesis and serotype specificity in *Vibrio cholerae* O1 by transposon mutagenesis using Tn5 and Tn*2680*. *Mol. Gen. Genet.* 218, pp. 367-370, 1989.

22. Bhaskaran, K. and R.H. Gorrill. A study of antigenic variation in *Vibrio cholerae*. *J. Gen. Microbiol.* 16, pp. 721-729, 1957.

23. Fallarino A., C. Mavrangelos, U.H. Stroeher, and P.A. Manning. Identification of additional genes required for O-antigen biosynthesis in *Vibrio cholerae* O1. *J. Bacteriol.* 179, pp. 2147-2153, 1997.

24. Stroeher, U.H., L.E. Karageorgos, R. Morona, and P.A. Manning. In *Vibrio cholerae* serogroup O1, *rfaD* is closely linked to the *rfb* operon. *Gene* 155, pp. 67-72, 1995.

25. Mäkelä, P.H. and B.A.D. Stocker. Genetics of lipopolysaccharide, in *Handbook of Endotoxin* Vol. 1., Rietschel, E.T. Ed., Elsevier, Amsterdam, pp. 59-137, 1984.

26. Austin, E.A., J.F. Graves, L.A. Hite, C.T. Parker, and C.A. Schnaitman. Genetic analysis of lipopolysaccharide core biosynthesis of *Escherichia coli* K-12: insertion mutagenesis of the *rfa* locus. *J. Bacteriol.* 172, pp. 5312-5325, 1990.

27. Hobbs, M. and P.R. Reeves. The JUMPstart sequence: a 39bp element common to several polysaccharide gene clusters. *Mol. Microbiol.* 12, pp. 855, 1994.

28. Arakawa, Y., R. Wacharotayanku, T. Nagatsuka, H. Ito, N. Kato, and M. Ohta. Genomic organization of the *Klebsiella pneumoniae cps* region responsible for serotype K2 capsular polysaccharide synthesis in the virulent strain Chedid. *J. Bacteriol.* 177, pp. 1788-1796, 1995.

29. Bailey, M.J.A., C. Hughes, and V. Koronakis. RfaH and *ops* element, components of a novel system controlling bacterial transcription elongation. *Mol. Microbiol.* 26, pp. 845-852, 1997.

30. Stroeher, U.H., L.E. Karageorgos, M.H. Brown, R. Morona, and P.A. Manning. A putative pathway for perosamine biosynthesis is the first function encoded within the *rfb* region of *Vibrio cholerae* O1. *Gene* 166, pp. 33-42, 1995.

31. Piggot, N.H., I.W. Sutherland, and T.R. Jarman. Enzymes involved in the biosynthesis of alginate by *Pseudomonas aeruginosa*. *Eur. J. Appl. Microbiol. Biotechnol.* 13, pp. 179-183, 1981.

32. Pugashetti, B.K., L. Vadas, H.S. Prihar, and D.S. Feingold. GDP-mannose dehydrogenase and biosynthesis of alginate-like polysaccharide in a mucoid strain of *Pseudomonas aeruginosa*. *J. Bacteriol.* 153, pp. 1107-1110, 1983.

33. Sa'-Correia, I., A. Darzins, S.-K. Wang, A. Berry, and A.M. Chakrabarty. Alginate biosynthetic enzymes in mucoid and nonmucoid *Pseudomonas aeruginosa*: overproduction of phosphomannose isomerase, phosphomannomutase, and GDP-mannose pyrophosphorylase by overexpression of the phosphomannose isomerase *(pmi)* gene. *J. Bacteriol.* 169, pp. 3224-3231, 1987.

34. Darzins, A., B. Frantz, R.I. Vangas, and A.M. Chakrabarty. Nucleotide sequence analysis of the phosphomannose isomerase gene *(pmi)* of *Pseudomonas aeruginosa* and comparison with the corresponding *Escherichia coli manA*. *Gene* 42, pp. 293-302, 1986.

35. May, T.B., D. Shinabarger, A. Boyd, and A.M. Chakrabarty. Identification of amino acid residues involved in the activity of phosphomannose isomerase-guanosine 5'-diphospho-D-mannose pyrophosphorylase. A bifunctional enzyme in the alginate biosynthetic pathway of *Pseudomonas aeruginosa*. *J. Biol. Chem.* 269, pp. 4822-4827, 1994.

36. Haishima, Y., S. Kondo, and K. Hisatsune. The occurrence of alpha (1–2) linked N-acetylperosamine-homopolymer in lipopolysaccharides of non-O1 *Vibrio cholerae* possessing an antigenic factor in common with O1 *V. cholerae*. *Microbiol. Immunol.* 34, pp. 1049-1054, 1990.

37. Yamasaki, S., K. Hoshino, T. Shimizu, S. Garg, T. Shimada, S. Ho, R.K. Bhadra, G.B. Nair, and Y. Takeda. Comparative analysis of the gene responsible for lipopolysaccharide synthesis *Vibrio cholerae* O1 and O139 and those of non-O1 non-O139 *Vibrio cholerae*, in *The Thirty-Second Joint Conference U.S.–Japan Cooperative Medical Science Program Cholera and Related Diarrheal Diseases Panel 1996*, U.S.–Japan Cooperative Medical Science Program.

38. Raetz, C.R.H. Biochemistry of endotoxins. *Annu. Rev. Biochem.* 59, pp. 129-1700, 1990.

39. Higgins, C.F., M.P. Gallagher, S.C. Hyde, M.L. Mimmack, and S.R. Pearce. Periplasmic binding protein-dependent transport systems: the membrane associated components. *Philos. Trans. R. Soc. London Ser. B.* 326, pp. 353-365, 1990.

40. Manning, P.A., U.H. Stroeher, and R. Morona. Molecular basis for O-antigen biosynthesis in *Vibrio cholerae* O1: Ogawa–Inaba switching, in, *Vibrio cholerae and Cholera: Molecular to Global Perspectives*, Wachsmuth, I.K., Blake, P.A., and Olsvik, Ø., (Eds.) American Society for Microbiology. Washington, D.C., pp. 77-94, 1994.

41. Manning, P.A., U.H. Stroher, L.E. Karageorgos, and R. Morona. Putative O-antigen transport genes within the *rfb* region of *Vibrio cholerae* O1 are homologous to those of capsule transport. *Gene* 158, pp. 1-7, 1995.

42. Zhang, L., A. al-Hendy, P. Toivanen, and M. Skurnik. Genetic organisation and sequence of the *rfb* gene cluster of *Yersinia enterocolitica* serotype O:3: similarities to the dTDP-rhamnose biosynthesis pathway of *Salmonella* and to the bacterial polysaccharide export systems. *Mol. Microbiol.* 9, pp. 309-321, 1993.

43. Hashimoto, Y., N. Li, H. Yokoyama, and T. Ezaki. Complete nucleotide sequence and molecular characterisation of ViaB region encoding Vi antigen in *Salmonella typhi*. *J. Bacteriol.* 175, pp. 4456-4465, 1993.

44. Frosch, M., C. Weisgerber, and T.F. Meyer. Molecular characterisation and expression in *Escherichia coli* of the gene complex encoding the polysaccharide capsule of *Neisseria meningitidis* groupB. *Proc. Natl. Acad. Sci. U.S.A.* 86, pp. 1169-1673, 1989.

45. Pavelka, Jr., M.S., L.F. Wright, and R.P.R. Silver. Identification of two genes *kpsM* and *kpsT*, in region 3 of the polysialic acid gene cluster of *Escherichia coli* K1. *J. Bacteriol.* 173, pp. 4603-4610, 1991.

46. Smith, A.N., G.J. Boulnois, and I.S. Roberts. Molecular analysis of the *Escherichia coli kps* locus: identification and characterisation of an inner-membrane capsular polysaccharide transport system. *Mol. Microbiol.* 4, pp. 1863-1869, 1990.

47. Kroll, J.S., B. Loynds, L.N. Brophy, and E.R. Moxon. The *bex* locus in encapsulated *Haemophilus influenzae*: a chromosomal region involved in capsule polysaccharide export. *Mol. Microbiol.* 4, pp. 1853-1862, 1990.

48. Walker, J.E., M. Saraste, M.L. Runswick, and N.J. Gray. Distantly related sequences in the alpha- and beta-subunits of ATP synthase, myosin, kinases and other ATP-requiring enzymes and a common nucleotide binding fold. *EMBO J.* 1, pp. 9435-9451, 1982.

49. Macpherson, D.F., R. Morona, D.W. Beger, K.C. Cheah, and P.A. Manning. Genetic analysis of the rfb region of *Shigella flexneri* encoding the Y-serotype O-antigen specificity. *Mol. Microbiol.* 5, pp. 1491-1499, 1991.

49a. Macpherson, D.F., P.A. Manning, and R. Morona. Genetic analysis of *rfbX* of *Shigella flexneri. Gene*, 155, 9-17, 1995.

50. Morona, R., U.H. Stroeher, L.E. Karageorgos, M.H. Brown, and P.A. Manning. A putative pathway for biosynthesis of the O-antigen component, 3-deoxy-ʟ-*glycero*-tetronic acid, based on the sequence of the *Vibrio cholerae* O1 *rfb* region. *Gene* 166, pp. 19-31, 1995.

51. Bairoch, A. PROSITE: a dictionary of sites and patterns in proteins. *Nucleic Acids Res.* 20, pp. 2013-2018, 1992.

52. Conway, T. and L.O. Ingram. Similarity of *Escherichia coli* propanediol oxidoreductase (*fucO*) product and an unusual alcohol dehydrogenase from *Zymomonas mobilis* and *Saccharomyces cerevisiae. J. Bacteriol.* 171, pp. 3754-3759, 1989.

53. Kessler, D., I. Leibrecht, and J. Knappe. Pyruvate-formate-lyase-deactivase and acetyl-CoA reductase activity of *Escherichia coli* resides on a polymeric protein particle encoded by *adhE. FEBS Lett.* 281, pp. 59-63, 1991.

54. Hale, R.S. and P.F. Leadlay. Oligonucleotide probes for bacterial acylcarrier protein genes. *Biochemie* 6, pp. 835-839, 1985.

55. Chang, S.I. and G. Hammes. Amino acid sequences of substrate-binding sites in chicken liver fatty acid synthase. *Biochemistry* 27, pp. 4753-4760, 1988.

56. Morona, R., M.S. Matthews, J.K. Morona, and M.H. Brown. Regions of the cloned *Vibrio cholerae rfb* genes needed to determine the Ogawa form of the O-antigen. *Mol. Gen. Genet.* 224, pp. 405-412, 1990.

57. Fields, P., D.N. Cameron, and P.A. Manning. The molecular basis for Ogawa–Inaba serotype shift in *Vibrio cholerae* O1, in *The Twenty-Ninth Joint Conference U.S.–Japan Cooperative Medical Science Program Cholera and Related Diarrheal Diseases Panel, 1993.* U.S.–Japan Cooperative Medical Science Program, pp. 63-65, 1993.

58. Jedani, K., U.H. Stroeher, and P.A. Manning, submitted for publication.

59. Brade, H. Occurrence of 2-keto-deoxyotonic 5-phosphate in lipopolysaccharide of *Vibrio cholerae* Ogawa and Inaba. *J. Bacteriol.* 161, pp. 795-798, 1985.

60. Neoh, S.H. and D. Rowley. Protection of infact mice against cholera by antibodies to three antigens of *Vibrio cholerae. J. Infect. Dis.* 126, pp. 41-47, 1972.

61. Iredell, J.R., U.H. Stroeher, H.M. Ward, C.J. Thomas, and P.A. Manning. Virulence of *rfb* mutants, defective in the synthesis of the O-antigen of the lipopolysaccharide of *Vibrio cholerae* O1. *FEMS Immunol. Med. Microbiol.* 20, pp. 45-52 1998.

62. Iredell, J.R. and P.A. Manning. Outer membrane translocation arrest of the TcpA pilin subunit in *rfb* mutants of *Vibrio cholerae* O1 strain 569B. *J. Bacteriol.* 179, pp. 2038-2046, 1997.

63. Waldor, M.K. and J.J. Mekalanos. Emergence of a new cholera pandemic: molecular analysis of virulence determinants in *Vibrio cholerae* O139 and development of a live vaccine prototype. *J. Inf. Dis.* 170, pp. 278-283, 1994.

64. Fazeli, A., S.R. Attridge, U.H. Stroeher, and P.A. Manning, submitted for publication.

65. Davis, M.A., R.W. Simons, and N. Kleckner. Tn*10* protects itself at two levels from fortuitous activation by external promoters. *Cell* 43, pp. 379-387, 1985.

66. Schulz, V.P. and W.S. Reznikoff. Translation initiation of IS*50*R read-through transcript. *J. Mol. Biol.* 221, pp.65-80, 1991.

67. Zhao, S., C.H. Sandt, G. Feuler, D.A. Valzany, J.A. Gray, and C.W. Hill. *Rhs* elements of *Escherichia coli* K-12: Complex composites of shared and unique components that have different evolutionary history. *J. Bacteriol.* 175, pp. 2799-2808, 1993.

68. Hill, C.W., C.H. Sandt, and D.A. Vlazny. *Rhs* elements of *Escherichia coli*: a family of genetic composites each encoding a large mosaic protein. *Mol. Microbiol.* 12, pp. 865-871, 1994.

69. Gustafson, C.E., S. Chun, and T.J. Trust. Mutagenesis of the paracrystalline surface protein array of *Aeromonas salmonicida* by endogenous insertion sequence. *J. Mol. Biol.* 237, pp. 452-463, 1994.

70. Xiang, S.-H., M. Hobbs, and P.R. Reeves. Molecular analysis of the *rfb* gene cluster of a group D2 *Salmonella enterica* strain: evidence for its origin from an insertion sequence mediated recombination event between group E and D1 strains. *J. Bacteriol.* 175, pp. 4357-4365, 1994.

71. Stroeher, U.H., G. Parasivam, B.K. Dredge, and P.A. Manning. Novel *Vibrio cholerae* O139 genes involved in lipoloysacchride biosynthesis. *J. Bacteriol.* 179, pp. 2740-2747 1997.

72. Stroeher, U.H., G. Parasivam, A. Fallerino, B.K. Dredge, K. Jedani, and P.A. Manning. LPS and capsule genes of *Vibrio cholerae* O1 and O139, in *Cytokines, Cholera and the Gut*, Keusch, G. and M. Kawakami (Eds.) IOS Press, Burke, VA, pp. 235-242, 1997.

74. Karaolis, D.K., R. Lan, and P.R. Reeves. Molecular evolution of the seventh pandemic clone of *Vibrio cholerae* and its relationship to other pandemic and epidemic *V. cholerae* isolates. *J. Bacteriol.* 176, pp. 6199-6206, 1994.

75. Johnson, J.A., C.A. Salles, P. Panigrahi, M.J. Albert, A.C. Wright, R.J. Johnson, and J.G. Morris Jr., *Vibrio cholerae* O139 synonym Bengal is closely related to *Vibrio cholerae* El Tor but has important differences. *Infect. Immun.* 62, pp. 2108-2110, 1994.

76. Morris, Jr., J.G. and the Cholera Laboratory Task Force. *Vibrio cholerae* O139 Bengal, in *Vibrio cholerae and Cholera: Molecular to Global Perspectives*. Wachsmuth, K., P.A. Blake, and Ø. Olsvik (Eds.). American Society for Microbiology. Washington, D.C., pp. 95-102, 1994.

77. Faruque, S.M., A. R. M. A. Alim, S.K. Roy, F. Khan, B.G. Nair, R.B. Sack, and M.J. Albert. Molecular analysis of rRNA and cholerae toxin genes carried by the new epidemic strain of toxigenic *Vibrio cholerae* O139 synonym Bengal. *J. Clin. Microbiol.* 32, pp. 1050-1053, 1994.

78. Waldor, M.K., R. Colwell, and J.J. Mekalanos. The *Vibrio cholerae* O139 serogroup antigen includes an O-polysaccharide capsule and lipopolysaccharide virulence determinant. *Proc. Natl. Acad. Sci. U.S.A.*, 91, pp. 11388-11392, 1994.

79. Johnson, J.A., P. Panigrahi, and J.G. Morris, Jr. Non-O1 *Vibrio cholerae* NRT36S produces a polysaccharide capsule that determines colony morphology, serum resistance and virulence in mice. *Infect. Immun.* 60, pp. 864-869, 1992.

80. Comstock, L.E., D. Maneval, Jr., P. Panigrahi, A. Joseph, M.M. Levine, J.B. Kaper, J.G. Morris Jr., and J.A. Johnson. Capsule and O antigen in *Vibrio cholerae* O139 Bengal associated with a genetic region not present on *Vibrio cholerae* O1. *Infect. Immun.* 63, pp. 317-323, 1995.

81. Bik, E.M., A.E. Bunschoten, R.D. Gouw, and F.R. Mooi. Genesis of the novel epidemic *Vibrio cholerae* O139 strain: evidence for horizontal transfer of genes involved in polysaccharide synthesis. *EMBO J.* 14, pp. 209-216, 1995.

82. Weintraub, A., personal communication.

83. Bik, E.M., A.E. Bunschoten, R.J.L. Willems, A.C.Y. Chang, and F.R. Mooi. Genetic organization and functional analysis of the *otn* DNA essential for cell-wall polysaccharide synthesis in *Vibrio cholerae* O139. *Mol. Microbiol.* 20, pp. 799-811, 1996.

84. Bastin, D.A., L.K. Romana, and P.R. Reeves. Molecular cloning and expression in *Escherichia coil* of the *rfb* gene cluster determining the O-antigen of an *E. coli* O111 strain. *Mol. Microbiol.* 5, pp. 2223-2231, 1991.

85. Comstock, L.E., J.A. Johnson, J.M. Michalski, J.G. Morris Jr., and J.B. Kaper. Cloning and sequencing of a region encoding a surface polysaccharide of *Vibrio cholerae* O139 and characterisation of the insertion site in the chromosome of *Vibrio cholerae* O1. *Mol. Microbiol.* 19, pp. 815-826, 1996.

86. Hisastune, K., S. Kondo, Y. Isshiki, T. Iguch, Y. Kawamata, and T. Shimada. O-antigenic lipopolysaccharide of *Vibrio cholerae* O139 Bengal, a new epidemic strain for recent cholera in the Indian subcontinent. *Biochem. Biophys. Res. Commun.* 196, pp. 1309-1315, 1993.

87. Preston, L.M., Q, Xu, J.A. Johnson, A. Joseph, D.R. Maneval Jr., K. Husain, G.P. Reddy, C.A. Bush, and J.G. Morris Jr. Preliminary structure determination of the capsular polysaccharide of *Vibrio cholerae* O139 Bengal AI1837. *J. Bacteriol.* 177, pp. 835-838, 1995.

88. Knirel, Y.A., L. Paredes, P.-E. Jansson, A. Weintraub, and M.J. Albert. Structure of the capsular polysaccharide of *Vibrio cholerae* O139 synonym Bengal containing D-galactose 4,6-cyclophosphate. *Eur. J. Biochem.* 232, pp. 391-396, 1995.

89. Weintraub, A., G. Widmalm, P.-E. Jansson, M. Jannson, K. Hultenby, and M.J. Albert. *Vibrio cholerae* O139 Bengal possesses a capsular polysaccharide which may confer increased virulence. *Microb. Pathogen.* 16, pp. 235-241, 1994.

90. Bastin, D.A. and P.R. Reeves. Sequence analysis of the O antigen gene (*rfb*) cluster of *Escherichia coli* O111. *Gene* 164, pp. 17-23, 1995.

91. Dodgson, C., P. Amor, and C. Whitfield. Distribution of the *rol* gene encoding the regulator of lipopolysaccharide O-chain length in *Escherichia coli* and its influence on the expression of group I capsule K antigen. *J. Bacteriol.* 178, pp. 1895-1902, 1996.

92. Mooi, F. R. and E.M. Bik. The evolution of epidemic *Vibrio cholerae* strains. *Trends Microbiol.* 4, pp.161-165, 1997.

93. Nandy, R.K., T.K. Sengupta, S. Mukhopadhyay, and A.C. Ghose. A comparative study of the properties of *Vibrio cholerae* O139, O1 and other non-O1 strains. *FEMS Microbiol. Lett.* 136, pp. 175-180, 1996.

chapter seven

Common themes in the genetics of streptococcal capsular polysaccharides

Janet Yother

Contents

The significance of streptococcal capsules 161
Capsular polysaccharide structures and basic genetics 162
    Common genes and functions among the streptococcal
        capsule loci...163
    Transcription and mutation analyses 165
    Insertion elements, repetitive sequences, and cryptic genes..................167
The *S. pyogenes* hyaluronic acid and *S. pneumoniae* type 3 capsule loci....168
Group B type III, *S. pneumoniae* types 14 and 19F, *S. thermophilus*, and
    L. lactis capsules 172
Regulation 174
Virulence..175
The special case of *S. pneumoniae*: genetic exchange and
    capsule diversity 176
Future directions...178
Acknowledgments...178
References..179

The significance of streptococcal capsules

The streptococci represent a diverse group of Gram-positive bacteria that includes both medically and industrially important members. Many species elaborate capsular polysaccharides, while others may produce extracellular

0-8493-0021-5/99/$0.00+$.50
© 1999 by CRC Press LLC

polysaccharides that do not form distinct capsules. Studies of capsular polysaccharides of the streptococci have been central to the advancement of genetics, pathogenesis, and immunology. Griffith's 1928 description of the transformation of pneumococcal capsular polysaccharides in the mouse was the first observation of a mechanism of bacterial gene transfer.[1] Attempts to define the "transforming principle" ultimately led Avery, MacLeod, and McCarty to identify DNA as the hereditary material.[2] Other classic studies with *S. pneumoniae* demonstrated the essential nature of the capsule in virulence, antiphagocytosis, and immunity.[3-6] Its significance is further reflected in the fact that virulence properties of the organism, including infectivity in animals and the ability to resist phagocytosis, are related in part to the specific capsular polysaccharide that is produced (see Kelly et al. [Reference 7] and references therein). Likewise, invasive neonatal infections due to *Streptococcus agalactiae* (group B streptococci) are predominantly caused by strains representing a single capsular serotype (III),[8] and encapsulated strains of *Streptococcus pyogenes* (group A streptococci) occur more frequently among isolates obtained from invasive and acute rheumatic fever infections than from uncomplicated pharyngitis.[9] In the industrially important *Streptococcus thermophilus* and *Lactococcus lactis*, synthesis of an exopolysaccharide is critical for enhancing the texture of fermented milk products.

Recent years have seen an explosion in information concerning the genetic basis of capsule expression in the streptococci. Since the first molecular characterization of the *S. agalactiae* serotype III capsule locus in 1987,[10] the capsule loci of *S. pyogenes* and of multiple *S. pneumoniae* serotypes have been characterized, as have the related exopolysaccharide loci in *S. thermophilus* and *L. lactis*. All are chromosomally encoded, except in *L. lactis* where the genes are located on a mobilizable plasmid.[11] Many common features have emerged from these studies, and it is from this perspective that the recent findings in this field are reviewed. The foundations for the current work were laid in classic genetic studies, and much of the credit for our present knowledge must go to earlier workers who envisioned many of the basic concepts that have now been proven with the aid of modern molecular biology. The recent advances will be considered first, and will then be placed in context with the earlier work.

Capsular polysaccharide structures and basic genetics

The capsules of the pathogenic streptococci — *S. pyogenes*, *S. agalactiae*, and *S. pneumoniae* — have both structural and genetic similarities (Figure 7.1 and 7.2). *S. pyogenes* produces a hyaluronic acid capsule that is structurally similar to the type 3 capsule of *S. pneumoniae*. As will be described in detail below, the genetic loci and the general mechanisms associated with synthesis of these two polysaccharides are also closely related. The two organisms represent the extremes, however, in terms of diversity of capsular polysaccharides. Whereas hyaluronic acid (HA) is the only capsular structure known to be produced by *S. pyogenes*, 90 serologically distinct capsular polysaccharides have been identified among *S. pneumoniae* isolates.[12] The structures of approximately half of these polysaccharides have been determined, and they can be as simple as two sugars in a linear repeating structure (type 3, for example), or as complex as multiple sugars, linkages, and side chains (type 12, for example).[13] In *S. agalactiae*, nine capsular serotypes have been recognized.[14] Each of these structures is linear, with short side chains. Most contain Gal, Glc, and GlcNAc, and all have a terminal sialic acid residue in the side chain. In addition, the *S. agalactiae* type III polysaccharide is identical to that of the type 14 *S. pneumoniae*, except for the lack of sialic acid residues in the latter. Genetic studies have revealed similarity between the loci, and in fact the type 14 *S. pneumoniae* genes were first identified by hybridization with the cloned *S. agalactiae* sequences.[15] The capsular polysaccharides and capsule loci of other streptococci also show similarity to those just described, and indeed many common sequences have been found among the capsule genes in each of the streptococci (Figure 7.2).

Common genes and functions among the streptococcal capsule loci

Characterization of the *S. pneumoniae* type 3 locus revealed sequences that were present only in type 3 strains, as well as sequences that were present in strains of apparently all types.[16,17] The "common" sequence located downstream of the "type 3-specific" genes proved to be *plpA* (also referred to as *aliA*), a previously identified oligopeptide permease.[18] *plpA* has been identified through hybridization, linkage, and/or sequence analysis in strains of all serotypes examined.[18-22] It is not apparent that it has any role in capsule synthesis, as mutations have not been found to alter capsule production.[23] Although it has been suggested that *plpA* represents a truncated version of *aliA*,[21] these two genes are identical, and it is only in the unusual case of the type 3 locus that *plpA* is known to be truncated.[18,19]

Garcia et al. identified common sequences located upstream of the *S. pneumoniae* type 3-specific genes,[17] and Guidolin et al. characterized these in the sequence analysis of the type 19F locus.[24] They found that these sequences share homology with genes associated with the *S. agalactiae* type III capsule locus and are also present in other *S. pneumoniae* serotypes. Since that time, sequence analyses have confirmed the presence of these common genes upstream of multiple *S. pneumoniae* capsule loci, as well as those of other streptococci, lactococci, and with a lesser degree of homology, the staphylococci (Figure 7.2). For most, the organization of the genes is the same. In *L. lactis*, however, their order is variant.[11] For all, putative functions are thus far based strictly on sequence homologies and have not yet been experimentally confirmed. The homology comparisons have been described in detail by Guidolin et al.,[24] and subsequently by others.[11,25-29] They are briefly presented here, using the *S. pneumoniae* designations that have been adopted for many of the loci. All of the streptococcal sequences have levels of homology similar to those indicated here for *S. pneumoniae*. They share

S. pyogenes
  →4)-β-D-Glc$p$A-(1→3)-β-D-Glc$p$NAc-(1→

S. pneumoniae type 3
  →3)-β-D-Glc$p$A-(1→4)-β-D-Glc$p$-(1→

S. pneumoniae type 19F
  →4)-β-D-Man$p$NAc-(1→4)-α-D-Glc$p$-(1→2)-α-L-Rha$p$-(1-PO$_4^-$)

S. pneumoniae type 1
  →3)-α-D-AATp-(1→4)-α-D-Gal$p$A-(1→3)-α-D-Gal$p$A-(1→

S. pneumoniae type 14
  →6)-β-D-Glc$p$NAc-(1→3)-β-D-Gal$p$-(1→4)-β-D-Glc$p$-(1→
               4
               1
          β-D-Gal$p$

S. agalactiae type III
  →4-β-D-Glc$p$-(1→6)-β-D-Glc$p$NAc-(1→3)-β-D-Gal$p$-(1→
                          4
                          1
                     β-D-Gal$p$
                          3
                          2
                  α-D-Neu$p$NAc   (sialic acid)

S. thermophilus
  →3)-β-D-Gal$p$-(1→3)-β-D-Glc$p$-(1→3)-α-D-Gal$p$NAc-(1→
                          6
                          1
                     α-D-Gal$p$

L. lactis
                     α-L-Rha$p$
                          1
                          2
  →4)-β-D-Glc$p$-(1→4)-β-D-Gal$p$-(1→4)-β-D-Glc$p$-(1→
                          3
                       (PO$_4$)
                          1
                     α-D-Gal$p$

Figure 7.1 Capsule and exopolysaccharide structures. AAT, 2-acetamido-4-amino-2,4,6-trideoxy-D-Gal; Gal, galactose; GalA, galacturonic acid; GalNAc, N-acetylgalactosamine; Glc, glucose; GlcA, glucuronic acid; GlcNAc, N-acetylglucosamine; ManNAc, N-acetylmannosamine; NeuNAc, N-acetylneuraminic acid (sialic acid); Rha, rhamnose.

approximately 40 to 60% homology with each other, except for *L. lactis*, where approximately 25 to 45% homology occurs. At the DNA level, identity is approximately 60% among the streptococcal sequences, except with *L. lactis*, where it is 40%. CpsA has homology (28% identity) with LytR, a transcriptional regulator of autolysin expression in *Bacillus subtilis*. The only known homology of CpsB is to similar genes in the capsule loci of other Gram-positive bacteria. CpsC is homologous to the N-terminal third of ExoP from *Rhizobium meliloti* (23% identity), whereas CpsD is homologous to the C-terminal third of this same protein (30% identity). In ExoP, the N-terminal domain functions in polysaccharide chain length determination, and the C-terminal domain may have a regulatory role in polysaccharide polymerization.[24,30] Based on hydropathy profiles and similarity to other proteins, both CpsA and CpsC may be membrane associated.

The *cpsA* and *cpsB* sequences have been identified in virtually all *S. pneumoniae* serotypes examined, whereas *cpsC* and *cpsD* do not always occur.[21,24,31] In *S. pneumoniae*, the *dexB* sequence (homologous to a glucan 1,6-α-glucosidase of *S. mutans*) has been found upstream of the *cpsABCD* region in all serotypes for which sequence data have been reported, and it has been detected by hybridization analyses in all other serotypes examined.[21,22,24,25,32,33] The *S. pneumoniae* type 3 locus presents a somewhat unusual case, in that the common upstream sequences are present, but, like the downstream *plpA*, the *cpsA, B,* and *D* homologs are mutated and appear not to encode functional proteins.[19,25,34] In addition, the type 3 locus contains an additional truncated open reading frame (ORF5) that is located between the last common gene and the first type 3-specific biosynthetic gene. An apparently intact copy of this sequence is present in other serotypes, but it is not linked to the capsule locus. The unlinked copy is also present in type 3 strains.[18,19]

Transcription and mutation analyses

In most cases, it appears that transcription of the upstream common sequences initiates at a promoter located upstream of *cpsA* and its respective homologs, as depicted in Figure 7.2. Near consensus σ^{70}-like promoter sequences have been identified 30 to 50 bp upstream of the CpsA (or homolog) translation start in the sequences for *S. pneumoniae* types 1, 14, and 19F, *S. thermophilus*, *S. agalactiae* type III, and further upstream in *L. lactis*.[11,22,24,29,32,35,36] Transcription start sites have been confirmed in *S. pneumoniae* type 1, *S. thermophilus*, *S. agalactiae* type III, and *L. lactis*.[11,22,29,36] The only potential transcription termination sequences identified lie 3' to the biosynthetic genes, and other potential promoters have not been identified in the tightly clustered type-specific biosynthetic sequences, except in the case of *S. thermophilus*.[29,32,35] These observations suggest that the upstream common genes and the biosynthetic genes may be transcribed as a unit in these capsule loci. Northern analyses have confirmed this expectation for the *L. lactis* locus.[11]

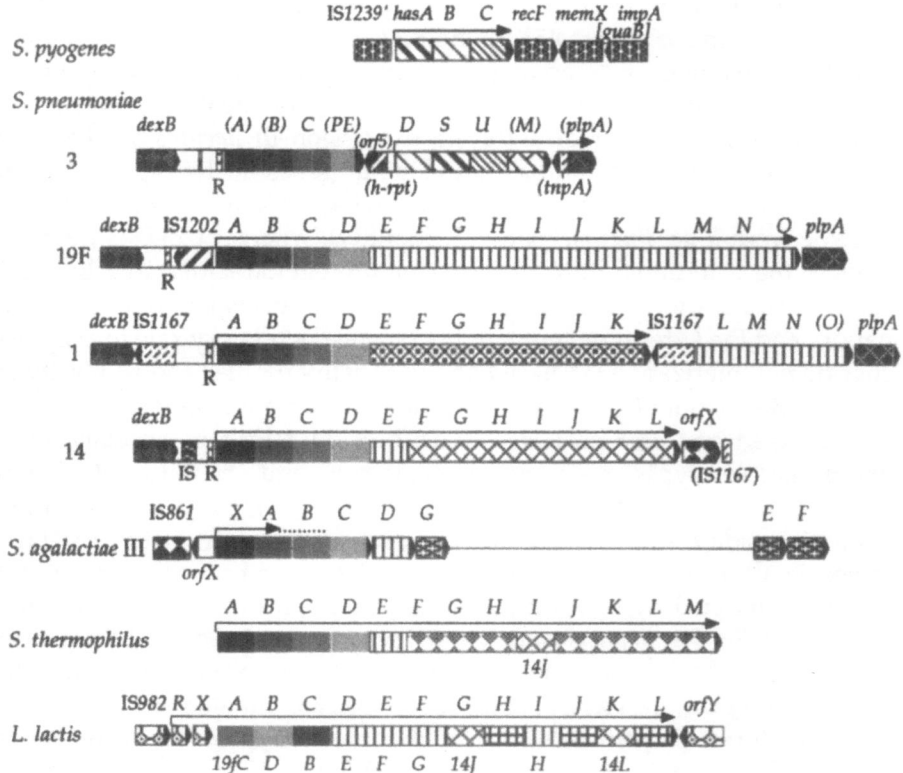

Figure 7.2 Capsule and exopolysaccharide loci. The shaded upstream common sequences (*ABCD*, in most cases) are virtually identical, both at the DNA and protein levels, in the *S. pneumoniae* serotypes, as is *plpA*. In the other loci, the predicted proteins of the upstream sequences are 25 to 60% homologous to those of *S. pneumoniae*. Other sequences indicated by the same pattern are homologous but not necessarily identical, as noted in the text. Their homology is with the sequence with which they are aligned, unless otherwise indicated. Genes below the *L. lactis* map indicate the type 19F homologs, except as indicated. Sequences in parentheses are mutated when compared to their respective homologs. Arrows indicate the direction and length of known or anticipated transcriptional units. R, 115 bp repetitive element. The overall GC contents of the capsule/exopolysaccharide loci (28 to 36%) are generally lower than that of the streptococcal chromosomes (38 to 40%). Maps are based on References 11, 19, 21, 22, 25, 28, 29, 32, 34, 52, and others cited in the text.

Insertion mutations in the *cpsABCD* region result in loss of capsule production in *S. pneumoniae* types 14[27,32] and 19F,[24] and in *S. agalactiae*.[36-38] As most of these mutations may affect expression of downstream genes, definitive proof of a requirement for each of these sequences awaits the construction of nonpolar mutations. Insertion mutations in *cpsD* of *S. pneumoniae* types 14 and 19F eliminate capsule production but do not result in loss of the enzymatic activity encoded by the next gene downstream, indicating a role for *cpsD* itself.[21,27] Mutations in the regions upstream of the *cpsABCD* sequences do not alter capsule production.

In contrast to the loci just described, the biosynthetic genes of the *S. pneumoniae* type 3 and *S. pyogenes* capsule loci are transcribed independent of flanking upstream sequences (see below for further details). In *S. pneumoniae* type 3, Northern analyses have provided no evidence for transcription in the region between *dexB* and *cps3D*, the first type 3-specific biosynthetic gene,[25] but data from *cat* fusions suggests there may be a low level of transcription.[19] Mutations in this region do not affect capsule production.[18,19,25]

Insertion elements, repetitive sequences, and cryptic genes

Insertion elements represent another commonly observed feature of the streptococcal capsule loci, although there is little relation between the elements identified. IS861, a 1.4 kb sequence containing two open reading frames, one with homology to the transposases of IS150 and IS3, was identified upstream of some *S. agalactiae* type III loci.[37,39,40] Different type III isolates were found to have either nine copies of the IS element in a highly conserved pattern, to have one or two copies that were always located in the same *Eco*RI restriction fragments, or to have no copies. Examination of other serotypes revealed only a single hybridizing restriction fragment in a type II isolate. Similarly, a 1.2 kb sequence related to IS1239 (IS1239') is located approximately 50 to 140 nt upstream of the *hasA* promoter in some strains of *S. pyogenes*.[41-43] Its presence does not appear to affect capsule production, as many strains lack the element, others have only a 12 to 100 bp insert that is related to IS1239,[42] and it can be deleted without obvious effect.[43] In *L. lactis*, a homolog of IS982 lies upstream of the capsule locus.[11]

The *S. pneumoniae* capsule loci abound with insertion elements, repetitive sequences, and genes with no apparent function in capsule production. In type 19F, IS1202 is located 87 bp upstream of *cpsA* but appears to have no role in capsule production, as insertion mutations are without effect in this regard.[33] Although this IS element is not present in the type 3 locus, its putative recognition sequence has been identified in a similar location.[25] The type 3 locus contains two other sequences suggestive of IS elements.[19,34] The first, *tnpA*, is inserted between *cps3M*, the last type 3-specific gene, and *plpA*, the first downstream common sequence. All three of these sequences are truncated genes: *tnpA* represents only an internal fragment, *cps3M* is 3' deleted, and *plpA* is 5' deleted. *tnpA* shares homology with a number of putative transposases from other IS elements, and has about 50% homology with the IS1167 sequence that is found in multiple copies in the *S. pneumoniae* chromosome. Hybridization analyses suggest that *tnpA* may also be located near the type 2 and 6B capsule loci.[19] Just upstream of *cps3D*, the first type 3-specific gene, is another internal gene fragment with homology to the IS-like H-rpt sequences of *Escherichia coli*, *Salmonella enterica*, and *Vibrio cholerae*.

These sequences have been found in association with polysaccharide bio-synthetic loci that exhibit numerous truncations, rearrangements, and other mutations in close proximity to the element.[34] As already noted, eight genes in the type 3 locus are mutated when compared to their respective homologs. In addition, a remnant of *cpsE*, a biosynthetic gene found in types 14, 19F, *S. agalactiae* type III, *S. thermophilus*, and *L. lactis*, is also present.[19] In the type 1 locus, copies of IS*1167* are located both upstream of the common sequences and downstream of the last type 1 biosynthetic gene. Interestingly, four genes nearly identical to the type 19F biosynthetic genes *cps19fLMNO* were iden-tified between IS*1167* and *plpA*.[22] In type 19F, these genes encode products involved in the biosynthesis of rhamnose,[21] a sugar found in the type 19F but not the type 1 capsule. In the type 1 locus, *cpsO* is mutated, and mutations in the other genes do not alter type 1 production.[22] A partial copy of IS*1167* is located downstream of the type 14 biosynthetic genes, where it interrupts a sequence with homology to teichoic acid synthesis enzymes. This ORF appears to have no role in either capsule or teichoic acid synthesis. Another truncated sequence with homology to putative transposases found in IS elements from *Synechocytis* and *Rhizobium meliloti* is located downstream of *dexB* in the type 14 locus.[32] In addition to these various IS-like elements, a highly conserved 115 bp sequence is located upstream of *cpsA* in the type 1, 3, 14, and 19F loci. This same sequence is found upstream of the *S. pneumoniae* genes encoding penicillin binding protein 3, neuraminidases A and B, and hyaluronidase. It is unrelated to the pneumococcal box element, and its significance is unknown.[22,32] The presence of IS elements and genes that are apparently vestigial in nature suggests the possibility that many of the *S. pneumoniae* capsule loci may have arisen as a result of aberrant transforma-tion and transposition events. Such events may have resulted in remnants of genes remaining in these loci, and may have played a role in the generation of new capsule types.[18,19,22,34]

The S. pyogenes *hyaluronic acid and* S. pneumoniae *type 3 capsule loci*

These two capsules are considered together because of their similarity at both the genetic and structural levels. Three genes comprise the hyaluronic acid (HA) capsule locus. *hasA* encodes the hyaluronan synthase (HAS), nec-essary for synthesis of the (GlcA-GlcNAc)$_n$ polysaccharide; *hasB* encodes a UDP-Glc dehydrogenase, which converts UDP-Glc to UDP-GlcA, one of the precursors of the polysaccharide; and *hasC* encodes a Glc-1-P uridylyltrans-ferase (UDP-Glc pyrophosphorylase), which converts Glc-1-P to UDP-Glc (Figure 7.3). HasA shares homology with polysaccharide synthases from a number of organisms, including NodC from *Rhizobium*, the *S. pneumoniae* type 3 synthase (Cps3S), chitin synthase, and a family of eukaryotic hyalu-ronan synthases.[18,44-47] HasB has homology with the GDP-mannose dehydro-genase from *Pseudomonas aeruginosa* (AlgD) and the *S. pneumoniae* type 3

(Cps3D) and bovine UDP-Glc dehydrogenases.[18,48,49] The uridylyltransferase, HasC, shares homology with enzymes from a number of bacteria, including *S. pneumoniae* type 3 (Cps3U), *Bacillus subtilis* (GtaB), and *E. coli* (GalU).[50] Expression of the cloned *hasB* and *hasC* genes in *E. coli* has been used to confirm the enzymatic activities expected from the sequence homologies.[44,49,50] DeAngelis and Weigel have shown that expression of *hasA* and *hasB* in either *E. coli* or *Enterococcus faecalis* is sufficient to direct HA synthesis, and to form capsules in the latter. In *E. coli* strains containing endogenous UDP-Glc dehydrogenase activity, the expression of *hasA* alone results in HA production.[44,51] No genes expected to be involved in the synthesis of UDP-GlcNAc, the other necessary precursor nucleotide sugar, have been identified in the *has* locus. Because GlcNAc is also a component of peptidoglycan, synthesis of the polysaccharide presumably utilizes cellular pools of the precursor nucleotide sugar in both streptococci and *E. coli*. UDP-Glc pools in *E. coli* likewise provide the source for this precursor, and the lack of *hasC* in some encapsulated group A streptococci indicates that the same situation can occur here.[44,52]

The hyaluronan synthase appears to possess the glycosyltransferase activities necessary to make both the GlcA-β-(1>3)-GlcNAc and the GlcNAc-β-(1>4)-GlcA linkages. As noted above, the expression of *hasA* alone is sufficient to synthesize polysaccharide in *E. coli*. In *E. coli* strains that do not make UDP-Glc dehydrogenase, the synthase is produced from the cloned *hasA*, but no polysaccharide is made. Isolated membranes from these cells can, however, direct HA synthesis when provided with the precursor molecules.[53] Unlike more complex polysaccharides, hyaluronic acid synthesis does not initiate on or involve lipid intermediates.[54] In *S. pyogenes*, the polysaccharide is apparently translocated out of the cell as it is synthesized, suggesting that the synthase also serves as the transporter molecule.[45,47] No other putative transporters have been identified in sequence or mutation analyses. Based on observations with the eukaryotic hyaluronan synthase, the mechanism of polysaccharide synthesis has been postulated to involve growth at the reducing end of the molecule.[55,56] This hypothesis has not been confirmed, and other evidence from studies with the *S. pyogenes* enzyme suggests that growth occurs at the nonreducing end.[57]

The genes of the *has* locus are transcribed as a single operon from a promoter located immediately upstream of *hasA*. The observation that insertions in *hasA* resulted in loss of capsule production, HA synthase activity, and UDP-Glc dehydrogenase activity first suggested the operon arrangement.[49,58] Further studies identified the transcription start site 43 bp upstream of the apparent GTG translational start codon.[46] A putative transcription terminator was identified downstream of *hasC*, and a single 4.1 kb transcript containing *hasABC* was found in Northern analyses.[41,50] Regions further upstream of *hasA* appear not to be necessary for expression of the *has* genes, as many strains have insertion sequences located within 140 bp of the translational start.[42,43] Using *cat* fusions to assess promoter activity, Alberti et al.

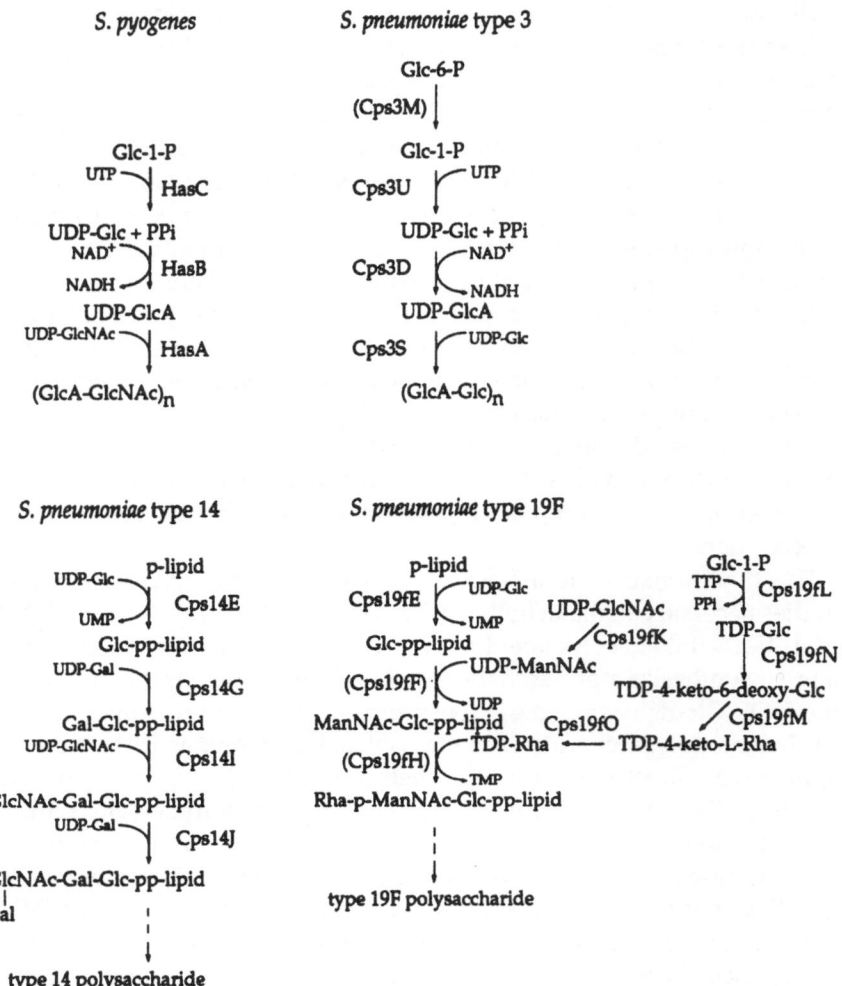

Figure 7.3 Biosynthetic pathways. Functions have been experimentally demonstrated, except where the gene product is enclosed in parentheses. Additional proteins are necessary to polymerize and export the type 14 and 19F polysaccharides. Pathways are based on References 18, 21, 32, and others cited in the text.

showed that clones containing only 12 bp of the region upstream of the *hasA* −35 site exhibited the same level of activity in both group A streptococci and *E. coli* as did clones containing more than 500 bp of the upstream sequence. The complete −35/−10 region was, however, necessary for maximum levels of transcription.[43]

In an arrangement very similar to that just described, the *S. pneumoniae* type 3 locus is comprised of four genes: *cps3D*, a UDP-Glc dehydrogenase; *cps3S*, the type 3 synthase; *cps3U*, a Glc-1-P uridylyltransferase; and *cps3M*, a phosphomutase homolog predicted to convert Glc-6-P to Glc-1-P.[16,18,19] (The first three genes have also been referred to as *capA*, *capB*, and *capC*, respectively.[25]) The homologies of the UDP-Glc dehydrogenase, the type 3 synthase, and the uridylyltransferase with the *S. pyogenes* enzymes and other proteins have been noted above. Mutant analyses in *S. pneumoniae* and enzymatic activities of the cloned products expressed in *E. coli* have been used to confirm the functions of the UDP-Glc dehydrogenase, synthase, and uridylyltransferase.[16,18,25,48,59,60] Expression of the type 3 synthase in *E. coli* resulted in production of type 3 polysaccharide, indicating that it, like the hyaluronan synthase, is capable of synthesizing the polysaccharide in the absence of other pneumococcal proteins. Introduction of the cloned synthase into pneumococcal strains of other serotypes which synthesize GlcA resulted in binary encapsulated strains that produced both the recipient and type 3 polysaccharides.[60] Like the hyaluronan synthase, the type 3 synthase is processive, and current evidence indicates that growth is at the nonreducing end of the molecule.[61] The only type 3 gene for which the expected biochemical function remains unproven is *cps3M*. The amino acid sequence predicted from this gene is C-terminal truncated with respect to other phosphomutases, but the putative substrate-binding, Mg^{2+}-binding, and active sites of these enzymes are retained in Cps3M.[19] Phosphomutase activity has not been observed with the cloned product expressed in *E. coli*,[62] but other data suggest that the protein is functional (see below).

Like the *has* operon, the genes of the type 3 locus are transcribed as a unit. Mutation analyses initially demonstrated that insertions located between *cps3D* and *cps3S* that did not disrupt either open reading frame resulted in the loss of capsule production and synthase activity. In contrast, point mutations in *cps3D* eliminated capsule production and UDP-Glc dehydrogenase activity, but had no effect on synthase activity.[18] That *cps3D* and *cps3S* form part of an operon was confirmed in Northern analyses, and the transcription start site was shown to lie in the predicted promoter region 58 bp upstream of the Cps3D translation start.[25] Insertion mutations located approximately 300 bp upstream of the *cps3D* promoter have no effect on capsule production, indicating that regions further upstream do not act in *cis* to affect capsule gene expression under the conditions examined.[18] The transcript that contains *cps3D* and *cps3S* is, in fact, comprised of *cps3DSUM-tnpA-plpA*.[19] Mutations between *cps3S* and *cps3U*, or anywhere in the *cps3UM-tnpA-plpA* region, have no apparent effect on capsule production.[16,18,19]

For the synthesis of both the hyaluronic acid and type 3 capsules, the only genes absolutely required are those encoding the dehydrogenase (*hasB* or *cps3D*) and the synthase (*hasA* or *cps3S*). No requirements for the uridylyltransferase (*hasC* or *cps3U*) or for the phosphomutase homolog (*cps3M*) in capsule production have been demonstrated. It has been postulated that these genes are unnecessary because the functions they encode duplicate those necessary for normal cellular functions. Our recent data show, however, that insertions in *cps3U* or *cps3M* reduce mouse virulence and result in growth defects that are manifest on certain media as growth predominantly as streptococci rather than diplococci, and the frequent occurrence of abnormal cellular shapes.[63] Similar effects are not seen with insertions that merely separate the *cps3UM-tnpA-plpA* region from the promoter upstream of *cps3D*, suggesting that other promoters are active. In addition, the Cps3U and Cps3M mutants do not produce reduced amounts of capsule.[19] These observations may result from the fact that UDP-Glc, the product of Glc-1-P uridylyltransferase (Cps3U) activity, is also an essential component of teichoic acids and, in some systems, may serve as an intracellular signal that influences gene expression.[64] Under specific environmental conditions, reduced levels of UDP-glucose in the mutants could result in altered levels of teichoic acids or other necessary cellular components, resulting in apparent growth defects. A similar effect in the animal environment could be the cause of the reduction in virulence. Thus, the presence of *cps3U* in the *S. pneumoniae* type 3 locus, and possibly of *hasC* in the hyaluronic acid locus, may reflect the need to maintain adequate levels of UDP-glucose for essential cellular functions, under specific environmental conditions.

Group B type III, S. pneumoniae *types 14 and* 19F, S. thermophilus, *and* L. lactis *capsules*

The genetic loci for these capsular polysaccharides share not only common upstream sequences but also some biosynthetic genes and a basic mechanism of capsule synthesis (Figure 7.3). As noted earlier, the *S. pneumoniae* type 14 and *S. agalactiae* type III polysaccharides are identical, except for the presence of a sialic acid on the side chain of the latter. In addition, all of these polysaccharides share some similarities in their sugar content (Figure 7.1). A CpsE homolog (CpsD in *S. agalactiae* and *L. lactis*) is found in all of the loci, and is a glycosyltransferase. In *S. agalactiae*, membranes from CpsD mutants were reduced in the ability to incorporate ^{14}C from UDP-Gal into a lipid intermediate, suggesting a galactosyltransferase activity.[38] Similar experiments with membranes from the corresponding CpsE mutants of type 14 *S. pneumoniae*, or from the cloned Cps14E *E. coli* product, demonstrated that Glc rather than Gal was transferred in this system.[15,27] The *S. pneumoniae* and *S. agalactiae* sequences share only partial identity, possibly accounting for differences in specificity. Alternatively, an epimerase activity in *S. pneumoniae* converts UDP-Gal to UDP-Glc, which can then be transferred to the lipid carrier. A similar activity in *S. agalactiae* could account for the apparent incorporation of Gal.[27] The *S. pneumoniae* type 19F CpsE is virtually identical to that of type 14 and appears to have the same activity,[27,35] as does the corresponding protein from *L. lactis*.[11] In both *S. pneumoniae* serotypes, the transfer of Glc to a lipid carrier is the first step in polysaccharide biosynthesis, and type 14 mutants specifically lacking CpsE accumulate no intermediate products.[27] The initiation of polysaccharide synthesis by transfer of a sugar to a lipid intermediate is common to each of the systems mentioned here, but contrasts with that just described for hyaluronic acid synthesis in *S. pyogenes* and type 3 synthesis in *S. pneumoniae*,[31] where lipid intermediates are not involved. As was noted above, a remnant of *cpsE* remains in the *S.*

pneumoniae type 3 locus. In numerous other *S. pneumoniae* serotypes in which Glc is part of the capsular polysaccharide structure, homologs of *cpsE* have been detected, and a glycosyltransferase activity that initiates synthesis by transfer of Glc to a lipid carrier has been observed.[31]

Using membranes from *S. pneumoniae* mutants and *E. coli* strains containing the cloned genes, Kolkman et al. have characterized the products necessary for synthesis of the remainder of the type 14 repeating tetrasaccharide unit (Figure 7.3). The second step, the addition of Gal to lipid-linked Glc, is mediated by a β-1,4-galactosyltransferase encoded by *cps14G*. Membranes from *S. pneumoniae* mutants containing insertions in either *cps14F* or *cps14G* incorporated Glc but not Gal, and deletion analyses with the cloned genes in *E. coli* demonstrated that Cps14G was responsible for the activity. The *cps14F* deletion mutants were reduced in galactosyltransferase activity, possibly suggesting an enhancing role for Cps14F.[27] Membranes from *E. coli* strains containing both *cps14EFG* and *cps14I* incorporated GlcNAc in the presence of UDP-Glc and UDP-Gal, and characterization of the product confirmed that Cps14I is an *N*-acetylglucosaminyltransferase that adds the third sugar to the saccharide.[32] The addition of Gal, the fourth sugar, requires a GlcNAc acceptor and is mediated by the product of *cps14J*.[32] Functions for the other genes in the type 14 locus have not been experimentally demonstrated, but, based on sequence homologies, their products may be involved in polymerization (Cps14H) and transport (Cps14L) of the polysaccharide. The function of *cps14K* is unknown, but mutants express reduced amounts of capsule on the cell surface.[27,32]

In *S. agalactiae* type III, *cpsF* encodes the enzyme necessary for activation of *N*-acetylneuraminic acid (NeuNAc, sialic acid) to CMP-NeuNAc. *S. agalactiae cpsF* mutants synthesize an asialo capsular polysaccharide, lack CMP-NeuNAc synthase activity, accumulate intracellular NeuNAc, and react with antiserum against the *S. pneumoniae* type 14 polysaccharide.[65,66] Enzymatic activity was demonstrated from the cloned gene in *E. coli*, and expression of *cpsF* restored capsule production in an *E. coli* mutant lacking CMP-NeuNAc synthase activity.[67] This same enzymatic activity has been demonstrated in *S. agalactiae* strains representing types Ia, Ib, and II.[68] From sequence analyses, other genes in the type III locus may encode enzymes involved in acetylation (*cpsE*) and *N*-acetylglucosaminyltransferase activity (*cpsG*).[28] A 30 kb region containing the capsule locus is highly conserved among strains of types Ia, Ib, Ic, II, and III.[10,37]

Evidence for enzymatic activities for gene products of the *S. pneumoniae* type 19F locus comes from complementation of specific *E. coli* mutants with the cloned genes. Cps19fK has homology to a UDP-GlcNAc-2-epimerase (RffE) of *E. coli* that is involved in the synthesis of UDP-ManNAc, a component of the enterobacterial common antigen. RffE mutants are resistant to phage N4 infection, but complementation with Cps19fK restores sensitivity.[21] The predicted *cps19fLMNO* products are homologous to proteins (RfbBDAC) involved in O-antigen-related rhamnose biosynthesis in *Shigella flexneri*.

Expression of *cps19fLMNO* in an *E. coli* strain containing an *rfb* locus that lacks *rfbBDAC* resulted in production of the O-antigen. The proteins are thus expected to be a Glc-1-P thymidylyltransferase (Cps19fL), a dTDP-4-keto-6-deoxyglucose-3,5-epimerase (Cps19fM), a dTDP-Glc-4,6-dehydratase (Cps19fN), and a dTDP-L-Rha-synthase (Cps19fO).[21] Putative functions for the other gene products are based on homology and suggest a UDP-*N*-acetyl-D-mannosamine transferase (Cps19fF), a rhamnosyltransferase (Cps19fH), a polysaccharide polymerase (Cps19fI), and a trisaccharide transporter (Cps19fJ). A possible function for Cps19fG is not apparent, but it has homology with a lipopolysaccharide-associated gene (*licD*) of *Haemophilus influenzae*.[21,24] The type 19B polysaccharide is immunologically cross-reactive with the type 19F, but its structure is more complex.[13] In the type 19B genetic locus, 5 genes replace the centrally located *cps19fI* and *cps19fJ*.[35] Homologies with other proteins suggest roles in polymerization of the polysaccharide (Cps19bI), addition of rhamnose in the side chain (CpsbQ), and transport of the polysaccharide repeat unit (Cps19bJ).

Genes encoding enzymes necessary for the synthesis of many of the precursor sugars are frequently not found in the capsule loci. These sugars have other cellular functions and are expected to be provided from other pathways. For example, 2-acetamido-4-amino-2,4,6-trideoxy-D-Gal (AAT, type 1 capsule) is a component of the *S. pneumoniae* teichoic acids, and UDP-Glc (types 14, 19, and *S. thermophilus*), UDP-Gal, and UDP-GlcNAc (type 14 and *S. thermophilus*) have various cellular functions, including roles in teichoic acid and/or cell wall metabolism.[21,22,29,32]

Regulation

In *S. pyogenes*, HA synthase activity (HasA) and capsule are lost during stationary phase.[69] Northern and primer extension analyses showed that the *has* locus is not transcribed during this time.[41] Transcripts from both mid- and late-exponential phases of growth had a similar half-life (approximately one minute), indicating that the level of control is most likely transcriptional.[42] As noted in the section describing the *has* locus, optimum transcription during exponential phase does not require sequences more than 12 bp upstream of the *hasA* promoter. Taken together, the results suggest that repression of capsule expression may occur during stationary phase. Among clinical isolates, which can produce greatly differing amounts of capsule, the *has* locus is conserved.[41,70] It is not, however, transcribed in nonencapsulated strains.[41] Although a number of mechanisms may be responsible for this observation, transcription analyses of two strains producing either low or high amounts of capsule identified a stronger promoter in the latter.[43] Sequence differences were found in the promoter regions and were consistent among different strains of the same M-type (M18, high capsule; M3, low capsule). Hybrid promoters exhibited altered strengths. Whether the increased promoter strength was directly responsible for the difference in capsule production was not determined. Recently, mutations in *mga* (multi-gene activator) of an M1 isolate were found to cause loss of capsule production and *has* transcription, indicating that activation of capsule expression may occur in some strains. Deletion of *mga* did not, however, affect capsule expression in an M49 isolate.[71]

Among the other streptococci, less is known regarding the genetic basis for regulation of capsule expression. Like clinical isolates of *S. pyogenes*, those of type 3 *S. pneumoniae* produce widely variant amounts of capsular polysaccharide. Comparisons of the type 3 loci from multiple *S. pneumoniae* isolates using RFLP, PCR, and sequence analyses have found them to be virtually identical. All retain *cps3U* and *cps3M*, as well as the truncated sequences that surround the locus.[19] The basis for the difference in capsule production has not been determined. Transcription of the type 3 capsule genes ceases during late exponential phase, but complete loss of capsule does not occur.[72] In *S. agalactiae* type III, the amount of capsular polysaccharide is reduced by late exponential phase, and longer doubling times (11 vs. <2 h) result in more than tenfold less polysaccharide.[73]

Virulence

The capsules of *S. pneumoniae*, *S. pyogenes*, and *S. agalactiae* have long been known or suspected to be important in virulence. With knowledge of the genetic basis of capsule expression has come the ability to generate defined, isogenic mutants for the purpose of further defining the roles of these capsules in virulence. In *S. pyogenes*, mutations in the *has* locus result in loss of capsule and corresponding alterations in a number of virulence phenotypes. It has been noted that such mutants are more efficiently phagocytized, do not grow in blood, and have reduced virulence in several mouse models of infection, including intraperitoneal challenge, intranasal colonization, and pneumonia following intratracheal infection.[70,74,75] Enhanced phagocytosis of the mutants is not due to increased deposition of complement but may rather be due to increased exposure of bound C3b to phagocytic receptors, a situation analogous to that observed in the staphylococci.[76] Capsule mutants and stationary phase cells, which would not be expressing capsule, are more readily internalized by keratinocytes and epithelial cells.[71,77,78] Whether these latter observations are reflective of decreased virulence, or whether they instead suggest that the ability to regulate capsule expression is an important factor in the various stages of infection, has not been resolved.

Mutants of *S. agalactiae* type III that either express no capsule (*cpsD*) or that fail to add sialic acid to the side chain (*cpsF*) are avirulent in a neonatal rat model of infection, are readily phagocytized, and invade lung endothelial cells more effectively than the encapsulated parent.[10,38,65,66,79-81] In contrast to *S. pyogenes*, the lack of capsule results in increased accumulation of C3, and the effect is also seen in mutants lacking only the sialic acid moiety.[82]

In *S. pneumoniae*, nonencapsulated mutants derived by transposon mutagenesis were significantly reduced in virulence. The insertion itself was not responsible for the loss of capsule production, however, as co-transformation experiments showed it to be only 36% linked to the acapsular phenotype.[83]

The insertion has subsequently been mapped to a site distant to where the known capsule genes are located,[84] and other possible alterations in these mutants are not known. Specific mutations in *cps3D* and *cps3S*, which result in loss of capsule production, do result in avirulence in mice.[62] As described above, mutations in *cps3U* and *cps3M* reduce virulence, but perhaps not as a result of decreases in capsule production. The construction of isogenic strains expressing different capsular polysaccharides was used to determine whether the type of capsule produced is the determining factor in *S. pneumoniae* virulence.[7] Perhaps not unexpectedly, the effect of the type 3 capsule varied with the genetic background in which it was expressed: a highly virulent type 5 recipient became avirulent, a relatively avirulent type 6B recipient was enhanced in virulence, and a virulent type 2 recipient was unchanged.

The special case of S. pneumoniae: *genetic exchange and capsule diversity*

Although recent work has revealed the underlying molecular basis for capsule expression in the streptococci, the groundwork for these studies was laid decades ago. (See Reference 85 for a more extensive review of the earlier work in *S. pneumoniae*.) In particular, the demonstration of a common genetic organization of the capsule loci that is shared among all *S. pneumoniae* strains is confirmation of a classic and elegant series of genetic experiments that began in 1928 with Griffith's demonstration of transformation of capsule types during infection in the mouse. Later studies demonstrated transformation under laboratory culture conditions and suggested that a wide variety of donor and recipient combinations were possible.[86,87] In 1951, Ephrussi-Taylor postulated, and Austrian et al. later elaborated on the idea, that capsule genes for different serotypes might occupy identical sites in the chromosome, and that transformation of capsule type would result from replacement of the entire set of recipient capsule genes with those of the donor. DNA homology outside the multigenic type-specific region would provide the recombination sites necessary for the exchange.[88,89] Biochemical evidence for a replacement mechanism came from studies using strains that either lacked or contained uronic acids in their capsules. Type 14 and 18 capsules do not contain uronic acids, and accordingly do not possess UDP-Glc dehydrogenase or UDP-GlcA. In contrast, types 1 and 2 have these components, but they are lost upon transformation to type 14 or 18 encapsulation.[88] Each of these experiments supported the concept of capsule genes existing as exchangeable units, and clearly implied linkage of the genes encoding the biosynthetic enzymes. Recombination experiments confirmed these linkages and provided the first genetic maps of the capsule loci.[90,91] Further studies characterized the mechanisms and pathways for biosynthesis of the type 3 and other polysaccharides, and also identified some of the genes involved, including that encoding the type 3 UDP-Glc dehydrogenase.[92-95]

As described above, modern molecular genetic studies have borne out the hypothesized "cassette-like"[20,96] organization of the *S. pneumoniae* capsule loci, and have shown that the sequences upstream and downstream of the type-specific biosynthetic genes are homologous in apparently all pneumococcal capsular types. Molecular confirmation of the replacement of one capsule cassette by another has been provided through hybridization and mapping studies of isolates transformed to heterologous capsule types.[16] In these studies, type 2 recipients transformed to type 3 encapsulation were shown to contain the type 3-specific biosynthetic genes, whereas type 3 recipients transformed to type 2 encapsulation had lost the type 3-specific genes. Because of the truncations of many of the sequences in the type 3 locus, it was possible to demonstrate the presence of the type 3 downstream flanking region in the transformed type 2 recipients. Characterization of clinical isolates has also clearly shown that capsule loci are transferred in the environment, and that both the immediate common flanking sequences and more distant regions can participate in the recombination events.[97,98]

Austrian and Bernheimer described binary encapsulated strains of *S. pneumoniae* that, as a result of integration of a second set of capsule genes, expressed two different capsular polysaccharides.[88,99] These isolates were most often observed when a nonencapsulated mutant lacking UDP-Glc dehydrogenase activity was transformed with DNA of a heterologous type that also contained GlcA in its capsule. Complementation of the defect permitted expression of both polysaccharides. In general, the genes encoding the UDP-Glc dehydrogenases did not recombine, and recent studies have confirmed that only a low level of homology exists between these genes.[16,22] Stable binary strains resulted when integration of the second set of capsule genes occurred at a site unlinked to the recipient capsule locus, whereas integration at a closely linked site resulted in unstable binaries.[99] A mechanism involving a transposition-like event, where identical sequences flanking a capsule locus permit resolution and subsequent integration at homologous sites in a recipient chromosome, has been proposed as one mechanism for the generation of binary strains.[18] The finding of numerous transposase-like sequences in the capsule loci, and repetitive sequences that occur throughout the chromosome, has provided some support for this idea. In type 1, the capsule genes are flanked by copies of IS1167,[22] and multiple copies of this element are present in most strains of *S. pneumoniae*. *tnpA* in the type 3 locus shares about 50% homology with IS1167.[19] Strains of these two capsule types were frequently involved in the binary reaction. Munoz et al. have recently shown that plasmids containing *cap1HIJKIS1167* from the type 1 locus can integrate into nonencapsulated type 3 strains at sites distant to the type 3 genes, apparently via homologous recombination with other IS1167 sequences. UDP-Glc dehydrogenase expression from *cap1K* complemented the defect of the type 3 strain, resulting in production of type 3 capsule.[22] Whether similar integration events occur when chromosomal DNA from type 1 strains is used as the donor has not been determined, but the possibility of generating not only binary encapsulated strains but potentially new capsular polysaccharides as a result of transposition events is an intriguing one.

Future directions

The streptococcal capsular polysaccharides have provided an area of fruitful discovery for over a century. Studies from the last decade have brought forth molecular explanations for many of the earlier genetic and biochemical observations, and they have revealed an underlying genetic similarity among the capsule loci of many of the streptococci and related organisms. From those studies, we can begin to address many of the remaining unanswered questions. For example, the origin of diversity of the *S. pneumoniae* polysaccharides may lie, in part, in the ability of the organism to undergo natural transformation, but other mechanisms, such as recombination, mutation, transposition, and genetic rearrangements, are likely to play a major role. The similarities between the streptococcal capsule loci suggest a common ancestral origin for many of the genes and loci, and imply that a mechanism of genetic exchange is operational in all. The significance of the genetic exchange of capsule genes in *S. pneumoniae* is only beginning to be appreciated as molecular analyses of natural isolates are pursued. It is clear that the genetic switching of capsule types may have a significant impact on virulence and on vaccine considerations, because immunity to one capsular polysaccharide is likely to enrich for isolates which have acquired the ability to produce another.

The mechanisms of polysaccharide biosynthesis are also just beginning to be fully explored, as are the underlying genetic aspects of their expression and regulation. It is apparent that, for many of the polysaccharides, there are basic similarities in the pathways leading to synthesis of their repeating units. Examination of the role of the upstream sequences, which are found in almost all of the loci, is likely to reveal a common mechanism of polysaccharide export and genetic regulation. The regulation of capsule expression in the natural environment is expected to be complex. As studies described herein have revealed, there is likely to be an intimate relationship between capsule production and basic cellular metabolism. A full appreciation of the streptococcal capsular polysaccharides will thus involve a better understanding of their biochemistry, genetics, and their niche in overall cellular processes.

Acknowledgments

The significant contributions made to these studies by the present and former members of my laboratory, especially my students, are gratefully acknowledged. Our work is supported by Public Health Service Grants AI28457 and GM53017 from the National Institutes of Health.

References

1. Griffith, F., The significance of pneumococcal types, *J. Hygiene*, 27, 113, 1928.
2. Avery, O. T., MacLeod, C. M., and McCarty, M., Studies on the chemical nature of the substance inducing transformation of pneumococcal types. Induction of transformation by a desoxyribonucleic acid fraction isolated from pneumococcus Type III, *J. Exp. Med.*, 79, 137, 1944.
3. Avery, O. T. and Dubos, R., The protective action of a specific enzyme against type III pneumococcus infection in mice, *J. Exp. Med.*, 54, 73, 1931.
4. MacLeod, C. M., Hodges, R. G., Heildeberger, M., and Bernhard, W. G., Prevention of pneumococcal pneumonia by immunization with specific capsular polysaccharides., *J. Exp. Med.*, 82, 445, 1945.
5. Stryker, L. M., Variations in the pneumococcus induced by growth in immune serum, *J. Exp. Med.*, 24, 49, 1916.
6. Wood, W. B., Jr. and Smith, M. R., The inhibition of surface phagocytosis by the capsular "slime layer" of pneumococcus type III, *J. Exp. Med.*, 90, 85, 1949.
7. Kelly, T., Dillard, J. P., and Yother, J., Effect of genetic switching of capsular type on virulence of *Streptococcus pneumoniae*, *Infect. Immun.*, 62, 1813, 1994.
8. Baker, C. J. and Edwards, M. S., Group B streptococcal infections, in *Infectious Diseases of the Fetus and the Newborn*, J. S. Remington and J. O. Klein, (Eds.), W. B. Saunders Co., Philadelphia, 1990, 742.
9. Johnson, D. R., Stevens, D. L., and Kaplan, E. L., Epidemiologic analysis of group A streptococcal serotypes associated with severe systemic infections, rheumatic fever, or uncomplicated pharyngitis, *J. Infect. Dis.*, 166, 374, 1992.
10. Rubens, C. E., Wessels, M. R., Heggen, L. M., and Kasper, D. L., Transposon mutagenesis of type III group B Streptococcus: correlation of capsule expression with virulence, *Proc. Natl. Acad. Sci. U.S.A.*, 84, 7208, 1987.
11. van Kranenburg, R., Marugg, J. D., van Swam, I. I., Willem, N. J., and de Vos, W. M., Molecular characterization of the plasmid-encoded *eps* gene cluster essential for exopolysaccharide biosynthesis in *Lactococcus lactis*, *Mol. Microbiol.*, 24, 387, 1997.
12. Henrichsen, J., Six newly recognized types of *Streptococcus pneumoniae*, *J. Clin. Microbiol.*, 33, 2759, 1995.
13. van Dam, J. E., Fleer, A., and Snippe, H., Immunogenicity and immunochemistry of *Streptococcus pneumoniae* capsular polysaccharides, *Antonie Van Leeuwenhoek*, 58, 1, 1990.
14. Kogan, G., Uhrin, D., Brisson, J. R., Paoletti, L. C., Blodgett, A. E., Kasper, D. L., and Jennings, H. J., Structural and immunochemical characterization of the type VIII group B streptococcus capsular polysaccharide, *J. Biol. Chem.*, 271, 8786, 1996.
15. Kolkman, M. A., Morrison, D. A., Van Der Zeijst, B. A., and Nuijten, P. J., The capsule polysaccharide synthesis locus of *Streptococcus pneumoniae* serotype 14: Identification of the glycosyltransferase gene *cps14E*, *J. Bacteriol.*, 178, 3736, 1996.
16. Dillard, J. P. and Yother, J., Genetic and molecular characterization of capsular polysaccharide biosynthesis in *Streptococcus pneumoniae* type 3, *Mol. Microbiol.*, 12, 959, 1994.
17. Garcia, E., Garcia, P., and Lopez, R., Cloning and sequencing of a gene involved in the synthesis of the capsular polysaccharide of *Streptococcus pneumoniae* type 3, *Mol. Gen. Genet.*, 239, 188, 1993.

18. Dillard, J. P., Vandersea, M. W., and Yother, J., Characterization of the cassette containing genes for type 3 capsular polysaccharide biosynthesis in *Streptococcus pneumoniae, J. Exp. Med.,* 181, 973, 1995.
19. Caimano, M. J., Hardy, G. G., and Yother, J., Capsule genetics in *Streptococcus pneumoniae* and a possible role for transposition in the generation of the type 3 locus, *Microbial Drug Resist.,* 4, 11, 1998.
20. Dillard, J. P., Caimano, M., Kelly, T., and Yother, J., Capsules and cassettes: genetic organization of the capsule locus of *Streptococcus pneumoniae, Dev. Biol. Stand.,* 85, 261, 1995.
21. Morona, J. K., Morona, R., and Paton, J. C., Characterization of the locus encoding the *Streptococcus pneumoniae* type 19F capsular polysaccharide biosynthetic pathway, *Mol. Microbiol.,* 23, 751, 1997.
22. Munoz, R., Mollerach, M., Lopez, R., and Garcia, E., Molecular organization of the genes required for the synthesis of type 1 polysaccharide of *Streptococcus pneumoniae:* formation of binary encapsulated pneumococci and identification of cryptic dTDP-rhamnose biosynthesis genes, *Mol. Microbiol.,* 25, 79, 1997.
23. Kelly, T. and Yother, J., unpublished data.
24. Guidolin, A., Morona, J. K., Morona, R., Hansman, D., and Paton, J. C., Nucleotide sequence analysis of genes essential for capsular polysaccharide biosynthesis in *Streptococcus pneumoniae* type 19F, *Infect. Immun.,* 62, 5384, 1994.
25. Arrecubieta, C., Garcia, E., and Lopez, R., Sequence and transcriptional analysis of a DNA region involved in the production of capsular polysaccharide in *Streptococcus pneumoniae* type 3, *Gene,* 167, 1, 1995.
26. Griffin, A. M., Morris, V. J., and Gasson, M. J., The *cpsABCDE* genes involved in polysaccharide production in *Streptococcus salivarius* ssp. *thermophilus* strain NCBF 2393, *Gene,* 183, 23, 1996.
27. Kolkman, M. A., van der Zeijst, B. A., and Nuijten, P. J., Functional analysis of glycosyltransferases encoded by the capsular polysaccharide biosynthesis locus of *Streptococcus pneumoniae* serotype 14, *J. Biol. Chem.,* 272, 19502, 1997.
28. Rubens, C. E., Haft, R. F., and Wessels, M. R., Characterization of the capsular polysaccharide genes of group B streptococci, *Dev. Biol. Stand.,* 85, 237, 1995.
29. Stingele, F., Neeser, J. R., and Mollet, B., Identification and characterization of the *eps* (exopolysaccharide) gene cluster from *Streptococcus thermophilus* Sfi6, *J. Bacteriol.,* 178, 1680, 1996.
30. Becker, A., Niehaus, K., and Puhler, A., Low-molecular weight succinoglycan is predominantly produced by *Rhizobium meliloti* strains carrying a mutated ExoP protein characterized by a periplasmic N-terminal domain and a missing C-terminal domain, *Mol. Microbiol.,* 16, 191, 1995.
31. Kolkman, M. A., van der Zeijst, B. A., and Nuijten, P. J., Diversity of capsular polysaccharide synthesis gene clusters in *Streptococcus pneumoniae, J. Biochem.* (*Tokyo*), 123, 937, 1998.
32. Kolkman, M. A., Wakarchuk, W., Nuijten, P. J., and van der Zeijst, B. A., Capsular polysaccharide synthesis in *Streptococcus pneumoniae* serotype 14: molecular analysis of the complete *cps* locus and identification of genes encoding glycosyltransferases required for the biosynthesis of the tetrasaccharide subunit, *Mol. Microbiol.,* 26, 197, 1997.
33. Morona, J. K., Guidolin, A., Morona, R., Hansman, D., and Paton, J. C., Isolation, characterization, and nucleotide sequence of IS*1202*, an insertion sequence of *Streptococcus pneumoniae, J. Bacteriol.,* 176, 4437, 1994.

34. Yother, J., Ambrose, K. D., and Caimano, M. J., Association of a partial H-rpt element with the type 3 capsule locus of *Streptococcus pneumoniae, Mol. Microbiol.*, 25, 201, 1997.
35. Morona, J. K., Morona, R., and Paton, J. C., Molecular and genetic characterization of the capsule biosynthesis locus of *Streptococcus pneumoniae* type 19B, *J. Bacteriol.*, 179, 4953, 1997.
36. Yim, H. H., Nittayarin, A., and Rubens, C. E., Analysis of the capsule synthesis locus, a virulence factor in group B streptococci, *Adv. Exp. Med. Biol.*, 418, 995, 1997.
37. Kuypers, J. M., Heggen, L. M., and Rubens, C. E., Molecular analysis of a region of the group B streptococcus chromosome involved in type III capsule expression, *Infect. Immun.*, 57, 3058, 1989.
38. Rubens, C. E., Heggen, L. M., Haft, R. F., and Wessels, M. R., Identification of *cpsD*, a gene essential for type III capsule expression in group B streptococci, *Mol. Microbiol.*, 8, 843, 1993.
39. Rubens, C. E., Heggen, L. M., and Kuypers, J. M., IS861, a group B streptococcal insertion sequence related to IS150 and IS3 of *Escherichia coli, J. Bacteriol.*, 171, 5531, 1989.
40. Rubens, C. E., Kuypers, J. M., Heggen, L. M., Kasper, D. L., and Wessels, M. R., Molecular analysis of the Group B streptococcal capsule genes, in *Genetics and Molecular Biology of Streptococci, Lactococci, and Enterococci*, G. M. Dunny, P. P. Cleary, and L. L. McKay (Eds.), Am. Soc. Microbiol., Washington, D.C., 1991, 179.
41. Crater, D. L. and van de Rijn, I., Hyaluronic acid synthesis operon (*has*) expression in group A streptococci, *J. Biol. Chem.*, 270, 18452, 1995.
42. van de Rijn, I., Bernish, B., and Crater, D. L., Analysis of hyaluronic acid capsule expression in group A streptococci, *Adv. Exp. Med. Biol.*, 418, 965, 1997.
43. Alberti, S., Ashbaugh, C. D., and Wessels, M. R., Structure of the *has* operon promoter and regulation of hyaluronic acid capsule expression in group A streptococcus, *Mol. Microbiol.*, 28, 343, 1998.
44. DeAngelis, P. L., Papaconstantinou, J., and Weigel, P. H., Molecular cloning, identification, and sequence of the hyaluronan synthase gene from group A *Streptococcus pyogenes, J. Biol. Chem.*, 268, 19181, 1993.
45. DeAngelis, P. L., Yang, N., and Weigel, P. H., The *Streptococcus pyogenes* hyaluronan synthase: sequence comparison and conservation among various group A strains, *Biochem. Biophys. Res. Commun.*, 199, 1, 1994.
46. Dougherty, B. A. and van de Rijn, I., Molecular characterization of hasA from an operon required for hyaluronic acid synthesis in group A streptococci, *J. Biol. Chem.*, 269, 169, 1994.
47. Weigel, P. H., Hascall, V. C., and Tammi, M., Hyaluronan synthases, *J. Biol. Chem.*, 272, 13997, 1997.
48. Arrecubieta, C., Lopez, R., and Garcia, E., Molecular characterization of *cap3A*, a gene from the operon required for the synthesis of the capsule of *Streptococcus pneumoniae* type 3: sequencing of mutations responsible for the unencapsulated phenotype and localization of the capsular cluster on the pneumococcal chromosome, *J. Bacteriol.*, 176, 6375, 1994.
49. Dougherty, B. A. and van de Rijn, I., Molecular characterization of hasB from an operon required for hyaluronic acid synthesis in group A streptococci. Demonstration of UDP-glucose dehydrogenase activity, *J. Biol. Chem.*, 268, 7118, 1993.

50. Crater, D. L., Dougherty, B. A., and van de Rijn, I., Molecular characterization of *hasC* from an operon required for hyaluronic acid synthesis in group A streptococci. Demonstration of UDP-glucose pyrophosphorylase activity, *J. Biol. Chem.*, 270, 28676, 1995.

51. DeAngelis, P. L., Papaconstantinou, J., and Weigel, P. H., Isolation of a *Streptococcus pyogenes* gene locus that directs hyaluronan biosynthesis in acapsular mutants and in heterologous bacteria, *J. Biol. Chem.*, 268, 14568, 1993.

52. DeAngelis, P. L. and Weigel, P. H., Characterization of the recombinant hyaluronic acid synthase from *Streptococcus pyogenes*, *Dev. Biol. Stand.*, 85, 225, 1995.

53. DeAngelis, P. L. and Weigel, P. H., Immuncchemical confirmation of the primary structure of streptococcal hyaluronan synthase and synthesis of high molecular weight product by the recombinant enzyme, *Biochemistry*, 33, 9033, 1994.

54. Sugahara, K., Schwartz, N. B., and Dorfman, A., Biosynthesis of hyaluronic acid by streptococcus, *J. Biol. Chem.*, 254, 6252, 1979.

55. Prehm, P., Synthesis of hyaluronate in differentiated teratocarcinoma cells; mechanism of chain growth, *Biochem. J.*, 211, 191, 1983.

56. Prehm, P., Synthesis of hyaluronate in differentiated teratocarcinoma cells; characterization of the synthase, *Biochem. J.*, 211, 181, 1983.

57. Stoolmiller, A. C. and Dorfman, A., The biosynthesis of hyaluronic acid by streptococcus, *J. Biol. Chem.*, 244, 236, 1969.

58. Dougherty, B. A. and van de Rijn, I., Molecular characterization of a locus required for hyaluronic acid capsule production in group A streptococci, *J. Exp. Med.*, 175, 1291, 1992.

59. Arrecubieta, C., Garcia, E., and Lopez, R., Demonstration of UDP-glucose dehydrogenase activity in cell extracts of *Escherichia coli* expressing the pneumococcal *cap3A* gene required for the synthesis of type 3 capsular polysaccharide, *J. Bacteriol.*, 178, 2971, 1996.

60. Arrecubieta, C., Lopez, R., and Garcia, E., Type 3-specific synthase of *Streptococcus pneumoniae* (Cap3B) directs type 3 polysaccharide biosynthesis in *Escherichia coli* and in pneumococcal strains of different serotypes, *J. Exp. Med.*, 184, 449, 1996.

61. Cartee, R., Forsee, T., Schutzbach, J., and Yother, J., unpublished data.

62. Caimano, M. J. and Yother, J., unpublished data.

63. Yother, J., A.D. Magee, C.L. Ventura, M.J. Caimano, unpublished data.

64. Böhringer, J., Fischer, D., Mosler, G., and Hengge-Aronis, R., UDP-glucose is a potential intracellular signal molecule in control of expression of σ^s and σ^s-dependent genes in *Escherichia coli*, *J. Bacteriol.*, 177, 413, 1995.

65. Wessels, M. R., Rubens, C. E., Benedi, V. J., and Kasper, D. L., Definition of a bacterial virulence factor: sialylation of the group B streptococcal capsule, *Proc. Natl. Acad. Sci. U.S.A.*, 86, 8983, 1989.

66. Wessels, M. R., Haft, R. F., Heggen, L. M., and Rubens, C. E., Identification of a genetic locus essential for capsule sialylation in type III group B streptococci, *Infect. Immun.*, 60, 392, 1992.

67. Haft, R. F., Wessels, M. R., Mebane, M. F., Conaty, N., and Rubens, C. E., Characterization of *cpsF* and its product CMP-N-acetylneuraminic acid synthetase, a group B streptococcal enzyme that can function in K1 capsular polysaccharide biosynthesis in *Escherichia coli*, *Mol. Microbiol.*, 19, 555, 1996.

68. Haft, R. F. and Wessels, M. R., Characterization of CMP-N-acetylneuraminic acid synthetase of group B streptococci, *J. Bacteriol.*, 176, 7372, 1994.

69. van de Rijn, I., Streptococcal hyaluronic acid: proposed mechanisms of degradation and loss of synthesis during stationary phase, *J. Bacteriol.*, 156, 1059, 1983.

70. Wessels, M. R., Goldberg, J. B., Moses, A. E., and DiCesare, T. J., Effects on virulence of mutations in a locus essential for hyaluronic acid capsule expression in group A streptococci, *Infect. Immun.*, 62, 433, 1994.

71. Cleary, P. P., McLandsborough, L., Ikeda, L., Cue, D., Krawczak, J., and Lam, H., High-frequency intracellular infection and erythrogenic toxin A expression undergo phase variation in M1 group A streptococci, *Mol. Microbiol.*, 28, 157, 1998.

72. Hardy, G. G. and Yother, J., unpublished data.

73. Paoletti, L. C., Ross, R. A., and Johnson, K. D., Cell growth rate regulates expression of group B Streptococcus type III capsular polysaccharide, *Infect. Immun.*, 64, 1220, 1996.

74. Husmann, L. K., Yung, D. L., Hollingshead, S. K., and Scott, J. R., Role of putative virulence factors of *Streptococcus pyogenes* in mouse models of long-term throat colonization and pneumonia, *Infect. Immun.*, 65, 1422, 1997.

75. Wessels, M. R. and Bronze, M. S., Critical role of the group A streptococcal capsule in pharyngeal colonization and infection in mice, *Proc. Natl. Acad. Sci. U.S.A.*, 91, 12238, 1994.

76. Dale, J. B., Washburn, R. G., Marques, M. B., and Wessels, M. R., Hyaluronate capsule and surface M protein in resistance to opsonization of group A streptococci, *Infect. Immun.*, 64, 1495, 1996.

77. LaPenta, D., Rubens, C., Chi, E., and Cleary, P. P., Group A streptococci efficiently invade human respiratory epithelial cells, *Proc. Natl. Acad. Sci. U.S.A.*, 91, 12115, 1994.

78. Schrager, H. M., Rheinwald, J. G., and Wessels, M. R., Hyaluronic acid capsule and the role of streptococcal entry into keratinocytes in invasive skin infection, *J. Clin. Invest.*, 98, 1954, 1996.

79. Gibson, R. L., Lee, M. K., Soderland, C., Chi, E. Y., and Rubens, C. E., Group B streptococci invade endothelial cells: type III capsular polysaccharide attenuates invasion, *Infect. Immun.*, 61, 478, 1993.

80. Martin, T. R., Ruzinski, J. T., Rubens, C. E., Chi, E. Y., and Wilson, C. B., The effect of type-specific polysaccharide capsule on the clearance of group B streptococci from the lungs of infant and adult rats, *J. Infect. Dis.*, 165, 306, 1992.

81. Wessels, M. R., Benedi, V.-J., Kasper, D. L., Heggen, L. M., and Rubens, C. E., Type III capsule and virulence of group B streptococci, in *Genetics and Molecular Biology of Streptococci, Lactococci, and Enterococci*, G. M. Dunny, P. P. Cleary, and L. L. McKay (Eds.), Am. Soc. Microbiol., Washington, D.C., 1991.

82. Marques, M. B., Kasper, D. L., Pangburn, M. K., and Wessels, M. R., Prevention of C3 deposition by capsular polysaccharide is a virulence mechanism of type III group B streptococci, *Infect. Immun.*, 60, 3986, 1992.

83. Watson, D. and Musher, D. M., Interruption of capsule production in *Streptococcus pneumoniae* serotype 3 by insertion of transposon Tn916, *Infect. Immun.*, 58, 3135, 1990.

84. Watson, D. A., Musher, D. M., and Verhoef, J., Pneumococcal virulence factors and host immune responses to them, *Eur. J. Clin. Microbiol. Infect. Dis.*, 14, 479, 1995.

85. Mäkelä, P. and Stocker, B. A. D., Genetics of polysaccharide biosynthesis, *Ann. Rev. Genet.*, 3, 291, 1969.

86. Dawson, M. H. and Sia, R. H. P., *In vitro* transformation of pneumococcal types. I. A technique for inducing transformation of pneumococcal types *in vitro, J. Exp. Med.*, 54, 681, 1931.

87. Langvad-Nielson, A., Change of capsule in the pneumococcus, *Acta. Path. et Microbiol. Scand.*, 21, 362, 1944.

88. Austrian, R., Bernheimer, H. P., Smith, E. E. B., and Mills, G. T., Simultaneous production of two capsular polysaccharides by pneumococcus. II. The genetic and biochemical basis of binary capsulation., *J. Exp. Med.*, 110, 585, 1959.

89. Ephrussi-Taylor, H., Genetic aspects of transformations of pneumococci, *Cold Spr. Harb. Symp. Quant. Biol.*, 16, 445, 1951.

90. Ravin, A. W., Reciprocal capsular transformations of pneumococci, *J. Bacteriol.*, 77, 296, 1959.

91. Ravin, A. W., Linked mutations borne by deoxyribonucleic acid controlling the synthesis of capsular polysaccharide in pneumococcus, *Genetics*, 45, 1387, 1960.

92. Bernheimer, H. P., Wermundsen, I. E., and Austrian, R., Mutation in pneumococcus type 3 affecting multiple cistrons concerned with the synthesis of capsular polysaccharide, *J. Bacteriol.*, 96, 1099, 1968.

93. Jackson, S., Genetic aspects of capsule transformation in the pneumococcus, *Br. Med. Bull.*, 18, 24, 1962.

94. Mills, G. T. and Smith, E. B., Biosynthetic aspects of capsule formation in the pneumococcus, *Br. Med. Bull.*, 18, 27, 1962.

95. Smith, E. B., Mills, G. T., and Bernheimer, H. P., Biosynthesis of pneumococcal polysaccharides. I. Properties of the system synthesizing type III capsular polysaccharide, *J. Biol. Chem.*, 236, 2179, 1961.

96. Lacks, S., Mannarelli, B., Springhorn, S., and Greenberg, B., Genetic basis of the complementary *Dpn*I and *Dpn*II restriction systems of *S. pneumoniae*: an intracellular cassette mechanism, *Cell*, 46, 993, 1986.

97. Coffey, T. J., Enright, M. C., Daniels, M., Morona, J. K., Morona, R., Hryniewicz, W., Paton, J. C., and Spratt, B. G., Recombinational exchanges at the capsular polysaccharide biosynthetic locus lead to frequent serotype changes among natural isolates of *Streptococcus pneumoniae*, *Mol. Microbiol.*, 27, 73, 1998.

98. Nesin, M., Ramirez, M., and Tomasz, A., Capsular transformation of a multidrug-resistant *Streptococcus pneumoniae in vivo*, *J. Infect. Dis.*, 177, 707, 1998.

99. Bernheimer, H. P. and Wermundsen, I. E., Unstable binary capsulated transformants in pneumococcus, *J. Bacteriol.*, 98, 1073, 1969.

chapter eight

Capsular polysaccharides
of Staphylococcus aureus

Jean C. Lee and Chia Y. Lee

Contents

Introduction 185
Composition of the *S. aureus* capsular polysaccharides............................186
The serotype 1 gene cluster 187
    Sequence analysis of *cap1* 188
    Transcriptional analysis of *cap1* 188
    The *cap1* element...191
The serotype 5 and 8 gene clusters 191
    Sequence analysis of *cap5* and *cap8* 192
    Transcriptional analysis of *cap8*...195
    Functional analysis of *cap5H*...198
    Functional analysis of *cap5O*...199
    Functional analysis of *cap5G* and *cap5P* 199
Regulation of capsule expression...200
Future directions 202
Acknowledgments 202
References...203

Introduction

Staphylococcus aureus is a Gram-positive coccus that is recognized worldwide
as an important opportunistic bacterial pathogen. The organism asymptom-
atically colonizes the anterior nares of about 35% of normal healthy individuals. In addition, *S. aureus* can colonize the intestinal and vaginal mucosa,
and it may reside transiently on the skin surface (primarily colonized by

coagulase-negative staphylococci). If mucosal barriers are breached or host immune defenses are impaired, *S. aureus* may produce a diverse array of human and animal diseases. These range from rather mild skin infections, such as impetigo, folliculitis and furuncles, to more invasive diseases, such as infections originating from prosthetic devices, wound infections, osteomyelitis, and bacteremia with metastatic complications. Toxin-mediated diseases caused by *S. aureus* include food poisoning, toxic shock syndrome, and scalded skin syndrome. The bacterial components and secreted products that affect the pathogenesis of *S. aureus* infections are numerous and include surface-associated adhesins, exoenzymes, exotoxins, and capsular polysaccharide.

Capsule production by the staphylococcus was first described in 1931 by Gilbert.[1] Because capsule detection methods were crude (India ink negative staining, colony morphology on agar plates and in serum-soft agar, and lack of cell-associated clumping factor), only a few strains of *S. aureus* were recognized as capsule positive. These highly encapsulated strains, typified by strains M and Smith diffuse, resisted phagocytosis and were virulent in mice.[2-4] J.C. Lee et al.[5] showed that the capsule was responsible for the enhanced virulence of a mucoid strain by creating isogenic, transposon-induced mutants and demonstrating that the nonmucoid mutants were less virulent in mice than the mucoid parental strain. Lin et al.[6] confirmed these results in a lethality study in which mice were challenged with strain M or a mutant with capsule genes deleted.

With more sensitive serologic methods, capsular polysaccharides can now be detected on ~90% of *S. aureus* strains. Although 11 capsular serotypes have been described, most clinical isolates of *S. aureus* belong to capsular types 5 or 8.[7,8] Because these strains produce small amounts of capsular polysaccharide that cannot be visualized on the cells by negative staining, serotype 5 and 8 strains are called microencapsulated, as suggested by Wilkinson.[9] This term distinguishes them from the heavily encapsulated *S. aureus* strains belonging to serotypes 1 and 2, which are rarely isolated. The role that the type 5 and 8 capsules play in the pathogenesis of staphylococcal infections is dependent on the bacterial growth conditions and on the animal model of infection tested.

Composition of the S. aureus *capsular polysaccharides*

Capsules or microcapsules from at least 18 *S. aureus* strains have been characterized to some extent,[4] but polysaccharides purified from only four of the eleven capsule types have been biochemically characterized (Figure 8.1). Strain M expresses a type 1 capsule composed of taurine, 2-acetamido-2-deoxy-D-fucose, and 2-acetamido-2-deoxy-D-galacturonic acid in the molar ratios of 1:2:4.[10,11] Two other strains, Dp[7] and SA1 mucoid,[12] produce capsules that are serologically and biochemically similar to that produced by strain M. The Smith diffuse capsule (serotype 2) consists of equimolar amounts of 2-acetamido-2-deoxy-D-glucuronic acid and 2-acetamido-2-deoxy-L-alanyl

Type 1:   (-->4)-α-D-GalNAcA-(1 --> 4)-α- D-GalNAcA-(1 -->3)-α-D-FucNAc -(1-->)$_n$

   (A taurine residue is amide-linked to every fourth D-GalNAcA residue)

Type 2:   (-->4)-ß-D-GlcNAcA- (1-->4)-ß-D-GlcNAcA-(L-alanyl)-(1-->)$_n$

Type 5:   (-->4)-3-O-Ac-ß-D-ManNAcA-(1-->4)-α-L-FucNAc- (1-->3)-ß-D-FucNAc-(1-->)$_n$

Type 8:   (-->3)-4-O-Ac-ß-D-ManNAcA-(1-->3)-α-L-FucNAc-(1-->3)-β-D-FucNAc-(1-->)$_n$

Figure 8.1   Repeating unit structures of the *S. aureus* capsular polysaccharides. Only polysaccharides from serotypes 1, 2, 5, and 8 have been chemically characterized, although 11 serotypes have been described. (From Lee, J. C., Xu, S. L., Albus, A., and Livolsi, P. J., Genetic analysis of type 5 capsular polysaccharide expression by *Staphylococcus aureus*, *J. Bacteriol.*, 176, 4883, 1994.)

D-glucuronic acid with β-(1→4) linkages.[13] Type 5 and 8 microcapsular polysaccharides are structurally very similar to each other and to the capsule made by strain T, described by Wu and Park.[14] They are trisaccharide repeating units comprising 2-acetamido-2-deoxy-D-mannuronic acid linked to two 2-acetamido-2-deoxy-fucose residues.[15-17] The serotype 5 and 8 polysaccharides differ only in the linkages between the sugars and in the sites of O-acetylation of the mannosaminuronic acids, yet they are serologically distinct. The biosynthetic pathway leading to production of the *S. aureus* capsular polysaccharide has not been elucidated.

The serotype 1 gene cluster

The mucoid colony morphology associated with the expression of serotype 1 capsular polysaccharide (CP1) by *S. aureus* strain M is unstable, as demonstrated by a spontaneous rate of capsule loss of 0.01% when the bacterium was cultivated at 37°C.[18] The frequency of capsule loss increased to a range of 1 to 38% of the total bacterial cells when they were cultivated at 43°C. This instability suggests that the capsule genes might be phage- or plasmid-encoded. However, when strain M was cured of a prophage or its 19-kb plasmid, it remained mucoid, suggesting that the genes for capsule production were, in fact, chromosomal. In a later study, loss of mucoidy was found to be due to random mutations within the genes in the *cap1* locus, rather than a result of gene rearrangement.[19]

   To clone the *S. aureus* serotype 1 capsule gene cluster, C.Y. Lee[18] prepared a collection of nonmucoid mutants that were derived from strain M by cultivation at 43°C or by mutagenesis with ethyl methane-sulphonate. One nonencapsulated mutant was transduced with a phage lysate of *S. aureus* RN4220 carrying a genomic library from strain M. The isolation of several transductants with a mucoid colony morphology suggested that the recombinant plasmids carried chromosomal fragments that complemented the capsule defect in the nonencapsulated mutant strain. One of these DNA fragments was used as a probe against a strain M genomic library constructed in *Escherichia coli*. Several cosmids hybridized with the probe, and subclones from these cosmids were able to complement various CP1-negative mutants. An extensive series of complementation experiments localized the *cap1* genes to a 14.6-kb region of the strain M chromosome.[6]

Sequence analysis of cap1

Sequence analysis of the *cap1* gene cluster (Figure 8.2) revealed 13 closely linked open reading frames (ORFs) that were transcribed in one orientation (GenBank accession number U10927).[6] The results of complementation tests and site-specific mutation studies indicated that all 13 genes were essential for CP1 expression. The deduced amino acid sequences of Cap1A to Cap1M were compared with sequences in the databanks. The authors reported a high homology between CapL and VipA, moderate homology between CapI and VipB, and limited homology between CapM and VipC. VipA, VipB, and VipC are the structural genes specific for *Salmonella typhi* Vi capsular antigen, a homopolymer of GalNAcA. Because CP1 also contains GalNAcA, it is likely that Cap1L, Cap1I, and Cap1M are involved in the biosynthesis of this sugar. Cap1G showed homology to the NodL-LacA-CysE family of acetyltransferases. Although the site of O-acetylation of CP1 is unknown,[12] by analogy with the chemical structure of other *S. aureus* capsules, it is probably the GalNAcA residue.

Transcriptional analysis of cap1

C.Y. Lee's laboratory performed a detailed transcriptional analysis of the *cap1* genes.[20] Plasmids carrying segments of the 14.6-kb CP1 gene region were used to complement chemical mutants with lesions mapped to each of the 13 *cap1* genes. These genetic complementation tests indicated that there were six promoters within the *cap1* gene cluster located upstream of *cap1A*, *cap1E*, *cap1F*, *cap1G*, *cap1H*, and *cap1J*. Northern hybridization analyses using the 14.6-kb DNA fragment as a probe revealed several hybridizing bands, ranging from 14 to 0.3 kb in size. If the RNA was hybridized with internal fragments of individual genes as probes, the 14-kb band was detected by all the probes tested, thereby suggesting that the 13 *cap1* genes were cotranscribed into a single polycistronic message. When Ouyang and Lee deleted the promoter region upstream of *cap1A*, neither the 14-kb transcript nor the smaller bands could be detected, thus suggesting that the smaller bands were probably the degraded or processed products of the major transcript.

Since the genetic complementation tests indicated the presence of six internal promoters within the *cap1* cluster, Ouyang and Lee hypothesized that the amount of messenger RNA transcribed from the internal promoters was too little to be detected by Northern blotting.[20] As an alternative

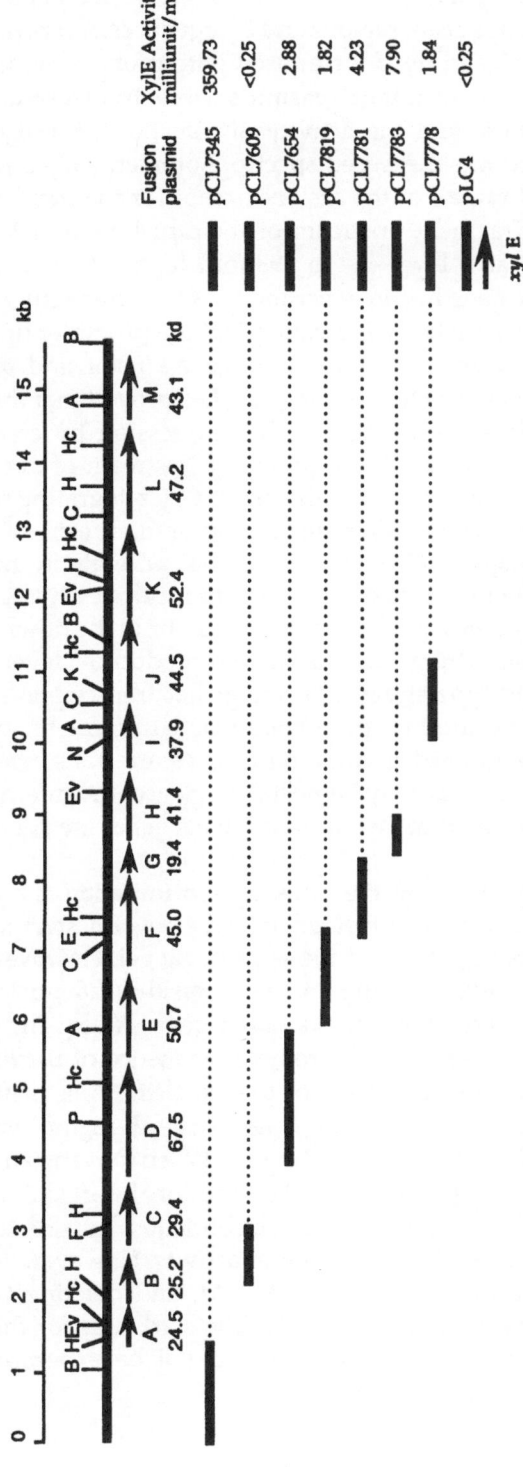

Figure 8.2 Genetic and restriction map of the *cap1* gene cluster. ORFs are indicated by arrows. Capital letters under arrows represent the corresponding genes. The predicted molecular mass of each predicted polypeptide is shown under each gene. Relative promoter activities within the *cap1* gene cluster were assessed by fusing the various potential promoter regions to the promoterless *xylE* reporter gene in a plasmid construct. A, *AccI*; B, *BglII*; C, *ClaI*; E, *EcoRI*; Ev, *EcoRV*; F, *FspI*; H, *HindIII*; Hc, *HincII*; K, *KpnI*; P, *PstI*; N, *NcoI*. (Adapted from Ouyang, S. and Lee, C. Y., Transcriptional analysis of type 1 capsule genes in *Staphylococcus aureus*, *Mol. Microbiol.*, 23, 473, 1997. With permission.)

approach, they tested the promoter activities of the *cap1* genes by creating fusions to a promoterless reporter gene, *xylE* from *Pseudomonas putida*. DNA fragments containing the upstream regions and 5′ sequences of *cap1A*, *cap1C*, *cap1E*, *cap1F*, *cap1G*, *cap1H*, and *cap1I* were fused with promoterless *xylE* in a plasmid (Figure 8.2). The resultant plasmids were transferred into a CP1-negative mutant strain, and the activity of catechol 2,3-dioxygenase encoded by the *xylE* gene was measured spectrophotometrically. Promoter activities were detected from all of the fusions except for the *cap1C* fusion. The fusion of the DNA fragment upstream of the *cap1A* gene yielded the maximum XylE activity, at a level 45- to 198-fold higher than the other fusions. The results of the gene fusion experiments were consistent with the results of the Northern blot analyses, indicating that the promoter upstream of *cap1A* is the primary promoter for the *cap1* gene cluster and that the internal promoters are much weaker. Ouyang and Lee proved that the internal promoters were sufficiently active for CP1 expression by creating a CP1-negative mutant strain in which the promoter upstream of *cap1A* was deleted.[20] CP1 expression was restored to the mutant by integrating a single copy of a DNA fragment comprising the promoter upstream of *cap1A* together with *cap1A* through *cap1E* at the phage L54a *attB* site in the chromosome (remote from the *cap1* locus). This construct physically separated the primary promoter together with the first five genes from the downstream *cap1* genes. The recombinant strain was mucoid and produced a level of CP1 similar to that of the wild-type strain, indicating that transcription of the downstream *cap1* genes by internal promoters was sufficient for capsule synthesis. The authors postulated that the *cap1* genes are transcribed as a single polycistronic message that may be unstable. Hence, the internal promoters may function to ensure that the promoter-distal genes are adequately expressed.

S1 nuclease mapping identified the transcription initiation site for the 14.6-kb transcript.[20] Transcription appeared to have several start sites as evidenced by a multibanding pattern on the sequencing gel. However, there was a prominent band corresponding to the A residue 26 nucleotides upstream from the ATG start codon of the *cap1A* gene. A sequence of 5′-TATAAT-3′ matching the consensus −10 promoter sequence of *Bacillus* and *E. coli* was found 5 nucleotides upstream of the start site. A sequence (5′-TTGCAA-3′) with two mismatches to the consensus −35 region was also found 18 nucleotides upstream from the −10 region. When Ouyang and Lee[20] constructed serial deletions upstream of the transcriptional start site containing the promoter region, deletion up to the nucleotide just outside the predicted −35 region did not affect the XylE reporter activity. However, deleting the −35 region reduced promoter activity ~30-fold, and deleting the −10 region abolished promoter activity completely. The start sites for the transcripts arising from the internal promoters could not be experimentally determined due to their weak activities.

The cap1 *element*

C.Y. Lee proposed that the CP1 gene cluster is associated with a discrete genetic element.[21] He made the observation that a 21.8-kb DNA fragment encompassing the *cap1* genes and ~7 kb of flanking DNA failed to hybridize to chromosomal DNA preparations from 14 *S. aureus* strains that did not elaborate CP1. Further characterization of a 53-kb region around the *cap1* genes revealed that at least 18 kb of DNA flanking the gene cluster were absent in the other staphylococcal strains tested. The "*cap1* element" was found to be 33.3 to 35.8 kb in size, extending from approximately 11 kb upstream of *cap1A* to about 7.5 kb downstream from *cap1M*. The *cap1* element is unlikely to be a compound transposon since repeat sequences were located on only one end of the genetic element. Other possibilities are that the *cap1* genes are located on a defective phage or on an integrated plasmid. The *cap1* locus mapped to the *Sma*I-G fragment on the physical map of NCTC 8325,[22] but at a different site on this fragment than the allelic *cap5* and *cap8* loci (described below).

One of the most interesting results revealed by the studies from C.Y. Lee's group was that strain M contained two *cap* gene clusters.[23] Sau and Lee created a *cap1* mutant strain with the entire *cap1* gene cluster deleted from the bacterial chromosome. Genomic DNA from this mutant was digested with restriction enzymes and probed with a DNA fragment carrying the cloned *cap8* genes. The positive signal obtained from the *cap1* mutant strain led the investigators to believe that strain M carried a second capsule gene cluster distinct from the *cap1* genes. They screened a cosmid library prepared from strain M with a *cap8* gene probe. The restriction map of one of the positive clones was similar, but not identical, to that of the *cap8* locus. These results suggest that strain M can synthesize two different capsular polysaccharides, and that the genes involved in their biosynthetic pathways are distinct but related.

The serotype 5 and 8 gene clusters

A genetic analysis of the capsule type 5 (CP5) genes was reported in 1991 when Albus et al.[24] described transposon mutagenesis experiments that resulted in the isolation of mutants altered in CP5 expression. Tn*918*, carrying the *tetM* gene, was introduced into strain Reynolds by filter mating, and a capsule-deficient transconjugate was recovered. Mutant JL236 produced levels of CP5 that were ~10% of the level expressed by the parental strain. The CP5-deficient mutant carried a single Tn*918* insert, and the transposon insertion was shown to be responsible for the capsule-deficient phenotype by transformation of the chromosomal mutation back into the parental strain, reconstituting the CP5-deficient phenotype.

The mutated region of the JL236 chromosome was targeted by selecting a DNA fragment from a chromosomal library of mutant JL236 that conferred tetracycline resistance on *E. coli*.[25] An ~27-kb *Eco*RI fragment containing the transposon insertion was cloned into a cosmid vector. In the absence of tetracycline selection, Tn*918* was spontaneously excised, thereby resulting in a plasmid containing 9.6 kb of *S. aureus* DNA flanking the Tn*918* insertion site. The 9.6-kb DNA fragment was used to screen a cosmid library prepared from the wild-type strain Reynolds. Positive colonies were identified by colony hybridization, and one cosmid clone (pJCL19) was mapped with restriction enzymes. Subclones of pJCL19 were able to complement EMS-derived, capsule-negative mutant strains in trans. By probing genomic DNA digests from a variety of *S. aureus* strains with DNA fragments from pJCL19, J. C. Lee's laboratory identified segments of the cosmid clone that were common among strains, segments that showed restriction fragment length polymorphism, and segments in the central region that hybridized only to DNA from strains of serotypes 2, 4, and 5.

Sau and C.Y. Lee[23] attempted to target the type 8 capsule (CP8) genes by their homology to the *cap1* genes. DNA from the serotype 8 strain Becker did not hybridize with the *cap1* genes under high-stringency, but did hybridize under low-stringency conditions. On the basis of this moderate level of homology, the investigators isolated a clone carrying the *cap8* genes from a chromosomal library made from strain Becker. Recombinant plasmids carrying strain Becker DNA were mapped with restriction enzymes and characterized for their ability to complement CP8-negative mutants derived by EMS treatment. The results of the complementation experiments showed that 18 CP8-negative mutants could be mapped to six complementation groups within an ~20.5-kb region of strain Becker chromosomal DNA. These results indicated that, like the *cap1* genes, the *cap8* genes were clustered together on the bacterial chromosome. Sau and Lee used different regions of the *cap8* gene cluster as probes in Southern analyses.[23] Whereas probes made from the ends of the gene cluster hybridized to DNA from all nine *S. aureus* strains tested, a centrally located DNA fragment hybridized to only four of nine strains tested.

Sequence analysis of cap5 and cap8

In a collaborative effort, the nucleotide sequences of the *cap5* and *cap8* gene clusters (GenBank accession numbers U81973 and U73374, respectively) were recently determined.[26] C.Y. Lee's laboratory sequenced the *cap8* gene cluster from strain Becker and identified 16 ORFs within a 17.5-kb region (Figure 8.3). The genes, designated *cap8A* through *cap8P*, are tightly clustered and transcribed in one orientation. T. J. Foster's laboratory sequenced a 7.1-kb DNA segment of the *cap5* gene cluster from strain Newman, and J. C. Lee's laboratory sequenced a 12.6-kb DNA segment of the *cap5* gene cluster from strain Reynolds. The region of overlapping sequence between the two serotype 5 strains was 1.55 kb in length and included most of *cap5F* and half of *cap5G*. Six nucleotide differences between the two strains were found in the overlapping segment, but only one of these resulted in a change in the

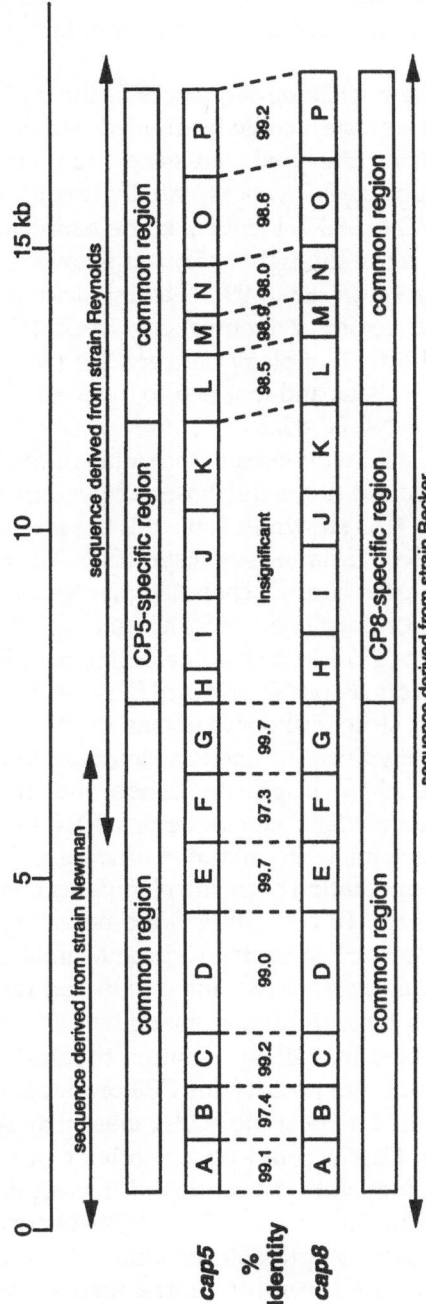

Figure 8.3 Comparison of *S. aureus cap5* and *cap8* gene clusters. The *cap5* sequence was derived from strains Newman and Reynolds, and the *cap8* sequence from strain Becker as shown. Gene designations are shown in boxes. Percentage identity indicates the amino acid identity of the deduced proteins of the two clusters. Both gene clusters are transcribed from left to right. (From Sau, S., Bhasin, N., Wann, E. R., Lee, J. C., Foster, T. J., and Lee, C. Y., The *Staphylococcus aureus* allelic genetic loci for serotype 5 and 8 capsule expression contain the type-specific genes flanked by common genes, *Microbiol.,* 143, 2395, 1997. With permission.)

amino acid sequence of the deduced protein. Amino acid 196 of Cap5G was shown to be serine in strain Reynolds and proline in strain Newman. At the equivalent position in Cap8G, the deduced amino acid was proline. Sequence analysis of the combined 18.1-kb region from the two CP5 strains revealed 16 contiguous ORFs that were transcribed in the same orientation and were named *cap5A* through *cap5P*.

Comparative analysis of the nucleotide sequences of the *cap5* and *cap8* loci revealed that each of the gene clusters could be divided into three regions (Figure 8.3). Region 1, comprising the predicted gene products of *cap5A* through *cap5G* and *cap8A* through *cap8G*, was essentially identical between the serotype 5 and 8 strains (97.3 to 99.7% identical at the amino acid level). Likewise, region 3, comprising *cap5L* through *cap5P*, was shown to be essentially identical to *cap8L* through *cap8P* (98 to 99.2% amino acid identity). In contrast, the centrally located region 2 (comprising *cap5(8)H*, *cap5(8)I*, *cap5(8)J*, and *cap5(8)K*) showed little homology between the two gene clusters. The *cap5* and *cap8* loci were allelic and mapped to the *Sma*I-G fragment on the physical map of *S. aureus* NCTC 8325.[27]

A comparison of the amino acid sequences of the putative *cap5* and *cap8* gene products with sequences found in the databases allowed us to predict functions for 15 of the 16 ORFs.[26] As shown in Table 8.1, the majority of the capsule genes appear to be involved in amino sugar synthesis. The remainder are likely involved in sugar transfer, capsule chain length regulation, transport, and polymerization. *cap5(8)A*, *cap5(8)B*, *cap5(8)C*, and *cap5(8)D* share a high degree of homology (60 to 72% identity at the amino acid level) and the same organization as *cap1A*, *cap1B*, *cap1C*, and *cap1D*, respectively. Based on amino acid homologies and structural comparisons of the polysaccharides, these genes are probably involved in functions common to CP1, CP5, and CP8 biosynthesis, such as chain length regulation and synthesis of FucNAc (Table 8.1). The function of CapC cannot be predicted because it is homologous to streptococcal proteins with unknown functions (Table 8.1).

Sau et al. showed by complementation of chemical mutants and by gene-specific mutagenesis that 11 of the 16 *cap8* ORFs were necessary for CP8 expression by *S. aureus* strain Becker.[28] In contrast, chromosomal mutations in five of the genes (*cap8A*, *cap8B*, *cap8C*, *cap8J*, and *cap8P*) had no effect on the level of CP8 expression. The *cap8A* mutation, made by a five-codon in-frame insertion, may have resulted in a silent mutation that did not affect CP8 expression. The *cap8B* mutant was positive for CP8 production, but the expressed polysaccharide exhibited a lower molecular mass than that made by the wild-type strain Becker. Cap8B could be a regulator of CP8 chain length, consistent with its homology with the conserved nucleotide-binding motif of the C-terminal half of *Rhizobium meliloti* ExoP.[26] The N-terminal half of ExoP has been implicated in chain-length determination of succinoglycan synthesis, whereas the C-terminal half seems to exert a regulatory function following nucleotide binding.

A mutant constructed with a 226-bp deletion within *cap8C* also expressed wild-type levels of CP8. The putative gene product of *cap8C* has a moderate degree of homology to EpsB of *Streptococcus thermophilus* and to Cps19fB of *S. pneumoniae*, neither of which has been functionally characterized. Therefore, neither a phenotype for the *capC* mutation nor a putative function for the gene product can be assigned at this time.

A *cap8J* mutant was also CP8 positive, but it did not react with a monoclonal antibody specific to O-acetylated CP8.[29] Cap8J shows homology to the C-terminal region of numerous bacterial acetyltransferases, including limited homology with Cap1G and Cap5H. Recent studies have revealed that a strain Reynolds *cap5H* mutant produced wild-type levels of O-deacetylated CP5 (discussed below).[30] By analogy, it is quite likely that a mutation in *cap8J* would not affect levels of CP8 production by strain Becker.

A mutant with a 69-bp deletion in *cap8P* still expressed CP8.[28] Likewise, a *cap5P* knockout mutant remained CP5-positive.[31] As discussed below, it is quite likely that another gene exists on the *S. aureus* chromosome that has a function similar to that of Cap5(8)P.

Transcriptional analysis of cap8

Sau et al. used Northern hybridizations to study the transcription of the *cap8* genes in strain Becker.[28] If a 17.2-kb DNA fragment containing almost the entire *cap8* gene cluster was used as a probe, a somewhat smeared ~17-kb transcript was detected. The Northern blot pattern was essentially the same if individual genes (*cap8A*, *cap8D*, *cap8I*, or *cap8P*) were used as probes. These results suggest that the 16 *cap8* genes are cotranscribed as a single polycistronic message from a promoter upstream of *cap8A* and that the smearing may be caused by degradation of the large transcript. However, the genetic complementation experiments performed in C.Y. Lee's laboratory indicated that there were numerous internal promoters within the *cap8* locus that might initiate transcription of the internal genes. They hypothesized that the amount of messenger RNA transcribed from the *cap8* internal promoters was too little to be detected by Northern blotting. Therefore, they fused several of the DNA fragments containing the potential promoter regions within *cap8* to the promoterless *xylE* reporter gene. The plasmid constructs were transferred to *S. aureus* Becker, and the *xylE* activities of the strains were measured spectrophotometrically. The experimental results indicated that a fusion containing the upstream region of *cap8A* was 10- to 25-fold more active than strains containing other fusions. These results confirm that *cap8* genes are transcribed primarily from the promoter upstream of *cap8A* and that the internal promoters are all weak.

Because C.Y. Lee's laboratory performed transcriptional analyses using Northern blots and *xylE* reporter gene fusions for both the *cap1* and *cap8* gene clusters, some comparisons can be made between the two systems.[20,28]

Table 8.1   Homology of *cap5* and *cap8* Putative Gene Products with Related Sequences in the Databases

ORF	Size	Homologous proteins	% Identity[a]	Function
cap5(8)A	222	*Staphylococcus aureus* Cap1A	63/167	Type 1 CP synthesis
		Streptococcus thermophilus EpsC	33/183	Probable chain-length regulator
		Streptococcus pneumoniae CpsC	32/183	Probable chain-length regulator
cap5(8)B	228	*S. aureus* Cap1B	62/228	CP1 synthesis
		S. pneumoniae CpsD	41/117	Probable chain-length regulator
		S. thermophilus EpsD	40/117	Probable chain-length regulator
		Rhizobium meliloti ExoP	30/113	Chain-length regulator
cap5(8)C	254	*S. aureus* Cap1C	59/254	CP1 synthesis
		S. thermophilus EpsB	31/165	Exopolysaccharide synthesis
		S. pneumoniae CpsB	31/165	Type 19F CP synthesis
cap5(8)D	607	*S. aureus* Cap1D	72/578	CP1 synthesis
		Yersinia enterocolitica TrsG	59/192	Lipopolysaccharide outer core synthesis
		Vibrio cholerae ORF11	62/114	O139 antigen synthesis
		Bordetella pertussis BplL	55/195	Lipopolysaccharide synthesis
cap5(8)E	342	*Methanococcus jannaschii* ProtD	43/239	CP synthesis
cap5(8)F	371	*Acholeplasma laidlawii*	58/369	Putative nucleotide-binding protein
cap5(8)G	374	*M. jannaschii* BplD homolog	46/71	Lipopolysaccharide synthesis
		B. pertussis BplD	38/76	Lipopolysaccharide synthesis
		Pseudomonas aeruginosa WbpI	40/80	Lipopolysaccharide synthesis
		Burkholderia solanacearum EpsC	33/68	Exopolysaccharide synthesis
		E. coli RffE	29/82	UDP-GlcNAc 2-epimerase
cap5H	208	*E. coli* Cat4	62/88	Chloramphenicol O-acetyltransferase

Gene	Length	Homolog	% identity[a]	Function
cap8H	360	None		
cap5I	369	None		
cap8I	464	None		
cap8J	185	R. meliloti NodL	54/58	O-acetyltransferase
		E. coli LacA	48/45	Thiogalactoside O-acetyltransferase
cap5K	394	None		
cap8K	412	None		
cap5(8)L	401	E. coli WcaI	39/41	Probable glycosyltransferase
cap5(8)M	185	S. enterica WbaP	65/41	Galactosyltransferase
		Xanthomonas campestris GumD	68/41	Galactosyltrasferase
		Streptococcus agalactiae CpsD	68/41	Galactosyltransferase
cap5(8)N	295	Salmonella typhimurium GalE	32/75	UDP-Glc 4-epimerase
		E. coli GalE	31/74	UDP-Glc 4-epimerase
cap5(8)O	420	M. jannaschii MJ0428	65/126	Probable UDP-ManNAc dehydrogenase
		B. solanacearum EpsD	53/208	Exopolysaccharide synthesis
		E. coli RffD	54/157	UDP-ManNAc dehydrogenase
		P. aeruginosa WbpA	31/184	Lipopolysaccharide biosynthesis
cap5(8)P	391	B. subtilis YvyH	58/373	Probable UDP-GlcNAc 2-epimerase
		S. pneumoniae cps19bK	55/194	Type 19b capsule synthesis
		S. pneumoniae cps19fK	56/194	Type 19f capsule synthesis
		B. solanacearum EpsC	48/210	Exopolysaccharide synthesis
		E. coli RffE	58/176	UDP-GlcNAc 2-epimerase

[a] Percentage identity of amino acid sequence/length of the homologous region.

Adapted from Sau, S., Bhasin, N., Wann, E. R., Lee, J. C., Foster, T. J., and Lee, C. Y., The *Staphylococcus aureus* allelic genetic loci for serotype 5 and 8 capsule expression contain the type-specific genes flanked by common genes, *Microbiol.*, 143, 2395, 1997. With permission.

Both capsule gene regions appeared to be transcribed from a primary promoter located at the beginning of the operon. However, the promoter activity of the primary *cap1* promoter was ~60-fold stronger than that of the primary *cap8* promoter. Likewise, the internal promoters of the *cap1* gene cluster were 10- to 50-fold stronger than those within the *cap8* gene cluster. Of note was the observation that the internal promoters of the *cap1* locus showed about the same activity as the *cap8* primary promoter.[20,28] These findings are consistent with a report indicating that *S. aureus* CP1 production exceeded CP8 production by a factor of ~60.[32] Promoter mapping experiments have revealed that, unlike the *cap1* promoter, the *cap8* promoter lacks a consensus −35 sequence.[46] This difference may explain why CP1 production by *S. aureus* greatly exceeds CP8 production.

Functional analysis of cap5H

The predicted amino acid sequence of *cap5H* encodes a protein of ~26 kDa with a high degree of homology to a family of bacterial O-acetyltransferase genes. A mutant of *S. aureus* strain Reynolds containing a Tn*918* insertion within *cap5H* produced wild-type levels of O-deacetylated CP5.[30] When provided in trans, *cap5H* complemented the gene defect in the O-deacetylated mutant. Southern blot analysis showed that genes similar to *cap5H* were present only in strains of *S. aureus* belonging to capsular serotypes 2, 4, and 5. The biological significance of CP5 O-acetylation was examined in an *in vitro* opsonophagocytic assay.[30] Both the parental and mutant strains were opsonized for phagocytic killing by CP5 antibodies and complement. However, the mutant was significantly more susceptible than the parent strain to opsonophagocytosis in the presence of teichoic acid antibodies and complement, teichoic acid antibodies alone, or complement alone. These differences may reflect greater exposure of the teichoic acid polymer on the bacterial surface in the absence of the O-acetyl substituents on the CP5 backbone. The biological differences between the two strains were further explored in a mouse model of staphylococcal infection.[30] The bacteria were inoculated intraperitoneally, and virulence was measured by the ability of the organisms to access the bloodstream and avoid immune clearance mechanisms. The parental strain Reynolds achieved higher concentrations in the bloodstream and was more efficient in provoking metastatic infection of the kidneys than the *cap5H* mutant. When *cap5H* was provided to the mutant in trans, it fully restored CP5 O-acetylation. The virulence of the complemented mutant strain closely approximated that of the parental strain. The data from both the *in vitro* and *in vivo* experiments suggest that the O-acetylated CP5 may be more proficient than the O-deacetylated polysaccharide in protecting the *Staphylococcus* from immune clearance. Whether the presence of O-acetyl groups on the polysaccharide affects the activation and deposition of complement components on the bacterial surface has not yet been explored.

Functional analysis of cap5O

The *cap5O* gene codes for a putative protein of 420 amino acids.[26] Cap5O shows homology to putative UDP-ManNAc dehydrogenases from *Methanococcus jannaschii* and *Burkholderia* (*Pseudomonas*) *solanacearum* and to RffD of *E. coli* (Table 8.1). The 420-amino acid RffD is a UDP-ManNAc dehydrogenase that oxidizes UDP-ManNAc to produce UDP-ManNAcA.[33] In direct sequence comparison, Cap5O is 68% similar and 45% identical to RffD across 399 amino acid residues. The *E. coli rffD* gene is involved in the biosynthesis of enterobacterial common antigen (ECA), a surface-associated glycolipid common to members of the Enterobacteriaceae family.[34] ECA has a trisaccharide repeating structure composed of GlcNAc, Fuc4NAc, and ManNAcA. The last sugar is also a component of *S. aureus* CP5 and CP8 (Figure 8.1). To determine whether the *S. aureus cap5O* gene could complement an *rffD* mutation in *E. coli*, Kiser and J.C. Lee subcloned the *S. aureus cap5O* gene into pUC19.[35] The recombinant plasmid was transformed into *E. coli* strain 21546, an ECA-negative mutant that has a Tn10 insertion in *rffD* and is deficient in UDP-ManNAc dehydrogenase activity.[33] By Western blot analysis with an ECA-specific monoclonal antibody, the *S. aureus cap5O* gene complemented the *rffD* mutation in strain 21546, thereby restoring ECA synthesis. The mutant strain transformed with the pUC19 vector alone or carrying *cap5P* or *cap5G* remained ECA negative. This experiment suggests that *cap5O* codes for a UDP-ManNAc dehydrogenase that oxidizes UDP-ManNAc to UDP-ManNAcA, a precursor molecule for CP5 biosynthesis. A nonpolar deletion in *cap5O*[36] or a chemically-induced mutation in *cap8O*[28] both resulted in an acapsular *S. aureus* phenotype.

Functional analysis of cap5G and cap5P

The *cap5G* and *cap5P* genes code for putative proteins of 374 and 391 amino acids, respectively,[26] and they show 29% overall identity. In direct sequence comparison, Cap5P shows substantial homology (68% similarity and 50% identity over 368 amino acids) with the *E.coli rffE* gene. The *E. coli rffE* gene encodes a UDP-GlcNAc 2-epimerase, an enzyme that catalyzes the conversion of UDP-GlcNAc to UDP-ManNAc during the biosynthesis of ECA in *E. coli*. Cap5G also shows homology, albeit less than that of Cap5P, to RffE (50% similarity and 30% identity over a 365-amino acid length). To determine whether *S. aureus cap5G*, *cap5P*, or both could complement an *rffE* mutation in *E. coli*, the genes were subcloned, alone or in tandem with *cap5O*, into pUC19.[35] The recombinant plasmids were introduced into *E. coli* strain 21566, an ECA-negative mutant that contains a Tn10 insertion in *rffD*, as well as an additional DNA insertion in *rffE*. Mutant 21566 is defective in both UDP-GlcNAc 2-epimerase and UDP-ManNAcA dehydrogenase activities and displays an ECA-negative phenotype.[33] The *cap5P* gene, but not *cap5G*, provided in trans, partially restored ECA synthesis to mutant 21566. When *cap5O* and *cap5P* were provided in tandem to strain 21566, ECA synthesis was restored fully. In contrast, a construct that included both *cap5O* and *cap5G* failed to restore ECA synthesis to mutant 21566. These experiments indicate that *cap5P*, but not *cap5G*, codes for a UDP-GlcNAc-2-epimerase. Recently, Bhasin et al. purified the *cap5P* gene product from *E. coli* and demonstrated its enzymatic function *in vitro*.[38] Because UDP-GlcNAc-2-epimerase catalyzes an early and critical step in the predicted CP5 biosynthetic pathway, inactivation of *cap5P* was expected to eliminate UDP-ManNAc synthesis, hence abrogating CP5 expression. However, both *cap5P* and *cap8P* mutants still express CP5 and CP8, respectively.[28,31] The linkage unit between ribitol teichoic acid and peptidoglycan in the cell wall of *S. aureus* contains ManNAc.[39] Thus, it is likely that *S. aureus* has another enzyme(s) encoded outside of the capsule gene cluster that mediates synthesis of UDP-ManNAc; identification of such a gene will be facilitated by the completion of the *S. aureus* genome project.

The function of the *cap5G* gene product in CP5 biosynthesis is uncertain. It is possible that Cap5G and Cap5P are both epimerases but with different substrates. We hypothesize that Cap5G serves as a 3-epimerase in the biosynthesis of UDP-L-FucNAc, the putative donor of FucNAc residues in CP5 and CP8 (see Figure 8.4). Nonetheless, the possibility remains that Cap5G has UDP-GlcNAc-2-epimerase activity in *S. aureus*.

Cap5G, Cap5O, and Cap5P lie in the common, flanking regions of the *cap5* gene cluster, and they are virtually identical to Cap8G, Cap8O, and Cap8P. These putative proteins are probably involved in the synthesis of ManNacA and FucNAc nucleotide precursors that are common to both CP5 and CP8. In contrast, *cap5H* lies in the central, serotype-specific region of the *cap5* cluster. This is consistent with its putative role as an O-acetyltransferase specific for the third carbon of ManNAcA. The corresponding gene in the *cap8* cluster is *cap8J*, which encodes a putative enzyme that O-acetylates the fourth carbon on CP8 ManNAcA. A likely biosynthetic pathway for CP5 is depicted in Figure 8.4. This pathway is consistent with the predicted functions of 15 of the 16 *cap5* genes and with the known structure of CP5.

Regulation of capsule expression

S. aureus capsule production is influenced by environmental and bacterial growth conditions, such as culture medium and the bacterial growth phase of the organism.[40,41] CP5 production is inhibited by high levels of yeast extract, alkaline growth conditions, CO_2, and anaerobiasis[40-42] but enhanced by growth of the bacterium in milk.[43] Growth of *S. aureus* under iron limitation and on solid medium both augmented the production of CP8.[32]

The expression of extracellular and cell-bound proteins by *S. aureus* is controlled by a regulatory locus called the accessory gene regulator (*agr*). The effect of an *agr* mutation on expression of CP5 by *S. aureus* Newman was investigated in different complex and synthetic media.[44] Compared with

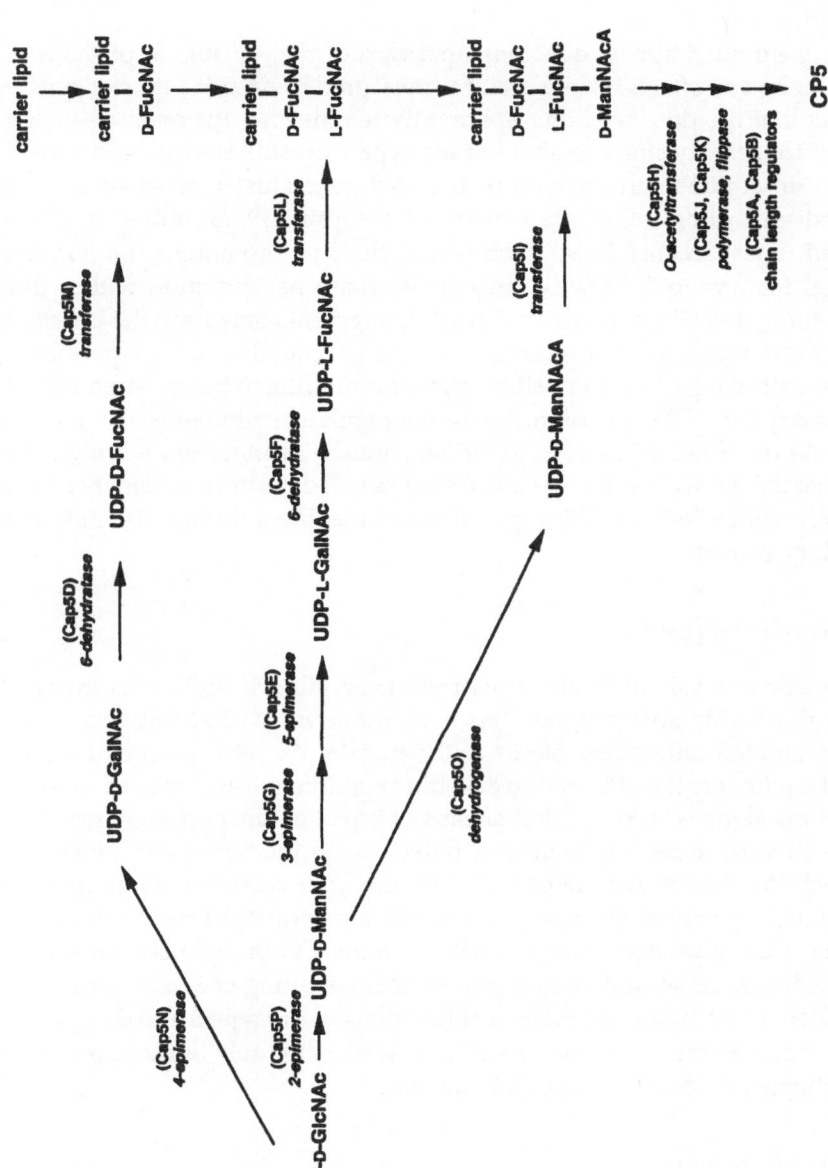

Figure 8.4 Proposed pathway for the biosynthesis of *S. aureus* CP5.

the wild-type strain, CP5 expression by the *agr* mutant was strongly reduced in some media and slightly reduced in others. The overall conclusion of the study was that CP5 production is positively controlled by *agr*. Another regulatory locus (*sar*) that interacts with *agr* has been described in *S. aureus*,[45] but its effect on capsule expression by the staphylococcus has not yet been reported.

As mentioned above, deletions upstream of the –35 region of the *cap1* promoter had no effects on its activity as measured by an *xylE* reporter gene.[20] The fact that the *cap1* promoter apparently requires no upstream cis-acting element for activity suggests that the serotype 1 capsule is expressed constitutively. In contrast, an analysis of the *cap8* gene cluster revealed several inverted and direct repeats upstream of the primary promoter. A 10-bp inverted repeat located 14 bp upstream of the *cap8* promoter is apparently required for promoter activity, since a chromosomal mutation within this repeat abrogated CP8 expression.[46] A DNA fragment containing the inverted repeat served as a protein binding site in a gel mobility shift experiment. These results suggest that a positive regulator binding to the inverted repeat is necessary for CP8 expression. A positive regulatory protein with a molecular mass of ~80 kDa has been identified, and N-terminal sequencing suggests that it is a novel protein.[46] Future studies will determine whether certain environmental effects of CP8 expression are mediated though the putative regulatory protein.

Future directions

The genetic analysis of *S. aureus* capsule expression is still in its infancy. Although considerable progress has been made in the last few years, we have assembled only a few pieces of the puzzle. We have proposed a biosynthetic pathway for CP5 (Figure 8.4), but experimental evidence to support most of the steps is lacking. Mechanisms of capsule transport and assembly at the cell surface remain to be elucidated, as do the regulatory processes involved. We do not even know whether the lipid carrier is undecaprenol phosphate, although it seems a likely candidate because of its involvement in both peptidoglycan and teichoic acid assembly. Continued research efforts in these directions should yield a greater understanding of *S. aureus* capsule expression. In addition, the biological functions of the type 5 and 8 capsular polysaccharides merit continued study, as well as the role of the capsule in the pathogenesis of staphylococcal infections.

Acknowledgments

This work was supported by NIH Grants AI29040 (to J.C.L) and AI37027 (to C.Y.L). We thank Navneet Bhasin, Kevin Kiser, Subrata Sau, and Ouyang Shu for their significant contributions to this work.

References

1. Gilbert, I., Dissociation in an encapsulated staphylococcus, *J. Bacteriol.*, 21, 157, 1931.
2. Cohn, Z., Determinants of infection in the peritoneal cavity. 1. Response to and fate of *Staphylococcus aureus* and *Staphylococcus albus* in the mouse, *Yale J. Biol. Med.*, 35, 12, 1962.
3. Melly, M., Duke, L., Liau, D.-F., and Hash, J., Biological properties of the encapsulated *Staphylococcus aureus* M. *Infect. Immun.*, 10, 389, 1974.
4. Wilkinson, B. J., Staphylococcal capsules and slime, in *Staphylococci and Staphylococcal Infections*, Vol. 2, Easmon, C. and Adlam, C., Eds., Academic Press, London, 1983, 481.
5. Lee, J. C., Betley, M. J., Hopkins, C. A., Perez, N.E., and Pier, G. B., Virulence studies, in mice, of transposon-induced mutants of *Staphylococcus aureus* differing in capsule size, *J. Infect. Dis.*, 156, 741, 1987.
6. Lin, W. S., Cunneen, T., and Lee, C. Y., Sequence analysis and molecular characterization of genes required for the biosynthesis of type 1 capsular polysaccharide in *Staphylococcus aureus*, *J. Bacteriol.*, 176, 7005, 1994.
7. Karakawa, W. W. and Vann, W. F., Capsular polysaccharides of *Staphylococcus aureus*, *Semin. Infect. Dis.*, 4, 285, 1982.
8. Sompolinsky, D., Samra, Z., Karakawa, W. W., Vann, W. F., Schneerson, R., and Malik, Z., Encapsulation and capsular types in isolates of *Staphylococcus aureus* from different sources and relationship to phage types, *J. Clin. Microbiol.*, 22, 828, 1985.
9. Wilkinson, J., The extracellular polysaccharides of bacteria, *Bacteriol. Rev.*, 22, 46, 1958.
10. Liau, D. F. and Hash, J. H., Structural analysis of the surface polysaccharide of *Staphylococcus aureus* M, *J. Bacteriol.*, 131, 194, 1977.
11. Murthy, S. V., Melly, M. A., Harris, T. M., Hellerqvist, C. G., and Hash, J. H., The repeating sequence of the capsular polysaccharide of *Staphylococcus aureus* M, *Carbohydr. Res.*, 117, 113, 1983.
12. Lee, J. C., Michon, F., Perez, N. E., Hopkins, C. A., and Pier, G. B., Chemical characterization and immunogenicity of capsular polysaccharide isolated from mucoid *Staphylococcus aureus*, *Infect. Immun.*, 55, 2191, 1987.
13. Hanessian, S. and Haskell, T., Structural studies on staphylococcal polysaccharide antigen, *J. Biol. Chem.*, 239, 2758, 1964.
14. Wu, T. and Park, J., Chemical characterization of a new surface antigenic polysaccharide from a mutant of *Staphylococcus aureus*, *J. Bacteriol.*, 108, 874, 1971.
15. Fournier, J. M., Vann, W. F., and Karakawa, W. W., Purification and characterization of *Staphylococcus aureus* type 8 capsular polysaccharide, *Infect. Immun.*, 45, 87, 1984.
16. Fournier, J. M., Hannon, K., Moreau, M., Karakawa, W. W., and Vann, W. F., Isolation of type 5 capsular polysaccharide from *Staphylococcus aureus*, *Ann. Inst. Pasteur Microbiol.*, 138, 561, 1987.
17. Moreau, M., Richards, J. C., Fournier, J. M., Byrd, R. A., Karakawa, W. W., and Vann, W. F., Structure of the type-5 capsular polysaccharide of *Staphylococcus aureus*, *Carbohydr. Res.*, 201, 285, 1990.
18. Lee, C. Y., Cloning of genes affecting capsule expression in *Staphylococcus aureus* strain M, *Mol. Microbiol.*, 6, 1515, 1992.

19. Lin, W. S. and Lee, C. Y., Instability of type 1 capsule production in *Staphylococcus aureus*, in *Abstracts of the 96th Annual Meeting of the American Society for Microbiology,* American Society for Microbiology, Washington, D.C., 1996, B236.

20. Ouyang, S. and Lee, C. Y., Transcriptional analysis of type 1 capsule genes in *Staphylococcus aureus, Mol. Microbiol.,* 23, 473, 1997.

21. Lee, C. Y., Association of staphylococcal type-1 capsule-encoding genes with a discrete genetic element, *Gene,* 167, 115, 1995.

22. Ouyang, S. and Lee, C. Y., unpublished data, 1996.

23. Sau, S. and Lee, C. Y., Cloning of type 8 capsule genes and analysis of gene clusters for the production of different capsular polysaccharides in *Staphylococcus aureus, J. Bacteriol.,* 178, 2118, 1996.

24. Albus, A., Arbeit, R. D., and Lee, J. C., Virulence of *Staphylococcus aureus* mutants altered in type 5 capsule production, *Infect. Immun.,* 59, 1008, 1991.

25. Lee, J. C., Xu, S. L., Albus, A., and Livolsi, P. J., Genetic analysis of type 5 capsular polysaccharide expression by *Staphylococcus aureus, J. Bacteriol.,* 176, 4883, 1994.

26. Sau, S., Bhasin, N., Wann, E. R., Lee, J. C., Foster, T. J., and Lee, C. Y., The *Staphylococcus aureus* allelic genetic loci for serotype 5 and 8 capsule expression contain the type-specific genes flanked by common genes, *Microbiol.,* 143, 2395, 1997.

27. Pattee, P. A., Lee, H.-C., and Bannantine, J. P., Genetic and physical mapping of the chromosome of *Staphylococcus aureus*, in *Molecular Biology of the Staphylococci,* Novick, R. P., Ed., VCH Publishers, Inc., New York, 1990, 41.

28. Sau, S., Sun, J., and Lee, C. Y., Molecular characterization and transcriptional analysis of type 8 capsule genes in *Staphylococcus aureus, J. Bacteriol.,* 179, 1614, 1997.

29. Sau, S., and Lee, C. Y., unpublished data, 1997.

30. Bhasin, N., Albus, A., Michon, F., Livolsi, P. J., Park, J.-S., and Lee, J. C., Identification of a gene essential for O-acetylation of the *Staphylococcus aureus* type 5 capsular polysaccharide, *Mol. Microbiol.,* 27, 9, 1998.

31. Bhasin, N., Kiser, K. B., Deng, L., and Lee, J. C., Purification and characterization of UDP-GlcNAc 2-epimerase encoded by the *cap5P* gene of *Staphylococcus aureus,* manuscript submitted.

32. Lee, J. C., Takeda, S., Livolsi, P. J., and Paoletti, L. C., Effects of in vitro and in vivo growth conditions on expression of type-8 capsular polysaccharide by *Staphylococcus aureus, Infect. Immun.,* 61, 1853, 1993.

33. Meier-Dieter, U., Starman, R., Barr, K., Mayer, H., and Rick, P. D., Biosynthesis of enterobacterial common antigen in *Escherichia coli, J. Biol. Chem.,* 265, 13490, 1990.

34. Kuhn, H.-M., Meier-Dieter, U., and Mayer, H., ECA, the enterobacterial common antigen, *FEMS Microbiol. Rev.,* 54, 195, 1988.

35. Kiser, K. B. and Lee, J. C., *Staphylococcus aureus cap5O* and *cap5P* genes functionally complement mutations affecting Enterobacterial common antigen biosynthesis in *Escherichia coli, J. Bacteriol.,* 180, 403, 1998.

36. Kiser, K. B. and Lee, J. C., unpublished data, 1998.

37. Marolda, C. L. and Valvano, M. A., Genetic analysis of the dTDP-rhamnose biosynthesis region of the *Escherichia coli* VW187 (O7:K1) *rfb* gene cluster: identification of functional homologs of *rfbB* and *rfbA* in the *rff* cluster and correct location of the *rffE* gene, *J. Bacteriol.,* 177, 5539, 1995.

38. Bhasin, N., Kiser, K. B., Deng, L., and Lee, J. C., Purification and biochemical characterization of the UDP-GlcNAc-2-epimerase encoded by the *cap5P* gene of *Staphylococcus aureus*, in *98th Annual Meeting of the American Society for Microbiology*, American Society for Microbiology, Washington, D.C., 1998, B415.

39. Kojima, N., Araki, Y., and Ito, E., Structure of the linkage units between ribitol teichoic acids and peptidoglycan, *J. Bacteriol.*, 161, 299, 1985.

40. Dassy, B., Stringfellow, W. T., Lieb, M., and Fournier, J. M., Production of type 5 capsular polysaccharide by *Staphylococcus aureus* grown in a semi-synthetic medium, *J. Gen. Microbiol.*, 137, 1155, 1991.

41. Stringfellow, W. T., Dassy, B., Lieb, M., and Fournier, J. M., *Staphylococcus aureus* growth and type 5 capsular polysaccharide production in synthetic media, *Appl. Environ. Microbiol.*, 57, 618, 1991.

42. Herbert, S., Worlitzsch, D., Dassy, B., Boutonnier, A., Fournier, J.-M., Bellon, G., Dalhoff, A., and Doring, G., Regulation of *Staphylococcus aureus* capsular polysaccharide type 5: CO_2 inhibition in vitro and in vivo, *J. Infect. Dis.*, 176, 431, 1997.

43. Sutra, L., Rainard, P., and Poutrel, B., Phagocytosis of mastitis isolates of *Staphylococcus aureus* and expression of type-5 capsular polysaccharide are influenced by growth in the presence of milk, *J. Clin. Microbiol.*, 28, 2253, 1990.

44. Dassy, B., Hogan, T., Foster, T. J., and Fournier, J. M., Involvement of the accessory gene regulator (*agr*) in expression of type-5 capsular polysaccharide by *Staphylococcus aureus*, *J. Gen. Microbiol.*, 139, 1301, 1993.

45. Heinrichs, J. H., Bayer, M. G., and Cheung, A. L., Characterization of the *sar* locus and its interaction with *agr* in *Staphylococcus aureus*, *J. Bacteriol.*, 178, 418, 1996.

46. Ouyang, S. and Lee, C. Y., Analysis of the promoter for the expression of type 8 capsular polysaccharide in *Staphylococcus aureus*, in *Abstracts of the 98th Annual Meeting of the American Society for Microbiology*, American Society for Microbiology, Washington, D.C., 1998, B416.

chapter nine

Arabinogalactan in mycobacteria: Structure, biosynthesis, and genetics

Michael McNeil

Contents

Introduction to mycobacteria...207
    Mycobacterial cell wall core..208
    Mycobacterial cell envelope carbohydrates............................208
    Drug development is a major rationale for studying mycobacterial
        cell surface carbohydrate biosynthesis.................................209
    The genome of *M. tuberculosis* has been sequenced 209
    Scope of the present review article...209
Structure, biosynthesis, and genetics of AG 210
    Structure 210
    Overall biosynthesis of AG 210
    Sugar donor biosynthesis and genetics.....................................210
    Biochemistry and genetics of the glycosyl transferases.........215
Conclusion 217
Acknowledgments...218
References..218

Introduction to mycobacteria

The genus *Mycobacterium* contains three important bacterial pathogens, *Mycobacterium tuberculosis*, *M. leprae*, and *M. avium*, and an important fast-growing nonpathogenic research species, *M. smegmatis*. Mycobacteria, although strictly speaking Gram-positive, are readily distinguished from other bacteria by their unique cell wall, which conforms neither to the classical Gram-positive nor Gram-negative wall but includes features of both.[1] The cell wall core refers to the covalently attached core of the cell wall; the cell envelope includes not only the cell wall core but also noncovalently associated molecules found at the mycobacterial surface.

Mycobacterial cell wall core

The mycobacterial cell wall core is present in essentially the same form in all mycobacterial species. It consists of three interconnected "macromolecules." The outermost of these are mycolic acids, unique 70- to 90-carbon branched fatty acids which form an outer lipid layer similar to, but differing from, the classical outer membrane of Gram-negative bacteria.[2] The mycolic acids are esterified to the middle component, "arabinogalactan" (AG), a polymer composed primarily of D-galactofuranosyl and D-arabinofuranosyl residues. AG is connected, via a linker disaccharide phosphate, to the 6 position of a muramic acid residue of the peptidoglycan. The peptidoglycan is the innermost of the three cell wall core macromolecules.

AG is an important polysaccharide in that it plays a key role in holding the lipid layer to the peptidoglycan layer in the cell wall core. Hence, AG is necessary for mycobacterial viability. For example, ethambutol inhibits biosynthesis of the arabinan component[3,4] of AG and as a result is an effective therapeutic.

Mycobacterial cell envelope carbohydrates

Carbohydrates in addition to AG are associated with the cell wall core. The phosphatidyl-based glycolipids include phosphatidyl inositol mannosides (PIMs),[5,6] mannosyl extended PIMs known as lipomannan (LM),[7] and finally, arabinosylated LM, lipoarabinomannan (LAM).[7-11] Although these carbohydrates are thought to be found primarily anchored in the plasma membrane via their phosphatidyl inositol reducing ends, it is possible that they are also found anchored to the mycolic acid layer of the cell wall core. LAM, LM, and PIMs are found in all mycobacterial species, although their structures vary somewhat depending on the species.

In addition, other glycolipids are anchored to the mycolic acids at the outer regions of the cell wall core, via fatty acid association. All species of mycobacteria have trehalose mycolates (trehalose acylated with one or two mycolic acids) and may also contain, depending on the mycobacterial species, lipooligosaccharides (LOS) based on additional glycosylation of acylated trehalose, phenolic glycolipids (PGL) based on glycosylation of an acylated derivative of phenol, and glycopeptidolipids (GPL) based on glycosylation of a complex lipopeptide. The structures of these glycolipids is well reviewed.[12,13] Finally, on the exterior of the cell are found the capsular polysaccharides, which include an α-1,4 glucan, an arabinomannan, and a mannan.[14-16] Because of the structural similarities between LAM and LM to capsular arabinomannan and mannan,[14-16] it is tempting to suggest that the extracellular arabinomannan and mannan are made by cleavage of the phosphatidyl inositol anchor unit from LAM and LM, respectively; however, this has not been demonstrated.

Drug development is a major rationale for studying mycobacterial cell surface carbohydrate biosynthesis

A major impetus for the study of the cell wall core molecule AG arises from the need for new drugs against *M. tuberculosis* and *M. avium*.[17-19] Two fundamental reasons to target AG are: (1) the fact that it appears to be essential for viability,[20] and (2) that three of the four sugars of which it is composed, D-Ara*f*, D-Gal*f*, and L-Rha*p*, are not found in humans. Thus, any of a score or more of enzymes involved in the formation of sugar donors and their polymerization are potential drug targets. The isolation and expression of the genes for these enzymes is a high research priority. Inhibitors of the resultant enzymes can be obtained by using "high throughput" screens and by enzyme characterization (ultimately X-ray analysis) and the subsequent design of "rational" inhibitors.

The PIMs may be necessary for viability as essential membrane components; in support of this PIM minus mutants of mycobacteria are not known. In addition, molecules such as LAM and GPL may be virulence factors. Thus, LAM is a potent down-regulator of functions involved in host cell mediated immunity,[21-23] and evokes a large array of cytokines.[24-28] Therefore, inhibition of LAM and other phosphatidyl inositol-based compounds may result in bacterial death by the host immune system and/or an amelioration of pathogenesis. The situation is less clear-cut with the cell wall envelope glycolipids, but GPLs which are found in *M. avium* may be virulence factors.[29-32]

The genome of M. tuberculosis has been sequenced

The genome of *M. tuberculosis* strain H37Rv has been sequenced and annotated by the Sanger Centre, and the sequence and BLAST searches are readily accessed from: http://www.sanger.ac.uk/Projects/M_tuberculosis/.

In addition, the clinical isolate *M. tuberculosis* CSU 93 is also nearly sequenced by The Institute for Genomic Research, and current information can be found at http://www.tigr.org/tigr_home/tdb/mdb/mdb.html.

Scope of the present review article

In this review, the structure, biosynthetic pathway, and genetics of the cell wall core polysaccharide, AG, will be described. The biosynthesis of LAM is now actively being studied,[33] but these studies are in the early stages and will not be presented here. For the most part, the biosynthesis of the remaining cell envelope glycoconjugates has not been studied, with the exception of the glycopeptidolipids for which some important genetic studies have been performed.[31,32,34,35]

Structure, biosynthesis, and genetics of AG

Structure

The structure of AG[1,36-39] is summarized in Figure 9.1. The polymer is readily divided into three structural regions. The first of these is a specialized linker region[39] located at the reducing end of the polymer and consists of →4)-α-L-Rha$p$-(1→3)-α-D-GlcNAc-(1→phosphate) where the phosphate is in turn attached to the peptidoglycan. Attached to this linker is a galactofuran[38] consisting of [→6)-β-D-Gal$f$-(1→5)-β-D-Gal$f$]$_{-14}$. Attached to the galactofuran are two or three chains of a D-arabinofuran[36,38] containing variously linked arabinofuranosyl residues arranged as illustrated in Figure 9.1. The mycolic acids are esterified to the 5 hydroxyl group of the terminal and penultimate arabinosyl residues.[37]

Overall biosynthesis of AG

The biosynthetic pathway of AG formation (Figure 9.2) is largely worked out,[40-46] although several key questions remain. The sugar donors of GlcNAc-1-phosphate and Rha are UDP-GlcNAc and dTDP-Rha, respectively; these are those commonly used by bacteria.[47] The galactofuranosyl residues are donated by UDP-Gal$f$[48] which is made by a ring contraction from UDP-Gal$p$ in a fashion similar to that recently shown for Gal$f$ residues in *E. coli*.[49,50] Interestingly the donor of D-ara$f$ residues is decaprenylphosphate-D-arabinose (DPA);[41,43,51] no role for a sugar nucleotide of D-ara has been established. DPA is made by a unique pathway starting with phosphoribosyl pyrophosphate.[45]

It is well established that the linker unit and galactofuran are synthesized on a lipid carrier as shown in Figure 9.2.[42] However, the order of some of the later events in AG synthesis is not yet clear. Thus it is not known if the arabinan is built up stepwise on the lipid linked galactan or whether the arabinan is formed on its own lipid carrier and transferred as a block to the galactan. Also, it is not known when the mycolic acids are added to the arabinan and when the transfer from the lipid carrier to peptidoglycan occurs.

Sugar donor biosynthesis and genetics

Generally speaking, the genes for the enzymes involved in sugar nucleotide formation have been identified; those involved in decaprenyl phosphate arabinose biosynthesis remain to be elucidated.

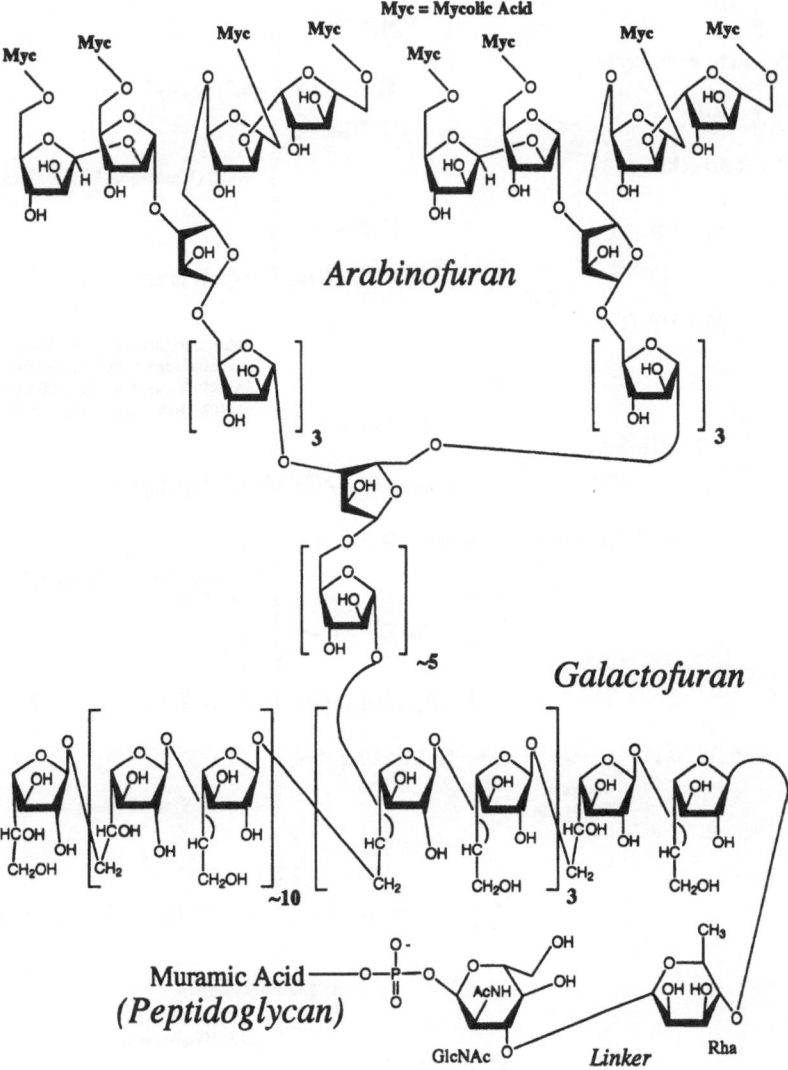

Figure 9.1 The structure of the mycobacterial cell wall arabinogalactan. The arabinan, galactofuran, and linker regions are indicated, as well as where the mycolic acids are attached and where the linker is attached to peptidoglycan. Although the 22 nonreducing structure of the arabinan is correct, it should be noted that the interior regions of the arabinan are not understood and that this 22 arabinosyl unit may be attached to additional arabinosyl residues rather than directly to the galactofuran chain as shown. Also it is expected that two or three complete arabinan chains are attached to the galactofuran rather than the single arabinan chain illustrated. Finally, the exact galactofuranosyl residues to which the arabinans are linked are unknown.

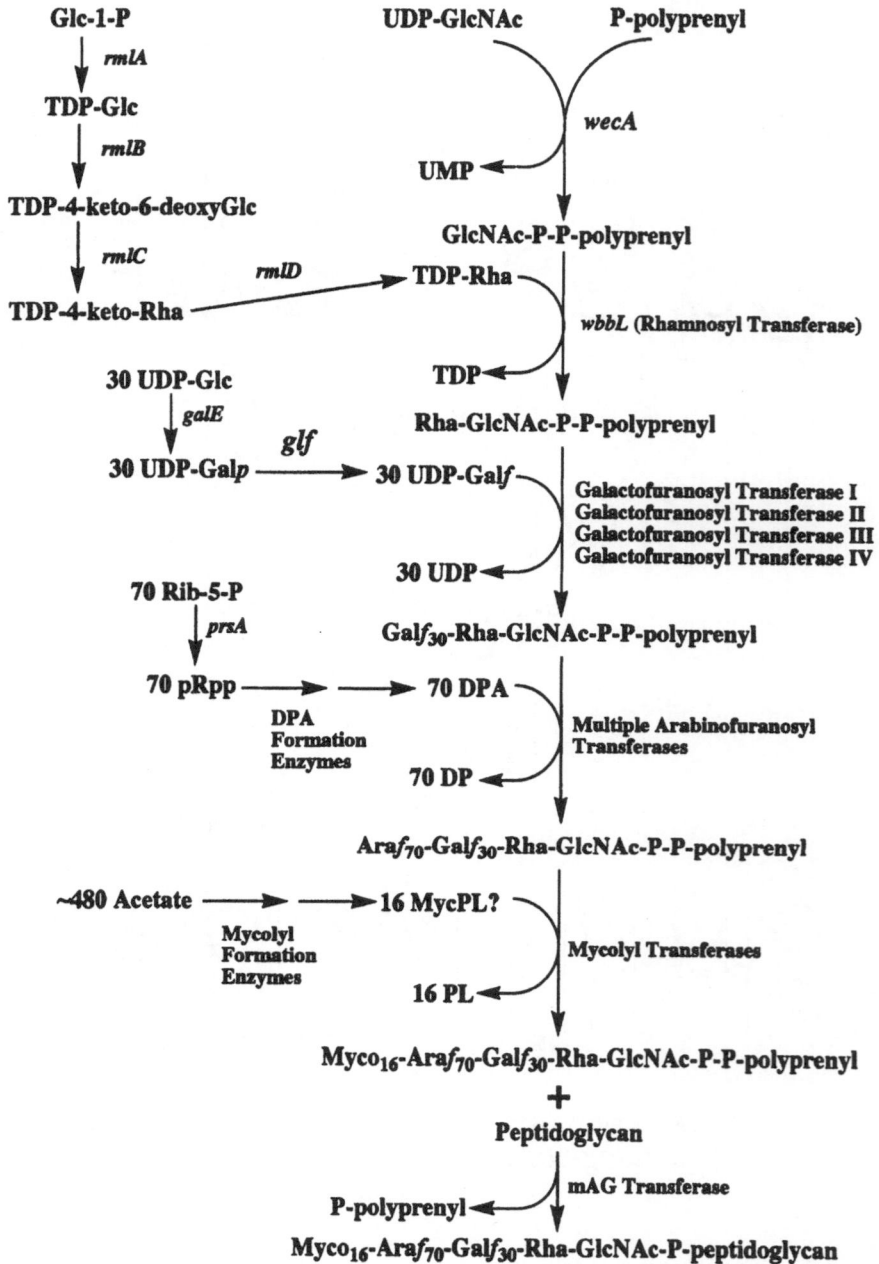

Figure 9.2   The pathway of cell wall arabinogalactan biogenesis. As noted in the text, the order of some of the later events has not been clarified.

dTDP-Rha

The genes for the four enzymes which convert glucose-1-phosphate to dTDP-Rha, *rmlA, rmlB, rmlC, rmlD* (previously known as *rfb A–D*[52]) are found clustered together in Gram-negative bacteria, usually in the order BDAC, where they are present for the synthesis of rhamnosyl residues found in O-antigen.[53] However, in mycobacteria these genes occur in three areas of the genome as discussed below.

A gene with homology to *rmlA*, which encodes for the first enzyme in the pathway for dTDP-Rha formation, α-D-glucose-1-phosphate thymidylyl-transferase, has been cloned and found to be separate from the other genes involved in dTDP-Rha formation.[44] This gene has been shown to encode for the expected enzyme, α-D-glucose-1-phosphate thymidylyltransferase (see Figure 9.2) by enzymatic assay.[44] In the Sanger genome sequence, this gene is Rv0334. Three open reading frames occur upstream that could possibly be linked to *rmlA* (stop codons are separated from the next start codons by about 25 nucleotides). These open reading frames (Rv0331, 0332, and 0333) are of unknown function, except RV0331 which shows homology to sulfide dehydrogenase and sulfide quinone reductase. In addition to Rv0034, another gene, Rv3264c, which encodes for a protein with a sequence homologous to RmlA of other bacteria (notated *rmlA2* at the present time in the Sanger sequence) is present in the *M. tuberculosis* genome. The Rv3264c gene is clustered with two other genes involved in rhamnose biosynthesis as discussed below. However, Rv3264c also shows homology to α-D-mannose-1-phosphate guanylyltransferase (*manB*), and experiments just conducted in the author's laboratory showed that the protein product of Rv4364c shows this GDP-mannose synthesis activity and not α-D-glucose-1-phosphate thymidylyltransferase activity. Why a gene involved in mannose biosynthesis is clustered with two genes involved in rhamnose biosynthesis is not known.

Dr. Tae-Yon Lee et al.,[54] in the midst of epidemiological studies, discovered *rmlB* and *rmlC*, by sequence homology, directly linked to each other in the order BC, but no other rhamnosyl genes were on either side of these. These genes are notated *rmlB* and *rmlC* in the Sanger sequence and are open reading frames Rv3465 and Rv3466, respectively. Both of these genes have recently been cloned in the author's laboratory[55] and shown, using enzyme assays, to encode for the expected enzymes, dTDP-D-glucose 4,6-dehydratase and dTDP-4-keto-6-deoxy-D-glucose 3,5 epimerase, respectively (Figure 9.2). The genome sequence also reveals several other genes with sequence homology to *rmlB* of other organisms. These genes are open reading frames Rv3468, Rv3634, and Rv3784, notated at present *rmlB3*, *rmlB2*, and *epiB*, respectively. Evidence is presented below that Rv3634 (now notated *rmlB2*) encodes for UDP-galactose epimerase (GalE); GalE and RmlB proteins do have substantial homology. The homology of the other two open reading frames with *rmlB* is significantly lower than that of Rv3465, and initial experiments in the author's laboratory have suggested that neither of these genes encodes for dTDP-D-glucose 4,6-dehydratase. Thus, tentatively, only one copy of *rmlB* is present in the genome (Rv3465).

A gene with strong homology to *rmlD* (Rv3266c, notated *rmlD* in the Sanger sequence), the final gene involved in dTDP-Rha biosynthesis, has been cloned in the author's laboratory. Recent enzyme assay of the expressed protein has shown that this gene does code for dTDP-rhamnose synthetase (RmlD, Figure 9.2). This gene is clustered with the aforementioned Rv3264c, *manB*, and with the rhamnosyl transferase gene (Rv3265c, *wbbL* described below) in the order *rmlD, wbbL, manB*.

Thus, the rhamnose biosynthetic genes are located in three different regions in the TB genome. There are two cluster regions [*rmlB, RmlC* in one cluster, and *rmlD, wbbL*, (and *manB*) in the second cluster] and *rmlA* separate from any other dTDP-Rha synthetic genes. None of these regions is near to the others. There also appears to be only one copy of each of the four genes, but this needs to be confirmed in the case of *rmlB*.

UDP-Gal*f*

The galactofuranosyl donor, UDP-Gal*f*, and the enzyme which forms it from UDP-Gal*p*, UDP-galactopyranose mutase, have been studied in some detail in *E. coli*[49,50] and *Klebsiella pneumoniae*.[56] This enzyme is encoded for by the gene *glf*;[49] in *E. coli* the function of the product of this gene was elegantly predicted by Reeves et al.[52] The enzymes have been purified and studied from both organisms. The *E. coli* enzyme has been shown to contain a bound FAD molecule,[49] and the *K. pneumoniae* enzyme is yellow, so it also is likely to contain a bound FAD.[56] In addition, the *K. pneumoniae* enzyme, as purified, requires either NADH or NADPH for activity.[56] Since the conversion of UDP-Gal*p* to UDP-Gal*f* does not require a net oxidation or reduction and since there is no obvious reason for an oxidation followed by reduction to accomplish the conversion, the role of these cofactors is unclear. It is also interesting to note that the equilibrium value of the UDP-Gal*f* to UDP-Gal*p* ratio, catalyzed by these enzymes, appears to be approximately 0.07.

The *M. tuberculosis* version of *glf* has been sequenced in the author's laboratory[48] and appears in the Sanger genome sequence (Rv3809). The gene has been expressed and shown to produce active enzyme.[48] Although not noted in the original publication,[48] this enzyme also requires NAD(P)H for optimal activity, and its sequence suggests that it also binds FAD.[48]

The early part of the pathway involves the formation of UDP-Gal*p* from UDP-Glc by UDP-galactopyranose epimerase (GalE); this enzyme from *M. smegmatis* was purified and *N*-terminally sequenced.[48] The resulting amino acid sequence allowed for tentative identification of the corresponding *M. tuberculosis* gene[48] (Rv3634, notated *rmlB2* as of this writing), which has an identical 24 amino acid *N*-terminal protein except for one amino acid change. It should be noted that there are three other genes in the genome with homology to *galE*; these are Rv0501 (notated *galE1*), Rv0536 (notated *galE2*), and Rv3784 (notated *epiB*).

The galactose salvage pathway begins with the conversion of galactose to galactose-1-phosphate[57] by the enzyme galactokinase (*galK*), which is present in *M. tuberculosis* as Rv0620 (Sanger sequence). The galactose-1-phosphate is converted to UDP-Gal*p* via the enzyme UDP glucose-galactose-1-phosphate uridylyltransferase (GalT); this activity has been shown to be present in *M. smegmatis*.[48] Interestingly, in the Sanger genome sequence

(Sanger Rv 0618 and Rv 0619) of *M. tuberculosis* H37Rv and also in the TIGR sequence of *M. tuberculosis* CSU 93, there appears to be a frame shift mutation in *galT*. These mutations imply that in *M. tuberculosis* GalT enzyme cannot be produced and thus that *M. tuberculosis* could not be grown on media with galactose as the sole carbon source and suggests that galactose may be toxic to *M. tuberculosis* due to the buildup of galactopyranose-1-phosphate.

Decaprenylphosphate arabinose

The donor of arabinosyl residues, decaprenylphosphate arabinose (DPA), was first suspected by the characterization of this compound[51] in mycobacteria and subsequently confirmed by forming radioactive arabinan from ^{14}C-DPA[41,43] in cell-free experiments. Polyprenylphosphate sugars are generally biosynthesized by transfer of a glycosyl residue from a sugar nucleotide to the phosphate moiety of a polyprenylphosphate.[47] Thus, in mycobacteria, decaprenylphosphate mannose is formed by GDP-Man reacting with a deca-prenylphosphate to form decaprenylphosphate mannose plus GDP.[58-60] However, no clearly characterized sugar nucleotide of arabinose has been found in mycobacteria. One report[3] suggested the possibility of GDP-Ara, and a second has suggested the possibility of UDP-Ara,[61] but the existence of either of these arabinofuranosyl donors has not yet been clearly established. In addition, DPA fails to react with nucleotide diphosphates to form a sugar nucleotide of arabinose[45] in contrast to the known reverse reaction of deca-prenylphosphate mannose with GDP to form GDP-Man.[59] Thus a novel pathway for the formation of DPA was searched for.

The arabinose carbon skeleton in DPA was first shown to arise from the pentose shunt.[46] This result led to the incubation of phospho[14C]ribosyl pyrophosphate (pRpp) with crude membranes prepared from *M. smegmatis*,[45] which yielded both DPA and decaprenylphosphate ribose (DPR). The pathway to DPR formation was readily deciphered and shown to occur by transfer of ribose-5-phosphate to decaprenyl phosphate to form decaprenylphosphate-5-phosphoribose (DP-5PR). DP-5PR is then dephosphorylated at the 5 position[45] to form mature DPR. The isolation of decaprenylphosphate-5-phosphoarabinose suggests that DPA is formed by a very similar pathway.[45] Clearly in the formation of DPA an epimerization must take place, and although it is expected that this occurs by epimerization of phosphoribosyl pyrophosphate to form 5-phosphoarabinosyl pyrophosphate, this has not yet been demonstrated. None of the genes encoding for the enzymes involved in the formation of DPA have been identified at this time, and such genes will be difficult to find merely by homology as D-Ara*f* is rarely found in nature.

Biochemistry and genetics of the glycosyl transferases

α-D-GlcNAc-1-phosphate transferase

The transferase that attaches a GlcNAc-phosphate to the decaprenyl phosphate is analogous to the product of the *wecA* (formerly *rfe*) gene in *E. coli*.

The biochemical evidence for an enzyme encoded for by a gene analogous to *wecA* in *M. smegmatis* is quite strong,[40,42] including the susceptibility of this transferase to tunicamycin.[42] Thus not surprisingly, a gene with homology to *E. coli wecA* was found in the Sanger genome sequence, Rv 1302 notated *rfe*.

Rhamnosyl transferase

The rhamnosyl transferase of *M. tuberculosis* was identified using an *E. coli* rhamnosyl transferase gene sequence. By coincidence, *E. coli* K12 O-antigen biosynthesis[53] requires dTDP-Rha:α-D-GlcNAc-pyrophosphate polyprenol α-3-L-rhamnosyl transferase (notated wbbL[52]) which forms the same product from the same substrates as the rhamnosyl transferase needed for mycobacterial AG production (Figure 9.2[42]). Thus, when an *M. tuberculosis* gene with sequence homologous to *E. coli wbbL* was found in the TB genome[62] (Sanger sequence, TIGR sequence), it was of interest to determine if this gene encoded for the mycobacterial version of *wbbL*. An *E. coli* with an insertion mutation in *wbbL*, and thus unable to form O-antigen,[53,63] was complemented by a plasmid containing the *M. tuberculosis wbbL* gene candidate (Rv 3265c), resulting in restoration of O-antigen formation.[64] This experiment demonstrated that the TB gene did, in fact, encode for the rhamnosyl transferase. The TB *wbbL* is linked with *rmlD* and *manB* as described above.

Galactofuranosyl transferases

As shown in Figure 9.2, four galactofuranosyl transferases are expected to be present in *M. tuberculosis* and other mycobacteria. Galactofuranosyl transferase I attaches a β-D-Galf residue to the 4 position of the rhamnosyl residue; galactofuranosyl transferase II attaches a β-D-Galf residue to the 5 position of the terminal β-D-Galf residue attached to the rhamnosyl residue; galactofuranosyl transferase III attaches a β-D-Galf residue to the 6 position of any β-D-Galf residue, which is itself attached to the 5 position of the penultimate Galf; and galactofuranosyl transferase IV attaches a β-D-Galf residue to position 5 of any β-D-Galf residue, which is itself attached to the 6 position of the penultimate Galf. Galactofuranosyl transferase activity can be readily detected[65] in *M. smegmatis* cell-free extracts. Whether the genes coding for these enzymes can be identified by homology with galactofuranosyl transferases, which form rather different end products such as those found in *E. coli* and *Klebsiella*, is unknown, but given both the different galactofuranosyl acceptors and products it is unlikely that a simple homology search will reveal these enzymes.

Arabinosyl transferases

Given the structural complexity of the D-arabinofuran present in AG (Figure 9.1), many different arabinosyl transferases must be involved in its formation. Arabinosyl transferase activity in a mixture of membranes and cell walls has been conclusively demonstrated,[41,43] and the product is of the size expected for mature arabinan.[43] Hence, this cell-free system probably contains many different active arabinosyl transferases. To assay for particular arabinosyl transferases, simple di- and trisaccharide hydrophobic acceptors have been synthesized and utilized successfully.[40,66]

Two genes, *embA* and *embB*, have been cloned from *M. avium* which are believed to code for arabinosyl transferases.[67] These genes confer resistance to ethambutol, a drug which inhibits arabinan formation.[3,4,68] This inhibition is observed when the genes are overexpressed in *M. smegmatis* using a multicopy plasmid vector. Cells containing this vector are not only resistant to ethambutol,[67] but, importantly, the ability of ethambutol to inhibit arabinosyl transferase assays in cell-free extracts prepared from these bacteria is diminished. The two gene products EmbA and EmbB are very similar to each other (45% identity) and not to any other nonmycobacterial protein sequences present in the databank. Analysis of the predicated amino acid sequences suggests the presence of 8 to 10 potential membrane-spanning domains in the proteins. The exact reaction catalyzed by either putative arabinosyl transferase has yet to be determined. In *M. tuberculosis* H37Rv genome, three genes homologous to *embA* and *embB* are found and in the order *embC* (Rv 3793), *embA* (Rv 3794), and *embB* (Rv 3795).

Mycolyl transferase

After synthesis from acetate, mycolic acids are present on a "carrier lipid" in the form of the mycolic acid being attached to the 6 position of the mannose in octahydroheptaprenylphosphate mannose.[69] The mycolic acids are then transferred to the terminal arabinosyl residues of AG either directly or via another glycolipid, trehalose monomycolate.[40] An interesting enzyme in this regard has recently been identified. The protein has long been known as a dominant antigen, the antigen 85 complex. This enzyme is assayed by its ability to catalyze the exchange of mycolic acids from trehalose monomycolate to free [14C]trehalose.[70] The exact biologically relevant reaction catalyzed by the enzyme is unknown, and thus it is not yet clear if it is involved in the mycolation of cell walls for sure. The mycolyltransferase exists in three closely related forms (antigen 85A, B, and C) which all show mycolyl transferase activity. The genes are not linked but are located at three different places on the *M. tuberculosis* genome.

Conclusion

The structure of AG is largely worked out (Figure 9.1), although the structure of the interior regions of the arabinan remain unknown. With respect to biosynthesis, the donors of both the usual and unusual sugars have been identified, and the basic pathways of polymerization are largely known, although some details remain to be elucidated. With both standard tools, and the genome sequencing efforts, most genes that share homology with known genes in other bacteria have been identified. Thus, all the genes involved in rhamnose biosynthesis, and in UDP-Gal*f* formation are known. Two important endeavors are now apparent. The first is to identify the remaining glycosyl transferase and DPA formation genes where no homology to genes identified in other bacteria exists. Important methods in this regard involve sophisticated homology techniques that might characterize, for example, Gal*f* binding homologies and thus allow gene candidates to be identified. However, it can be expected that in many cases such homologies may prove elusive. Perhaps more important will be affinity labeling techniques at the protein level, which might allow sufficient protein purification (perhaps merely by 2-dimension electrophoresis) for *N*-terminal sequence and recognition of the gene in the genome sequence.

A second endeavor is to determine which polysaccharide biosynthetic gene products are essential for bacterial growth. It is to be expected that any gene product required for AG biosynthesis will be essential unless another protein can catalyze the same reaction, but this needs to be experimentally demonstrated. Although gene replacement methods are available for *M. smegmatis*[71,72] and even in slow-growing mycobacteria,[73] such approaches are made even more difficult by the expectation of the genes being essential. Another approach involves the isolation and characterization of TS mutants, but it is difficult to target a specific gene in such an approach. Perhaps the most promising technique will be the use of conditional anti-sense mutagenesis using systems such as the one recently demonstrated by Parish and Stoker.[74] In this scenario, a plasmid with a regulated anti-sense orientation of the gene of interest can be introduced, the bacteria allowed to grow with the transcription of the anti-sense message turned off, and then bacterial growth monitored when the anti-sense message is induced.

Acknowledgments

The author gratefully acknowledges NIH–NIAID support (AI 33706) for the studies performed at Colorado State University, his many collaborators, and his colleagues at the Colorado State Microbiology Department Mycobacterial Research Laboratories.

References

1. McNeil, M. and P. J. Brennan. Structure, function, and biogenesis of the cell envelope of mycobacteria in relation to bacterial physiology, pathogenesis, and drug resistance; some thoughts and possibilities arising from recent structural information. *Res. Microbiol.* 142, pp. 451-463, 1991.
2. Brennan, P. J. and H. Nikaido. The envelope of mycobacteria. *Annu. Rev. Biochem.* 64, pp. 29-63, 1995.
3. Takayama, K. and J. O. Kilburn. Inhibition of synthesis of arabinogalactan by ethambutol in *Mycobacterium smegmatis*. *Antimicrob. Agents Chemother.* 33, pp. 1493-1499, 1989.

4. Deng, L., K. Mikusova, K. G. Robuck, M. Scherman, P. J. Brennan, and M. McNeil. Recognition of multiple effects of ethambutol on the metabolism of the mycobacterial cell envelope. *Antimicrob. Agents Chemother.* 39, pp. 694-701, 1995.

5. Lee, Y. C. and C. E. Ballou. Complete structures of the glycophospholipids of mycobacteria. *Biochemistry* 4, pp. 1395-1404, 1965.

6. Khoo, K. H., A. Dell, H. R. Morris, P. J. Brennan, and D. Chatterjee. Structural definition of acylated phosphatidylinositol mannosides from *Mycobacterium tuberculosis*: definition of a common anchor for lipomannan and lipoarabinomannan. *Glycobiology* 5, pp. 117-127, 1995.

7. Hunter, S. W., H. Gaylord, and P. J. Brennan. Structure and antigenicity of the phosphorylated lipopolysaccharide antigens from the leprosy and tubercle bacilli. *J. Biol. Chem.* 261, pp. 12345-12351, 1986.

8. Khoo, K.H., A. Dell, H. R. Morris, P. J. Brennan, and A. K. Chatterjee. Inositol phosphate capping of the nonreducing termini of lipoarabinomannan from rapidly growing strains of *Mycobacterium*. *J. Biol. Chem.* 270, pp. 12380-12389, 1995.

9. Khoo, K. H., A. Dell, H. R. Morris, P. J. Brennan, and D. Chatterjee. Inositol phosphate capping of the nonreducing termini of lipoarabinomannan from rapidly growing strains of Mycobacterium. *J. Biol. Chem.* 270, pp. 12380-12389, 1995.

10. Chatterjee, D., K. H. Khoo, M. R. McNeil, A. Dell, H. R. Morris, and P. J. Brennan. Structural definition of the non-reducing termini of mannose-capped LAM from *Mycobacterium tuberculosis* through selective enzymatic degradation and fast atom bombardment-mass spectrometry. *Glycobiology* 3, pp. 497-506, 1993.

11. Chatterjee, D., S. W. Hunter, M. McNeil, and P. J. Brennan. Lipoarabinomannan: multiglycosylated form of the mycobacterial mannosylphosphatidylinositols. *J. Biol. Chem.* 267, pp. 6228-6233, 1992.

12. Aspinall, G. O., D. Chatterjee, and P. J. Brennan. The variable surface glycolipids of mycobacteria: structures, synthesis of epitopes, and biological properties. [Review]. *Adv. Carbohydr. Chem. Biochem.* 51, pp. 169-242, 1995.

13. McNeil, M., D. Chatterjee, S. W. Hunter, and P. J. Brennan. Mycobacterial glycolipids: isolation, structures, antigenicity, and synthesis of neoantigens, in *Methods in Enzymology*, Ginsburg V. (Ed.), Academic Press. San Diego. 1989.

14. Ortalo-Magne, A., M. A. Dupont, A. Lemassu, A. B. Andersen, P. Gounon, and M. Daffe. Molecular composition of the outermost capsular material of the tubercle bacillus. *Microbiology* 141, pp. 1609-1620, 1995.

15. Lemassu, A., A. Ortalo-Magne, F. Bardou, G. Silve, M. A. Laneelle, and M. Daffe. Extracellular and surface-exposed polysaccharides of non-tuberculous mycobacteria. *Microbiology* 142, pp. 1513-1520, 1996.

16. Lemassu, A. and M. Daffe. Structural features of the exocellular polysaccharides of *Mycobacterium tuberculosis*. *Biochem. J.* 297, pp. t2)351-t2)357, 1994.

17. Cohen, M. L. Epidemiology of drug resistance: Implications for a post-antimicrobial era. *Science* 257, pp. 1050-1052, 1992.

18. Bloom, B. R. Foreword. *Tuberculosis*, First edition, Rom, W.N. and Garay, S. (Eds.), Little, Brown, Boston. 1996.

19. Moran, N. WHO issues another gloomy tuberculosis report. *Nature Med.* 2, p. 377, 1996.

20. McNeil, M. Targeted preclinical drug development for *Mycobacterium avium* complex: a biochemical approach, in *Mycobacterium avium*-Complex Infection, Korvick, J.A. and Benson, C.A. (Eds.), Marcel Dekker, New York. 1996.

21. Kaplan, G., R. R. Gandhi, D. E. Weinstein, W. R. Levis, M. E. Patarroyo, P. J. Brennan, and Z. A. Cohn. *Mycobacterium leprae* antigen-induced suppression of T cell proliferation *in vitro. J. Immunol.* 138, pp. 3028-3034, 1987.
22. Sibley, L. D., S. W. Hunter, P. J. Brennan, and J. L. Krahenbuhl. Mycobacterial lipoarabinomannan inhibits gamma interferon-mediated activation of macrophages. *Infect. Immun.* 56, pp. 1232-1236, 1988.
23. Chan, J., X. Fan, S. W. Hunter, P. J. Brennan, and B. R. Bloom. Lipoarabinomannan, a possible virulence factor involved in persistence of *Mycobacterium tuberculosis* within macrophages. *Infect. Immun.* 59, pp. 1755-1761, 1991.
24. Moreno, C., A. Mehlert, and J. Lamb. The inhibitory effects of mycobacterial lipoarabinomannan and polysaccharides upon polyclonal human T-cell proliferation. *Clin. Exp. Immunol.* 74, pp. 206-210, 1988.
25. Moreno, C., J. Taverne, A. Mehlert, C. A. W. Bate, R. J. Brealey, A. Meager, G. A. W. Rook, and J. H. L. Playfair. Lipoarabinomannan from *Mycobacterium tuberculosis* induces the production of tumor necrosis factor from human and murine macrophages. *Clin. Exp. Immunol.* 76, pp. 240-245, 1989.
26. Chatterjee, D., A. D. Roberts, K. Lowell, P. J. Brennan, and I. M. Orme. Structural basis of capacity of lipoarabinomannan to induce secretion of tumor necrosis factor. *Infect. Immun.* 60, pp. 1249-1253, 1992.
27. Adams, L. B., Y. Fukutomi, and J. L. Krahenbuhl. Regulation of murine macrophage effector functions by lipoarabinomannan from mycobacterial strains with different degrees of virulence. *Infect. Immun.* 61, pp. 4173-4181, 1993.
28. Barnes, P. F., D. Chatterjee, J. S. Abrams, S. Lu, E. Wang, M. Yamamura, P. J. Brennan, and R. L. Modlin. Cytokine production induced by *Mycobacterium tuberculosis* lipoarabinomannan: relationship to chemical structure. *J. Immunol.* 149, pp. 541-547, 1992.
29. Rastogi, N., K. S. Goh, and H. L. David. Enhancement of drug susceptibility of *Mycobacterium avium* by inhibitors of cell envelope synthesis. *Antimicrob. Agents Chemother.* 34, pp. 759-764, 1990.
30. Rastogi, N. Recent observations concerning structure and function relationships in the mycobacterial cell envelope: elaboration of a model in terms of mycobacterial pathogenicity, virulence and drug resistance. *Res. Microbiol.* 142, pp. 464-476, 1991.
31. Belisle, J. T., L. Pascopella, J. M. Inamine, P. J. Brennan, and W. R. Jacobs. Isolation and expression of a gene cluster responsible for the biosynthesis of the glycopeptidolipid antigens of *Mycobacterium avium. J. Bacteriol.* 173, pp. 6991-6997, 1991.
32. Belisle, J. T., K. Klaczkiewicz, P. J. Brennan, W. R. Jacobs, and J. M. Inamine. Rough morphological variants of *Mycobacterium avium*: characterization of genomic deletions resulting in the loss of glycopeptidolipid expression. *J. Biol. Chem.* 268, pp. 10517-10523, 1993.
33. Besra, G. S., C. B. Morehouse, C. M. Rittner, C. J. Waechter, and P. J. Brennan. Biosynthesis of mycobacterial lipoarabinomannan. *J. Biol. Chem.* 272, pp. 18460-18466, 1997.
34. Belisle, J. T., M. McNeil, D. Chatterjee, J. M. Inamine, and P. J. Brennan. Expression of the core lipopeptide of the glycopeptidolipid surface antigens in rough mutants of *Mycobacterium avium. J. Biol. Chem.* 268, pp. 10510-10516, 1993.
35. Mills, J. A., M. McNeil, J. T. Belisle, W. R. Jacobs, and P. J. Brennan. Genetic loci of *Mycobacterium avium ser2* gene cluster and their functions. *J. Bacteriol.* 176, pp. 4803-4808, 1994.

36. Besra, G. S., K.H. Khoo, M. McNeil, A. Dell, H. R. Morris, and P. J. Brennan. A new interpretation of the structure of the mycolyl-arabinogalactan complex of *Mycobacterium tuberculosis* as revealed through characterization of oligogly-cosylalditol fragments by fast-atom bombardment mass spectrometry and ^{1}H nuclear magnetic resonance spectroscopy. *Biochemistry* 34, pp. 4257-4266, 1995.

37. McNeil, M., M. Daffe, and P. J. Brennan. Location of the mycolyl ester substituent in the cell walls of mycobacteria. *J. Biol. Chem.* 266, pp. 13217-13223, 1991.

38. Daffe, M., P. J. Brennan, and M. McNeil. Predominant structural features of the cell wall arabinogalactan of *Mycobacterium tuberculosis* as revealed through characterization of oligoglycosyl alditol fragments by gas chromatography/mass spectrometry and by ^{1}H and ^{13}C-NMR analyses. *J. Biol. Chem.* 265, pp. 6734-6743, 1990.

39. McNeil, M., M. Daffe, and P. J. Brennan. Evidence for the nature of the link between the arabinogalactan and peptidoglycan components of mycobacterial cell walls. *J. Biol. Chem.* 265, pp. 18200-18206, 1990.

40. Besra, G. S. and P. J. Brennan. The mycobacterial cell envelope: a target for novel drugs against tuberculosis. *J. Pharm. Pharmacol.* 49, pp. 25-30, 1997.

41. Lee, R. E., K. Mikusova, P. J. Brennan, and G. S. Besra. Synthesis of the myco-bacterial arabinose donor β-D-Arabinofuranosyl-1-monophosphoryldecaprenol, development of a basic arabinosyl-transferase assay and identification of etham-butol as an arabinosyl transferase inhibitor. *J. Am. Chem. Soc.* 117, pp. 11829-11832, 1995.

42. Mikusova, K., M. Mikus, G. Besra, I. Hancock, and P. J. Brennan. Biosynthesis of the linkage region of the mycobacterial cell wall. *J. Biol. Chem.* 271, pp. 7820-7828, 1996.

43. Xin, Y., R. E. Lee, M. S. Scherman, K. H. Khoo, G. S. Besra, P. J. Brennan, and M. McNeil. Characterization of the *in vitro* synthesized arabinan of mycobac-terial cell walls. *Biochim. Biophys. Acta* 1335, pp. 231-234, 1997.

44. Ma, Y., J. A. Mills, J. T. Belisle, V. Vissa, M. Howell, K. Bowlin, M. S. Scherman, and M. McNeil. Determination of the pathway for rhamnose biosynthesis in mycobacteria: cloning, sequencing, and expression of the *Mycobacterium tu-berculosis* gene encoding α-D-glucose-1-phosphate thymidylyltransferase. *Microbiology* 143, pp. 937-945, 1997.

45. Scherman, M. S., L. Kalbe-Bournonville, D. Bush, Y. Xin, and M. McNeil. Poly-prenylphosphate-pentoses in mycobacteria are synthesized from phosphori-bose pyrophosphate. *J. Biol. Chem.* 271, pp. 29652-29658, 1996.

46. Scherman, M., A. Weston, K. Duncan, A. Whittington, R. Upton, L. Deng, R. Comber, J. D. Friedrich, and M. McNeil. The biosynthetic origin of the myco-bacterial cell wall arabinosyl residues. *J. Bacteriol.* 177, pp. 7125-7130, 1995.

47. Shibaev, V. N. Biosynthesis of bacterial polysaccharide chains composed of repeating units. [Review]. *Adv. Carbohydr. Chem. Biochem.* 44, pp. 277-339, 1986.

48. Weston, A., R. J. Stern, R. E. Lee, P. M. Nassau, D. Monsey, S. L. Martin, M. S. Scherman, G. S. Besra, K. Duncan, and M. R. McNeil. The biosynthetic origin of the mycobacterial cell wall galactofuranosyl residues. *Tubercle and Lung Disease* 78, pp. 123-131, 1998.

49. Nassau, P. M., S. L. Martin, R. E. Brown, A. Weston, D. Monsey, M. McNeil, and K. Duncan. Galactofuranose Biosynthesis in *Escherichia coli* K12: Identifi-cation and cloning of UDP-galactopyranose mutase. *J. Bacteriol.* 178, pp. 1047-1052, 1996.

50. Lee, R., D. Monsey, A. Weston, K. Duncan, C. Rithner, and M. McNeil. Enzymatic synthesis UDP-galactofuranose and assays for UDP-galactopyranosyl mutase based on HPLC and on linked enzymatic production of NADH. *Anal. Biochem.* 242, pp. 1-7, 1996.

51. Wolucka, B. A., M. R. McNeil, E. de Hoffmann, T. Chojnacki, and P. J. Brennan. Recognition of the lipid intermediate for arabinogalactan/arabinomannan biosynthesis and its relation to the mode of action of ethambutol on mycobacteria. *J. Biol. Chem.* 269, pp. 23328-23335, 1994.

52. Reeves, P. R., M. Hobbs, M. A. Valvano, M. Skurnik, C. Whitfield, D. Coplin, N. Kido, J. Klena, D. Maskell, C. R. Raetz, and P. D. Rick. Bacterial polysaccharide synthesis and gene nomenclature. *Trends Microbiol.* 4, pp. 495-503, 1996.

53. Stevenson, G., B. Neal, D. Liu, M. Hobbs, N. H. Packer, M. Batley, J. W. Redmond, L. Lindquist, and P. R. Reeves. Structure of the O antigen of *Escherichia coli* K-12 and the sequence of its *rfb* gene cluster. *J. Bacteriol.* 176, pp. 4144-4156, 1994.

54. Lee, T. Y., T. J. Lee, J. T. Belisle, P. J. Brennan, and S. K. Kim. A novel repeat sequence specific to *Mycobacterium tuberculosis* complex and its implications. *Tubercle and Lung Disease* 78, pp. 13-19, 1997.

55. Lee, T. Y., R. J. Stern, M. S. Scherman, and W. Yan. Unpublished data 1998.

56. Koplin, R., J. R. Brisson, and C. Whitfield. UDP-galactofuranose precursor required for formation of the lipopolysaccharide O antigen of *Klebsiella pneumoniae* serotype 1 is synthesized by the product of the $rfbD_{KPO1}$ gene. *J. Biol. Chem.* 272, pp. 4121-4128, 1997.

57. Szumilo, T. The occurrence of the Leloir pathway in non-pathogenic mycobacteria. *Acta Microbiol. Polonica* 30, pp. 327-336, 1981.

58. Yokoyama, K. and C. E. Ballou. Synthesis of α-1-6-mannooligosaccharide in *M. smegmatis*. *J. Biol. Chem.* 264, pp. 21621-21628, 1989.

59. Schultz, J. and A. D. Elbein. Biosynthesis of mannosyl- and glucosyl-phosphoryl polyprenols in *Mycobacterium smegmatis*. *Arch. Biochem. Biophys.* 160, pp. 311-322, 1974.

60. Takayama, K. and D. S. Goldman. Enzymatic synthesis of mannosyl-1-phosphoryldecaprenol by a cell-free system of *Mycobacterium tuberculosis*. *J. Biol. Chem.* 245, pp. 6251-6257, 1970.

61. Singh, S. and S. E. Hogan. Isolation and characterization of sugar nucleotides from *Mycobacterium smegmatis*. *Microbios.* 77, pp. 217-222, 1994.

62. Mills, J. A. Unpublished data 1997.

63. Liu, D. and P. R. Reeves. *Escherichia coli* K12 regains its O antigen. *Microbiology* 140, pp. 49-57, 1994.

64. Mills, J. A., K. Motichka, R. J. Stern, M. S. Scherman, and M. R. McNeil. Manuscript in preparation 1999.

65. Besra, G. S. and P. J. Brennan. The mycobacterial cell wall: biosynthesis of arabinogalactan and lipoarabinomannan. *Biochem. Soc. Trans.* 25, pp. 845-850, 1997.

66. Lee, R. E., P. J. Brennan, and G. S. Besra. Mycobacterial arabinan biosynthesis: the use of synthetic arabinoside acceptors in the development of an arabinosyl transfer assay. *Glycobiology* 7, pp. 1121-1128, 1997.

67. Belanger, A. E., G. S. Besra, M. E. Ford, K. Mikusova, J. Belisle, P. J. Brennan, and J. M. Inamine. Molecular characterization of the target of the antimycobacterial drug ethambutol. *Proc. Natl. Acad. Sci. U.S.A.* 93, pp. 11919-11924, 1996.

68. Mikusova, K., R. A. Slayden, G. S. Besra, and P. J. Brennan. Biogenesis of the mycobacterial cell wall and the site of action of ethambutol. *Antimicrob. Agents Chemother.* 39, pp. 2484-2489, 1995.
69. Besra, G. S., T. Sievert, R. E. Lee, R. A. Slayden, P. J. Brennan, and K. Takayama. Identification of the apparent carrier in mycolic acid synthesis. *Proc. Natl. Acad. Sci. U.S.A.* 91, pp. 12735-12739, 1995.
70. Belisle, J. T., V. D. Vissa, T. Sievert, K. Takayama, P. J. Brennan, and G. S. Besra. Role of the major antigen of *Mycobacterium tuberculosis* in cell wall biogenesis. *Science* 276, pp. 1420-1422, 1997.
71. Husson, R. N., B. E. James, and R. A. Young. Gene replacement and expression of foreign DNA in mycobacteria. *J. Bacteriol.* 172, pp. 519-524, 1990.
72. Sander, P., A. Meier, and E. C. Bottger. RspL+: A dominant selectable marker for gene replacement in mycobacteria. *Mol. Microbiol.* 16, pp. 991-1000, 1995.
73. Kalpana, G. V., B. R. Bloom, and W. R. Jacobs. Insertional mutagenesis and illegitimate recombination in mycobacteria. *Proc. Natl. Acad. Sci. U.S.A.* 88, pp. 5433-5437, 1991.
74. Parish, T. and N. G. Stoker. Development and use of a conditional antisense mutagenesis system in mycobacteria. *FEMS Microbiol. Lett.* 154, pp. 151-157, 1997.

Index

A

A antigen, 136
abe, 26, 41
Accessory gene regulator, 200
Acholeplasma laidlawii, 196
Acyltransferases, 27
adk, 39, 45
Aeromonas salmonicida, 145
Afipia felis, 65
Agrobacterium, 53
Agrobacterium radiobacter, 65
Agrobacterium rhizogenes, 65
Agrobacterium tumefaciens, 60, 65
  lipid-A, 65
alg genes, 4, 5, 6, 8
Alginate, 3
  biosynthesis, 4-5
    genes encoding, 4
  LPS synthesis and, 14-15
  role, 8
Alpha-chain biosynthesis, 114-116
Alpha-D-GlcNAc-1-phosphate transferase, 215-216
Anion exchange chromatography, high performance, 68-70
Anthocyanin, 75
Arabinogalactan, 208
  biosynthesis, 209, 210, 212
    enzymes for, 212-215
    genetics of, 212-215
  function, 208
  genetics, 209
  structure, 208, 210, 211
Arabinosyl transferase, 216-217
Azorhizobium, 53
Azorhizobium caulinodans, 65

B

Bacillus subtilis, 165, 169, 197
Bacterial resistance, 118
Bacterial virulence, 120-121

capsules and, 175
polysaccharides and, 46
Bacteriophages, 28-30
B antigen, 136
bas genes, 100
Beta-chain biosynthesis, 119
bexA genes, 94
Bordetella pertussis, 10, 196
Bradyrhizobium, 53
  glucans, 57
  O-chain polysaccharides, 70-71
Bradyrhizobium elkanii, 55
  core oligosaccharides, 68
  lipid-A, 64
Bradyrhizobium japonicum, 55
  lipid-A, 64
Bradyrhizobium lupini, 64
Brucella abortus, 65
Burkholderia solanacearum, 196

C

Capsular polysaccharides, 54, 139
  basic genetics, 162-168
    loci, 163, 165, 166
    transcription and mutation analyses, 165-167
  biosynthesis, 196
  gene clusters
    amino acid sequences of, 194
    element of, 191
    functional analysis of, 198-200
    mutant, 195
    regions of, 194
    sequence analysis of, 188, 192-195
    transcriptional analysis of, 188-190, 192-195
  glucose and, 173
  hyaluronic acid, 168-172
  identification, 163
  insertion elements, 167
  insertion mutations and, 166
  O-acetylation, 198

Staphylococcus aureus, 185-202
Streptococcus agalactiae, 164
Streptococcus pneumoniae, 13
Streptococcus pyogenes, 164
Streptococcus thermophilus, 164
structure, 162-168
type 19, 197
type 3 capsule, 168-172
Capsule(s), 92-93
    biology of, 95
    locus, 94-95
    production, genetics of, 94
    virulence and, 175
Carbohydrates
    biosynthesis, drug development and, 209
    cell envelope, 208-209
Chalcone synthase, 62
Chloramphenicol O-acetyltransferase, 196
Cholera, 133. *See also Vibrio cholerae*
Chromatography, high performance anion
        exchange, 68-70
cld, 99
Cloning
    O:8 O-antigen gene cluster, 40
Common antigen, 12-13
    genes, 12-13
    role, 13
Core oligosaccharides, 66-70
Core region, 66-70
    biosynthesis, 74, 113-114
    Bradyrhizobium elkanii, 68
    gene cluster
        outer, 33
    genetics, 11, 33, 113-114
    LPS, 67
    oligosaccharides, 66-70
    outer, 33
cps 14G, 173
cps genes, 165-168, 170-173
Cyclic glucans, 57
Cystic fibrosis
    LPS-rough phenotype and, 14

D

dct genes, 74
ddh genes, 26, 31, 39, 41, 44
Decaprenylphosphate arabinose, 215
Drug development, carbohydrate synthesis
        and, 209

dTDP-Glc-4,6,-dehydratase, 174
dTDP-4-keto-6-deoxyglucose-3,5-epime-
        rase, 174
dTDP-L-Rha-synthase, 174

E

emb genes, 217
Endotoxin, 13
Enterobacterial common antigene gene clus-
        ter, 28
Enterococcus faecalis, 169
Epimerases, 27
Erwinia amylovora, 60
Erwinia carotovora, 60
Erwinia stewartii, 60
Escherichia coli, 167, 168, 196, 197
    capsular exporters, 140
    cps 14G activity, 173
    heat-shock sigma factor, 4
    K-antigen, 59, 60
    lipid A, 13, 71
    polysaccharides
        core synthesis of, 74
        O-antigen portion, 8
        synthesis of, 168
    rfaD, 137
    UDP-galactopyranose mutase, 214
exo genes, 57
Exopolysaccharide
    biosynthesis, 196, 197
exsA, 57

F

fcl, 26
fimS, 4, 6
fix-23 gene region, 60
Flippase gene, 44
fumC, 8

G

Galactofuranosyl transferase, 216
Galactosyltransferase, 197
galE, 26, 32, 34, 39, 119, 121
    mutant, 120, 122
GDP-perosamine biosynthesis, 137, 139,
        140
Glc-1-P thymidylyltransferase, 174

glpM, 8
Glycolipids, 208
Glycopeptidolipids, 208
Glycosyltransferases, 27, 41-42, 197
    biochemistry and genetics, 215-217
glyS, 115
gmd, 12-13, 26, 39
Gonococcus(i), 111
    alpha-chain biosynthesis, 114-116
    lipooligosaccharide, 113-123
        high-frequency structure, 116-118
        sialylation of, 118-119
    opacity proteins, 120-121
    pili, 120
    resistance, 118
gsk, 39, 43, 45
Guanosine pyrophosphorylase, 137

H

Haemophilus influenzae, 10, 91, 174
    capsular exporters, 140
    capsule, 92-93
        biology of, 95
        locus, 94-95
        production, genetics of, 94
    lipooligosaccharide, 95-103
        biosynthesis of, 99-103
        genetic organization of, 101
        genetic regulation of, 101-103
        structure of, 95-99
has genes, 168-169
hemH, 34, 41, 45
Hexasaccharide, 30
High performance anion exchange chroma-
    tography, 68-70
htrB, 100
Hyaluronan synthase, 169
Hyaluronic acid, 162-163, 168-172
    biosynthesis, 171
Hyaluronidase, 168
Hydratase, 27

I

Insertion mutations, 166
IS*3,* 167
IS*150,* 167
IS*1016,* 95
IS*1167,* 167

IS*1202,* 167
IS*1229,* 167
IS*1358,* 144-145, 152
isn, 100

J

JUMPstart sequence, 44-45

K

K-antigens, 59-62
    biology of, 61-62
    genetics of, 60-61
    structures of, 59-60
kds genes, 100
2-keto-3deoxyoctulosonic acid, 95
Kinases, 27
kinB, 6
Klebsiella pneumoniae
    UDP-galactopyranose mutase, 214

L

Lactococcus lactis, 162
    capsule and exopolysaccharide structures,
        164
lex2, 100
lgt genes, 100, 114, 115
lic genes, 100, 101, 102
Ligase, 27
Lipid A, 13-14
    bacterial infections and, 95-96
    biosynthesis, 13-14, 71-73
    genes, 13-14
    rhizobial, 64-66
    role, 14
Lipoarabinomannan, 208
Lipo-chitin-oligosaccharides, 54
    signal molecules, 58
Lipomannan, 208
Lipooligosaccharides, 95-103
    advantages in pathogenesis, 120
    biosynthesis, 99-103
        alpha-chain, 114-116
        beta-chain, 119
        genes of, 99-101
        genes with secondary effects on, 119-
            120
        genetic organization of, 101

genetic regulation of, 101-103
gonococcal, 113-123. *See also* Gonococ-
     cus(i), lipooligosaccharide
Haemophilus influenzae, 95-103
high-frequency structure, 116-118
LPS *vs.,* 96
mucosal infection and, 122-123
mycobacterial, 208
neisserial, 112-113, 120
structure, 95-99
Lipopolysaccharides, 2
  biosynthesis, 14-15, 25-28, 196, 197
     genes for, 25-26
     genetics of, 73-74
     initiation of, 26
     pathways of, 27
  gene products, 27
  LOS *vs.,* 96
  O antigen portion, 8, 27. *See also* O anti-
     gen
  outer core biosynthesis, 196
  outer core gene cluster, 33
  rhizobial, 63-77. *See also Rhizobium,* li-
     popolysaccharides
  topology and export, 139
  Vibrio cholerae. See Vibrio cholerae
  Yersinia. See Yersinia, lipopolysaccharide
lps genes, 100
lpx genes, 13, 99, 100
lse, 41
lsg, 100

M

man genes, 26, 39, 41
Membrane proteins, 139-140
Meningococcus(i), 111
  opacity proteins, 121-122
Mesorhizobium, 53
Mesorhizobium loti, 64, 70
Methanococcus jannaschii, 196, 197
Microencapsulation, 186
msb, 100
mucA, 4
mucC, 4
Mucoid exopolysaccharide. *See* Alginate
Mucosal infection, 122-123
Mutagenesis, 73, 122
  opacity proteins and, 122
  transposon, 40, 175

virulence and, 144
Mycobacterium, 207
  cell envelope, 208-209
     drug development and, 209
  cell wall core, 208
Mycobacterium avium, 207
Mycobacterium leprae, 207
Mycobacterium smegmatis, 207, 214
Mycobacterium tuberculosis, 207
  CSU 93, 209
  genome, 209
  H37Rv, 209
  UDP-galactopyranose mutase, 214
Mycolyl transferase, 217

N

N-acetylneuraminic acid, 172
ndk, 8
Neisseria, 10, 112-113, 120
  capsular exporters, 140
Neisseria meningitidis, 140
Neuraminidases, 168
Nitrobacter hamburgensis, 65
Nitrobacter winogradskyi, 65
nol genes, 61

O

O-acetylation, 198
O-acetyltransferase, 197
O-antigen, 8
  biosynthesis, 8-10, 27
     genes for, 7-9
     site of, 27
  ligase, 27
  O:1. *See Vibrio cholerae,* O1 O-antigen
  O:3, 30, 31, 32, 35-38
     biosynthesis of, 35-37
     gene cluster promoters, 37-38
     related serotypes of, 38
     transport of, 37
  O:8, 31, 38-43
     biosynthesis, 38-40
        genes for, 38-40, 41
        temperature regulation of, 43
     gene cluster, 38-43
  polymerase, 27
  role, 10

O139 antigen, 196
Ogawa-Inaba serotype switching, 142-143
27-OHC28:0, 65
Oligosaccharides, 66-70
Oligotropha carboxydovorans, 65
O:3 O-antigen, 30, 31, 32, 35-38
  biosynthesis of, 35-37
  gene cluster promoters, 37-38
  related serotypes of, 38
  transport of, 37
O:8 O-antigen, 31, 38-43
  biosynthesis
    genes for, 38-40, 41
  gene cluster, 40-41
    cloning of, 40-41
  gene cluster of, 38-43
Open reading frames, 9, 165, 188
  OtnA, 147
  OtnB, 147
ops genes, 44-45, 100
Oxidoreductase, 147

P

Penicillin binding protein 3, 168
Perosamine biosynthesis, 137, 139, 140
Perosamine synthetase, 139
Perosamine transferase, 139
pFV100, 9, 10
pgm genes, 100, 119, 121
Phenolic glycolipids, 208
Phosphatidyl inositol mannosides, 208
Phosphomannose isomerase, 137
Phospho-mannose-mutase, 137
Phosphomutase, 171
pilC, 117
pilE, 120, 121
Pili, virulence and, 120-121
pilS, 121
pilS1, 120
pilS2, 120
Plant pathogens, 60
plpA, 163, 167
pLPS2, 9
Polymerase, 27, 42
Polysaccharide polymerase, 174
Polysaccharides. *See also* Lipopolysaccha-
    rides
  capsular, 54, 162-168
  core genes, 11

extracellular, 54
genes, 11, 151
O-antigen portion, 8
Pseudomonas aeruginosa, 2-3. *See also*
    Pseudomonas aeruginosa,
    polysaccharides
  genes involved production of, 6, 9
  type 14, 172
  type 19B, 172, 197
  type 19F, 172, 197
  type III, 172
  virulence and, 46
prt, 26
Pseudomonas aeruginosa, 2, 168, 196
  endotoxin, 13
  lipid A, 13-14
  mortality, 2
  polysaccharides, 2
    biosynthesis, 197
    common antigen, 12-13
    core, 10-11
    core genes, 11
    structure, 3
  resistance, 2
Pseudomonas solanacearum, 60
Pseudomonas syringae, 60

R

rfa, 100, 114, 137, 143-144
rfb, 99, 135, 136, 137, 138, 143
  mutants, 144
  V. cholerae capsule, 148-149
Rhamnosyltransferase, 174, 215-216
Rhizobial-legume symbiosis, 54
Rhizobial lipo-chitin-oligosaccharide signal
    molecules, 58
Rhizobium, 53
  glucans, 57
  K-antigens, 59-62
    biology of, 61-62
    genetics of, 60-61
    structures of, 59-60
  lipopolysaccharides, 63-77
    biosynthesis of, 73-74
    core oligosaccharides, 66-70
    lipid-A regions of, 64-66
    O-chain polysaccharides, 70-71
    structural variation, 74-77

structures of, 63
    during symbiotic nodule development,
        74-77
    polysaccharides, extracellular, 55-57
Rhizobium etli, 55
    lipid-A, 64, 71-73
    LPS core region, 67
Rhizobium leguminosarum, 55
    lipid-A, 64, 65, 71-73
Rhizobium meliloti, 168, 194, 196, 197
Rhizobium tropici, 70
Rhodopseudomonas palustris, 65
Rhodopseudomonas viridis, 65
rkp genes, 60, 62
rmlB, 35, 37
RosAB operon, 43
ros genes, 39, 43

S

Salmonella enterica, 167, 197
    rfaD gene, 137
Salmonella typhimurium, 197
    lipid A, 13
    polysaccharides
        core biosynthesis of, 74
        O-antigen portion, 8
Sanger genome sequence, 214
Shigella flexneri, 173
Shigella sonnei, 44
siaB, 100
Sialylation, 118-19, 122
Sialyltransferase, 118
Sinorhizobium, 53
    glucans, 57
    K-antigens, 59
Sinorhizobium fredii, 59
    lipid A, 64
    soy bean interactions, 61
Sinorhizobium meliloti, 55, 59
    lipid-A, 64
    mutants, 61-62
sodA, 8
Soy bean, 61
Staphylococcus aureus, 185
    accessory gene regulator, 200, 202
    adhesins, 186
    capsular polysaccharides, 185-202

biosynthesis of, 200-202
    composition of, 186-187
    gene clusters, 187-200
    O-acetylation, 198
    type 1, 187-191, 196
    type 5, 191, 192-194, 196, 198-202
    type 8, 192, 194-196, 202
    exoenzymes, 186
    exotoxins, 186
    microencapsulation, 186
    serotype 1 gene cluster, 187-191
    serotype 5 gene cluster, 191-200
    serotype 8 gene cluster, 191-200
    strain M, 187
Streptococcal capsules, 161-162
Streptococcus, capsule loci, 163, 168, 176
Streptococcus agalactiae, 162, 197
    capsular serotypes, 163
    capsule and exopolysaccharide structures,
        164
    mutants, 175
    type III polysaccharide, 172, 173, 175
    virulence, 175
Streptococcus mutans, 165
Streptococcus pneumoniae, 162, 195, 196
    biosynthetic genes, 167
    capsular polysaccharides, 13
    capsule and exopolysaccharide structures,
        164
    capsule loci, 163, 168
    genetic exchange and capsule diversity,
        176-178
    hyaluronic acid, 168-172
    loss of capsule production, 166
    mutants, 171, 173
    nonencapsulated mutants, 175
    type 19B polysaccharide, 172, 197
    type 3 capsule loci, 168-172
    type 19F polysaccharide, 172, 197
    type 14 polysaccharide, 172
    type 3-specific genes, 163
    virulence, 175
Streptococcus pyogenes, 162
    biosynthetic genes, 167
    capsule and exopolysaccharide structures,
        164
    HA synthase activity, 174
    hyaluronic acid, 162-163, 168-172
    virulence, 175

Streptococcus thermophilus, 162, 168, 174, 195, 196
   capsule and exopolysaccharide structures, 164
Sugar donor biosynthesis, 210-215
Symbiotic nodule development, 74-77
Synechocystis, 168
Synthase, 171

T

Tetronate biosynthesis, 141-142
Thiobacillus, 65
Thiogalactoside O-acetyltransferase, 197
tnp genes, 167
Toxin coregulated pilus, 144
Transcriptional analysis, 165-167
   cap8 genes, 195
Transforming principle, 162
Translocases, 27
Transmembrane conductance regulator, 12
Transposases, 144, 167
Transposon mutagenesis, 40, 175
Trehalose mycolates, 208
Trisaccharide transporter, 174
Type 19B polysaccharide, 172, 197
Type 3 capsule, 168-172
   biosynthesis, 171
Type 19F polysaccharide, 172
Type III polysaccharide, 172, 173
Type 14 polysaccharide, 172
tyv, 26

U

UDP-galactopyranose mutase, 214
   Escherichia coli, 214
UDP-Glc dehydrogenase, 168
   function, 171
UDP-GlcNAc 2-epimerase, 196, 197
UDP-GlcNAc 4-epimerase, 197
UDP-glucose, 172
UDP-ManNAc dehydrogenase, 197
UDP-*N*-acetyl-D-mannosamine transferase, 174
Uridylytransferase, 169, 170
   function, 171

ush genes, 39, 43, 45

V

Vibrio anguillarum, 143, 144
Vibrio cholerae, 133-152, 167, 196
   O1 O-antigen, 134-144
      detection of, 146
      GDP-perosamine biosynthesis in, 137, 139, 140
      gene products, 135
      genetics of, 136-137
      mutants, 144
      Ogawa-Inaba serotype switching, 142-143
      rfa genes linked to *rfb* operon in, 143-144
      rfb region of, 137, 138
      structure, 134, 136
      tetronate biosynthesis in, 141-142
      transport, 139-141
   O139 O-antigen, 134, 145
      biosynthesis of, 147, 150-151
      detection of, 146
      mutants, 144
      rfb region of, 138
      rfc homolog and, 150
   rfb region of, 137, 138
      insertion sequences associated with, 144-145
   serogroups, 134
Vibrio fluvalis, 144
Vibrionaceae, 151
Vibrio parahaemolyticus, 144
Virulence, 46, 120-121
   capsules and, 175
   polysaccharides and, 46
   streptococcal, 175-176

W

waa genes, 11
wba, 42
wbb genes, 26, 35, 36, 37
wbc genes, 26, 32, 33, 34, 39, 41
wbpL, 10
wbyA, 26

Web site, 46
wecA, 10
wzm, 26, 35, 36
wzt, 26, 35, 36, 37
wzx, 26, 33, 34, 39, 41, 44
wzy, 7, 9, 10, 39, 42
wzz, 7, 26, 42

X

Xanthomonas campestris, 60, 197

Y

Yersinia, 25
   capsular exporters, 140
   characteristics, 24
   gene names, 26
   lipopolysaccharide, 23-47
      bacteria and bacteriophages in study of,
         28-30
      biosynthesis ang genetics of, 33
      outer core gene cluster, 33
   O-antigen gene clusters of, 43-45
      location of, 45-46
   polysaccharide, 46

Yersinia aldovae, 24
Yersinia antiqua, 25
Yersinia bercovieri, 24
Yersinia enterocolitica, 140, 196
   serotypes, 24-25
      O:3 LPS, 30, 31, 32, 35-38
      O:8 LPS, 31, 38-43
Yersinia frederikseni, 24
Yersinia intermedia, 24
Yersinia kristenseni, 24
Yersinia mediaevalis, 25
Yersinia mollareti, 24
Yersinia orientalis, 25
Yersinia pesti, 24
   biovariants, 25
Yersinia pseudotuberculosis
   O-antigen gene clusters, 43-44
   serotypes, 25
      O:1a, 31
      O:2a, 31
      O:4a, 31
      O:5a, 31
Yersinia rohdei, 24
Yersinia rückeri, 24